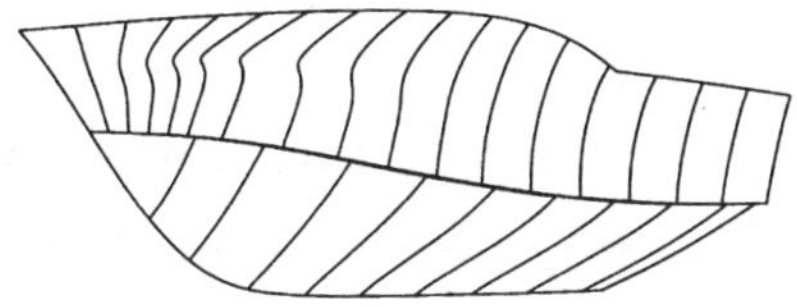

Designing Fair Curves and Surfaces

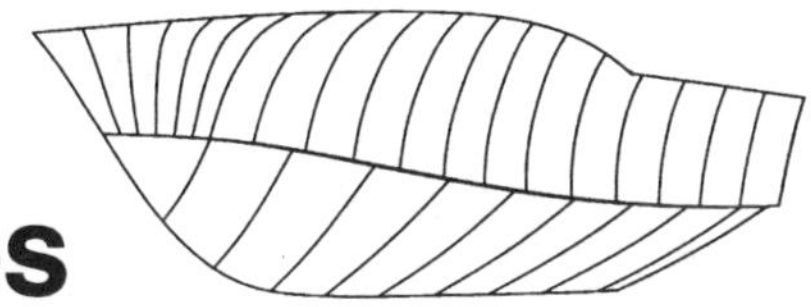

Geometric Design Publications

Editor
Gerald E. Farin
Arizona State University

Farin, Gerald E., editor, Geometric Modeling: Algorithms and New Trends (1987)
Farin, Gerald E., editor, NURBS for Curve and Surface Design (1991)
Barnhill, Robert E., editor, Geometry Processing for Design and Manufacturing (1992)
Hagen, Hans, editor, Curve and Surface Design (1992)
Hagen, Hans, editor, Topics in Surface Modeling (1992)
Goldman, Ronald N., and Lyche, Tom, editors, Knot Insertion and Deletion Algorithms for B-Spline Curves and Surfaces (1993)
Sapidis, Nickolas S., editor, Designing Fair Curves and Surfaces: Shape Quality in Geometric Modeling and Computer-Aided Design (1994)

Designing Fair Curves and Surfaces

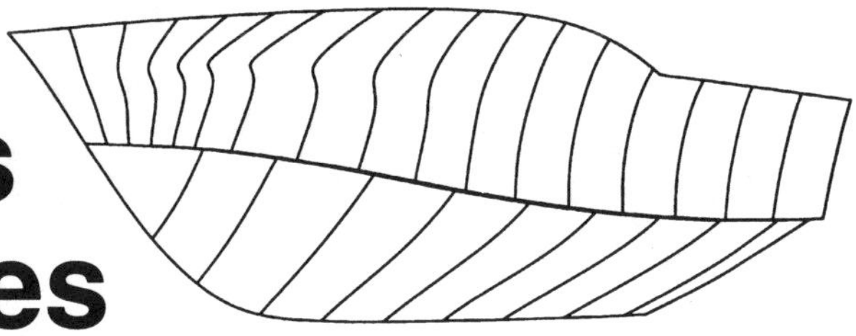

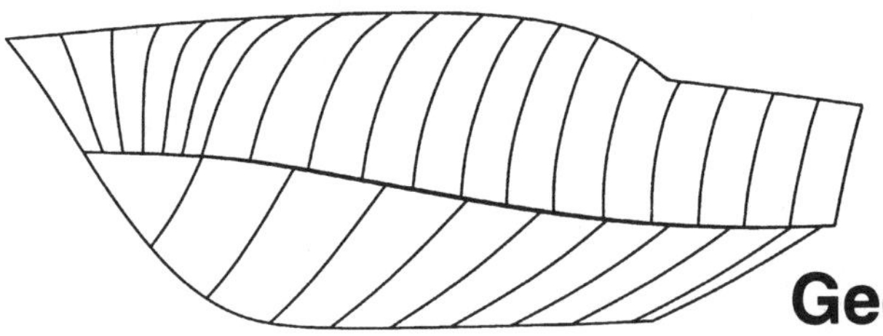

Shape Quality in Geometric Modeling and Computer-Aided Design

Edited by
Nickolas S. Sapidis
National Technical University of Athens

siam.

Philadelphia

Society for Industrial and Applied Mathematics

Library of Congress Cataloging-in-Publication Data

Designing fair curves and surfaces : shape quality in geometric
modeling and computer-aided design / edited by Nickloas S. Sapidis.
p. cm. — (Geometric design publications)
Includes bibliographical references and index.
ISBN 0-89871-332-3
1. Curves, Algebraic—Data processing. 2. Surfaces—Data
processing. 3. Computer-aided design. I. Sapidis, Nickolas S.
II. Series.
QA567.D47 1994
745.4'01'516352—dc20 94-26850

Cover art reprinted with permission from K. G. Pigounakis and P. D. Kaklis, created at the Ship-Design Laboratory of the National Technical University of Athens, Greece.

Sponsored by SIAM Activity Group on Geometric Design.

To Robert Barnhill
and
the Computer-Aided Geometric Design Research Group

List of Contributors

J.A. Ayers, Mathematics Department, General Motors Research Laboratory, Warren, MI 48090-9055.

Klaus-Peter Beier, Department of Naval Architecture and Marine Engineering, University of Michigan, 2600 Draper, NA & ME Bldg., Ann Arbor, MI 48109-2145.

Malcolm I. G. Bloor, Department of Applied Mathematical Studies, The University of Leeds, Leeds LS2 9JT, United Kingdom.

H.G. Burchard, Department of Mathematics, Oklahoma State University, Stillwater, OK 74078-0613.

Yifan Chen, Ford Motor Company, P.O. Box 2053/MD3135, Room 3135, SRL, Dearborn, MI 48121.

Matthias Eck, Department of Computer Science and Engineering, University of Washington, 423 Sieg Hall, FR-35, Seattle, WA 98195.

Mark Feldman, CAMAX Corporation, 7851 Metro Parkway, Minneapolis, MN 55425.

W.H. Frey, Mathematics Department, General Motors Research Laboratory, Warren, MI 48090-9055.

Tim Gallagher, 48 Nottinghill Road, Brighton, MA 02135.

Alexandros I. Ginnis, Department of Naval Architecture and Marine Engineering, National Technical University of Athens, Heroon Polytechneiou 9, Zografou 157 73, Athens, Greece.

Rainer Jaspert, Department of Mathematics, University of Science and Technology, AG 3, Schlossgartenstr. 7 D-64289, Darmstadt, Germany.

Alan K. Jones, Geometry and Optimization, Orgn. G-6413, M/S 7L-21, Boeing Computer Services, Seattle, WA 98124-0346.

Panagiotis D. Kaklis, Department of Naval Architecture and Marine Engineering, National Technical University of Athens, Heroon Polytechneiou 9, Zografou 157 73, Athens, Greece.

Henry P. Moreton, MS 6L-005, Silicon Graphics, 2011 N. Shoreline Blvd., Mountain View, CA 94039-7311.

Jörg Peters, Department of Computer Science, Purdue University, West Lafayette, IN 47907-1398.

Bruce Piper, Department of Mathematical Sciences, Rensselaer Polytechnic Institute, Troy, NY 12180.

Thomas Rando, 11 Bush Hill Drive, Niantic, CT 06357

Alyn Rockwood, Department of Computer Science and Engineering, Arizona State University, Tempe, AZ 85287.

John Roulier, Computer Science and Engineering Department, University of Connecticut, U155, Storrs, CT 06269-0001.

Nickolas S. Sapidis, Department of Naval Architecture and Marine Engineering, National Technical University of Athens, Heroon Polytechneiou 9, Zografou 157 73, Athens, Greece.

Carlo H. Séquin, Computer Science Division, Department of Electrical Engineering and Computer Science, University of California, Berkeley, CA 94720.

Michael J. Wilson, Department of Applied Mathematical Studies, The University of Leeds, Leeds LS2, United Kingdom.

Yan Zhao, Santa Teresa Laboratory, IBM Corporation, L53/F423, San Jose, CA 95141.

Contents

Preface

The present volume is a collection of papers focusing on the aesthetic aspects of geometric modeling, i.e., the problem of "fair" or "visually pleasing" curve/surface construction, which is of vital importance in many areas of work and especially in industrial design and styling. Current research deals with the issues of (i) mathematically defining "fairness" or "shape quality," (ii) developing new curve and surface schemes that guarantee fairness, and (iii) assisting a user in identifying shape aberrations in a surface model and removing them without destroying the principal shape characteristics of the model. The papers included in this book address the above issues for the cases of point sets, curves, and surfaces, and are highlighted below. Although the papers vary in terms of the problems considered and the solutions proposed, there is a common theme in quite a few of them. This common theme is the "principle of simplest shape" — an idea that is generally applied in the fine arts — which implies that "fair" shapes are always free of unessential features and simple in design (structure).

Part 1. The paper by Burchard, Ayers, Frey, and Sapidis deals with the problem of defining visual pleasantness for parametric curves, and proposes new definitions related to the "principle of simplest shape." Jones presents procedures, based on constraint optimization, that allow a user to prescribe various features of the curvature of a curve. The last two papers, one by Eck and Jaspert and one by Feldman, focus on fairing point sets, a problem that has received little attention from the research community, although it is very important in industrial applications (e.g., for processing digitized points or trim curves of surfaces). Eck and Jaspert use concepts from "difference geometry" to study the shape of a point set, while Feldman's algorithm is based on the mechanical model of a tight string.

Part 2. Roulier and Rando discuss existing and new fairness metrics for curves and surfaces, evaluate their effectiveness, and offer implementation strategies. The paper by Moreton and Séquin focuses on curvature variation as a measure of fairness and employs it to develop a new surface modeling technique. In the last paper in Part 2, Gallagher and Piper build on the "principle of simplest shape" and present an algorithm that constructs a composite surface interpolating discrete data by assembling spherical and cylindrical patches.

Part 3. Beier and Chen present a simplified model for calculating "reflection lines" on a surface and demonstrate its usefulness for fully interactive surface analysis and modification. The paper by Bloor and Wilson demonstrates that the partial differential equation method produces fair surfaces, and proposes techniques for interactive design of surfaces. Part 3 is concluded by the work of Ginnis, Kaklis, and Sapidis introducing a new family of polynomial splines that allow for direct manipulation of the curvature of a parametric interpolatory curve.

Part 4. The two papers included here reconsider some well-known design problems focusing on the fairness of the constructed surfaces. More specifically, Peters uses biquadratic and bicubic splines to develop a new technique for C^1 surface interpolation, while Zhao and Rockwood propose a convolution approach to produce "fair" solutions to the problems of N-sided patch design and vertex blending.

This volume evolved, in part, from presentations given at the Second SIAM Conference on Geometric Design held in Tempe, Arizona, in November 1991. Also, certain experts were invited to contribute papers. A total of twenty papers were submitted during the spring and summer of 1992, of which twelve were selected for publication on the basis of a peer-review process. Each paper was refereed by at least two reviewers. I would like to thank Robert E. Barnhill, the Organizing Committee, and the SIAM staff for their efforts in organizing the

Second SIAM/GD Conference, and for their help in initiating this book. I especially want to thank Gerald Farin for his continuous advice and encouragement in all stages of this project, and William Frey, Hans Hagen, and Ramon Sarraga for their suggestions and ideas. Also, I thank Susan Ciambrano for her cooperation on the paperwork and preparation of the manuscript. Finally, I would like to express my appreciation to the reviewers for their diligence; their names are listed below. This book was prepared while I was working in the Computer Science Department of General Motors Research Laboratories and it would not have been possible without the encouragement and support of Drs. Robert Tilove and Paul Besl, and the Head of the Department, Dr. George Dodd. I also thank the GM Design Staff, in particular Jeff Stevens and Tom Sanderson, for sponsoring my work during the last two years.

Nickolas S. Sapidis
Warren, MI

Referees

Antony DeRose	Gerald Farin	David Field	William Frey
Hans Hagen	Alan Jones	Panagiotis Kaklis	Michael Lounsbery
Weston Meyer	Henry Moreton	Jörg Peters	Bruce Piper
Steven Pruess	Alyn Rockwood	John Roulier	Ramon Sarraga
Carlo Séquin	Joe Warren	Andrew Worsey	

PART

1

Fairing Point Sets and Curves

CHAPTER

1

Approximation with Aesthetic Constraints

H.G. Burchard, J.A. Ayers, W.H. Frey, and N.S. Sapidis

1.1. Introduction

In this chapter, we address the problem of fitting a "fair" curve through or near a set of given points. "Fairness" in this case means that the curve must not only be "smooth" in some mathematical sense, but also that it must be pleasing to the eye. In other words, there are *aesthetic* considerations that constrain the curve-fitting operation. This problem arises in many areas of industrial design, whenever the appearance of a product is important for potential buyers. Prominent among such products are automobiles, and since our experience is primarily in the auto industry, we address the subject from this point of view. Most of the approach presented here was developed and implemented by two of us (Burchard and Ayers) at General Motors in the mid-1960s. But the ideas described should be useful to workers in all areas of computer-aided aesthetic geometric design.

In the traditional manual system for designing the exterior surfaces of automobiles, the "clay model" is the master source of information about the design [16]. It is accompanied by the so-called blackboard drawing, a full-size drawing on paper of the three standard views of a car, which contains various plane projections of the design lines and section lines of the present and past versions of the clay model. Information from the clay model is transferred to the blackboard drawing by measuring points on the model and plotting them on the drawing. The "boardman" then draws a smooth curve through or near these points. To complete the cycle, information from the blackboard drawing may be used to reshape the clay model by means of templates cut to the shape of lines on the drawing.

In this chapter we shall examine the role of the boardman more closely, and explore how mathematical curve fitting or curve design would have to work if it is to create mathematical curves as satisfactory to the designer as the ones produced manually by the boardman.

1.1.1. The curve-fitting problem. The work of the boardman is similar to that of other draftsmen who are faced with essentially the same problem at other stages of the design process [21, pp. 211–213]. Given a list of points P_i in an (x, y)-plane

$$P_i = (x_i, y_i), \qquad i = 1, \ldots, n,$$

he needs to construct a smooth curve passing through or near these points that captures the "design intent." These same points can also be used as input data for a curve-fitting computer program. But the boardman is actually given much more information than just the points P_i, and this information must also be made available for the computer program. How can this be done?

The boardman gets this additional information by studying the clay model, asking the designer about his intentions, and above all by using his own aesthetic judgement to tell him whether his line is "good looking" and consistent with other design lines on the car. Also based on this information is the boardman's decision on the deviations of the data points P_i. He may first attempt to fit the data within a standard tolerance, but he may revise this tolerance if one of the points turns out to be "wild" (i.e., has a much larger deviation)—which ordinarily would be due to either an error in measuring or plotting, or a depression in the clay.

But the most important decisions of the boardman as he strives for the best looking line occur when he detects "ogees" (S-shaped segments), "flat spots," "buckles," or "bumps" as he judges the overall shape of the curve. He may use expressions like "accelerating" or "decelerating," or determine whether a "bend," "break," or "corner" is "sharp" or "rounded." He may discard a line because it looks "bulgy" or "knobby," etc. To do his job properly, he must be able to decide, or know from some other source, if and where ogees, maxima, and minima of the curvature, corners, and flat segments are supposed to occur. This sensitivity to the visual properties and features of curves is standard practice among artists, as can be seen, for example, in a book by Nelms [21, pp. 31–34].

1.1.2. The principle of simplest shape. Generally speaking, the boardman will try to draw a curve in which ogees and other key features occur the minimum number of times within the design intent. We may say that he follows the *principle of simplest shape*, an idea that is generally applied in the fine arts. Beautiful objects, it is said, are uncluttered, i.e., free of unessential features and simple in design [3], [15]. To illustrate the principle of simplest shape, Fig. 1.1 shows a curve with *one* ogee and a curve with *many* ogees. The second curve is obviously *unsmooth* and would be undesirable on a car exterior. To lay the groundwork for exploring these ideas with more precision, in the next section we establish some notation and review some fundamental ideas from elementary differential geometry, emphasizing the concept of an intrinsic

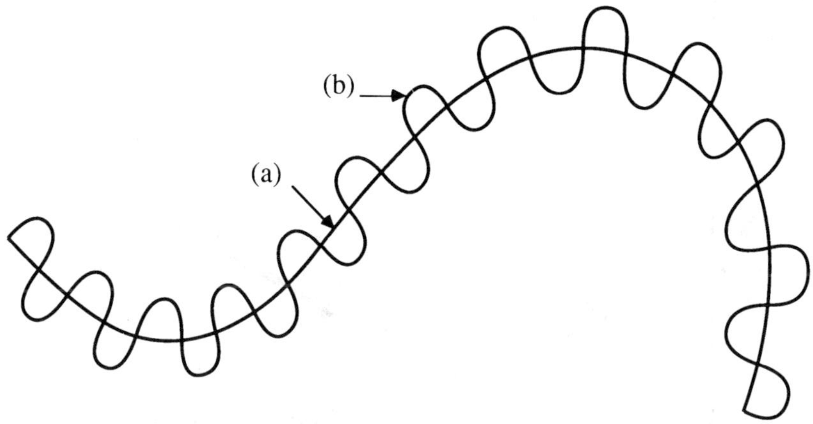

FIG. 1.1. (a) *A curve with one ogee.* (b) *A curve with many ogees.*

equation of a curve or a class of curves, i.e., a condition on the curvature. As we shall see, the *visual* features which concern the boardman are easily translated into certain corresponding *mathematical* properties of the intrinsic equation.

1.2. Differential Geometry of Plane Curves

In this section we briefly review certain fundamental concepts of the elementary differential geometry of plane curves and establish appropriate notation. The restriction to plane curves is in line with the boardman's work dealing with plane projections of design lines. The following considerations may be applied simultaneously to the several standard projections.

1.2.1. Parametric C^1-curves. We can represent plane curves in several different ways. However *parametric* representations have overwhelming advantages for investigating the aesthetic properties of curves. Our notation and terminology follow closely those of standard texts in differential geometry [19], [25]. We shall assume that any curve of interest is a *curve segment* lying in a Euclidean plane with a Cartesian (x, y)-coordinate system, as shown in Fig. 1.2. The parameter t is an independent real variable with domain in a bounded and closed interval $[a, b]$, and the curve is defined as a continuous vector function $\mathbf{c}(t) = (x(t), y(t))^T$. We assume throughout that the *tangent vector* $\mathbf{c}'(t)$ exists and is a continuous function of t. In other words, the curve is always of class C^1. Included among the C^1-curves are the C^k-curves $(k \geq 1)$, for which all derivatives for $i = 1, \ldots, k$ exist and are continuous.

1.2.2. Arc length. We can think of each point $\mathbf{c}(t)$ as the position at time t of a particle moving along the curve. *Arc length* $s = s(t)$, measured along the curve from the initial point $\mathbf{c}(a)$ of the curve to the point $\mathbf{c}(t)$, is the distance travelled by the particle since time $t = a$. With $\mathbf{c}'(t) = (x'(t), y'(t))^T$ we may

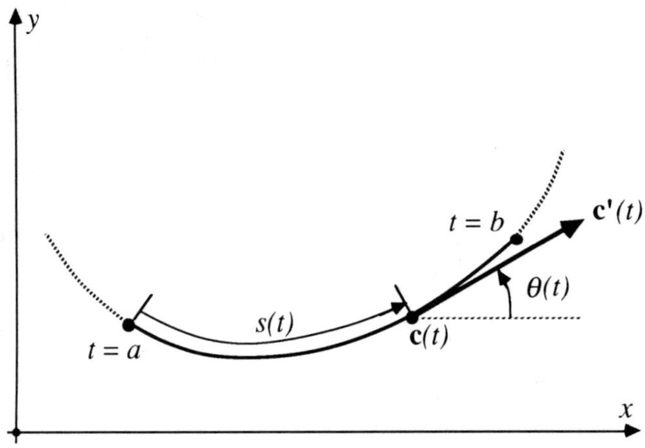

FIG. 1.2. *Defining a parametric C^1-curve.*

compute arc length as follows:

$$(1.1) \qquad s(t) - s_a = \int_a^t \frac{ds}{dt}\, dt = \int_a^t |\mathbf{c}'(t)|\, dt = \int_a^t \sqrt{[x'(t)]^2 + [y'(t)]^2}\, dt \ ,$$

where s_a is an arbitrary constant assigned as the arc length at $\mathbf{c}(a)$.

We shall assume that $s(t)$ increases as the particle moves along the curve in the direction of increasing t. In other words, in all the curves we shall examine, the *parametric speed* $ds/dt = |\mathbf{c}'(t)|$ will be positive. With this assumption, a parametric curve carries an *orientation* by which we can distinguish a *positive* direction of motion along the curve, in which t and s both increase, and a *negative direction*, in which t and s decrease.

It is a well-known result of differential geometry that if the constant s_a is given and the orientation of the curve is specified, then the *arc length parametrization* $\mathbf{c}(s)$ is the *unique* representation of the curve for which the parametric speed is constant and equal to unity [19]. The arc length at any point on the curve is independent of the (x, y)-coordinate system used and of the particular parameter t that is used to define the curve. We therefore say that the arc length parametrization is an *intrinsic* property of the curve and the arc length s is an intrinsic quantity. This is reflected in the fact that the arc length is a *visual* feature of a curve, one that the eye can see.

1.2.3. Tangent angle and curvature. The *tangent angle* $\theta(t)$ to a curve $\mathbf{c}(t)$ is the angle from the positive x-axis to the tangent vector $\mathbf{c}'(t)$, as shown in Fig. 1.2. Like the tangent vector, $\theta(t)$ is assumed to be a continuous function. The tangent angle is related to the components of $\mathbf{c}(t)$ by the differential equations

$$(1.2) \qquad \frac{dx(t)}{dt} = \cos\theta(t)\frac{ds}{dt} \qquad \text{and} \qquad \frac{dy(t)}{dt} = \sin\theta(t)\frac{ds}{dt} \ .$$

The *curvature* κ is defined to be the amount of bending of the curve at any point on the curve where $\theta(t)$ is differentiable. More precisely, κ is the derivative of the tangent angle θ with respect to arc length:

$$\kappa(s) = \frac{d\theta}{ds} = \frac{d\theta}{dt} \Big/ \frac{ds}{dt} \; . \tag{1.3}$$

Like the arc length, the curvature at any point of a curve is an intrinsic quantity, independent of the parametrization and the position or orientation in the Cartesian coordinate system. Similarly, the curvature and its reciprocal R, the *radius of curvature*, defined by

$$R(s) = \frac{1}{\kappa(s)} \; , \qquad \kappa \neq 0 \; , \tag{1.4}$$

are visual features of a curve.

Notice that definition (1.3) implies that *curvature is a signed quantity, positive* if the curve is turning to the left (increasing θ), *negative* if it is turning to the right (decreasing θ), when traversed in the positive direction. Equation (1.3) also reveals the intimate connection between the tangent angle θ and the curvature. In particular, if the function $\kappa(s)$ is given, then we can integrate (1.3) to obtain

$$\theta(s) - \theta(s_a) = \int_{s_a}^{s} \kappa(\sigma) \, d\sigma \; . \tag{1.5}$$

This shows that the net change in the tangent angle from $s = s_a$ to a generic point s on the curve depends only on the intrinsic quantities κ and s, and therefore $\theta(s) - \theta(s_a)$ is also an intrinsic quantity. Like the arc length and the curvature, it is a visual quantity depending only upon the shape of the curve.

1.2.4. Intrinsic equation. Another important result from differential geometry is that if a given function $\kappa(s)$ is piecewise continuous (has at most a finite number of jump discontinuities), then equations (1.2) and (1.3) can always be integrated to obtain the curve $\mathbf{c}(t)$ (up to a translation and a rotation) [25]. The equation

$$\kappa = \kappa(s) \tag{1.6}$$

is known as the *intrinsic equation* of $\mathbf{c}(t)$. This representation is of critical importance to our analysis since it completely defines the *shape* of a curve independent of its orientation or position within the (x, y)-coordinate system. Plots of κ versus s are now a widely used tool for designing and modifying plane curves [11], [17], [20].

We can also talk about the *intrinsic equation of a class of curves.* To illustrate, consider the equation

$$\frac{d^2\kappa(s)}{ds^2} = 0 \ . \tag{1.7}$$

This equation is not of the form (1.6), but integrating it twice gives

$$\kappa(s) = cs + d \ , \tag{1.8}$$

where c and d are arbitrary constants. For fixed c and d, (1.8) is the intrinsic equation of a curve, namely an Euler spiral. Thus the differential equation (1.7) characterizes *all* Euler spirals intrinsically, and is therefore the intrinsic equation of all Euler spirals. In the following sections we shall explore conditions still more general in form, for example, the condition

$$\frac{d^2\kappa(s)}{ds^2} \geq 0 \ . \tag{1.9}$$

While this is an inequality rather than an equation, it is an *intrinsic condition* characterizing a particular class of curves (which includes all Euler spirals).

1.3. Aesthetics and Monotonicity of Curvature

1.3.1. Visual properties and the intrinsic equation. There is an essential fact inherent in our discussion that must be brought out: all of the features about which the boardman must make decisions are mathematical properties of the intrinsic equation (1.6). This is clear a priori, since a plane curve is completely specified by its intrinsic equation.

Moreover, translating the *visual* properties of a curve into *mathematical* properties of its intrinsic equation is relatively *straightforward*, since κ and s are of a visual nature themselves. In fact, those visual properties of a curve that affect the aesthetic qualities of concern to the boardman and designers correspond to exact *metric* properties, and these in turn translate into specific mathematical equivalents: an *ogee* is a point where κ changes sign. In between ogees the curve is either *convex* or *concave*, or κ has *constant sign*, i.e., $\kappa \geq 0$ or $\kappa \leq 0$. A *flat spot* is a minimum of $|\kappa|$. A *buckle*, *bump*, *corner*, or *break* is a maximum of $|\kappa|$. In between maxima and minima of $|\kappa|$ the curve is either *accelerating* or *decelerating*, which translates into $|\kappa|$ being either *monotone increasing* or *decreasing*.

Fig. 1.3 illustrates these last concepts. The curve in Fig. 1.3(a) is "purely accelerating," i.e., the sign of κ is constant and $|\kappa|$ increases (from left to right) over the whole curve. By contrast, the curve in Fig. 1.3(b), still with constant sign of κ, has several maxima and minima of $|\kappa|$, giving the curve an unsmooth

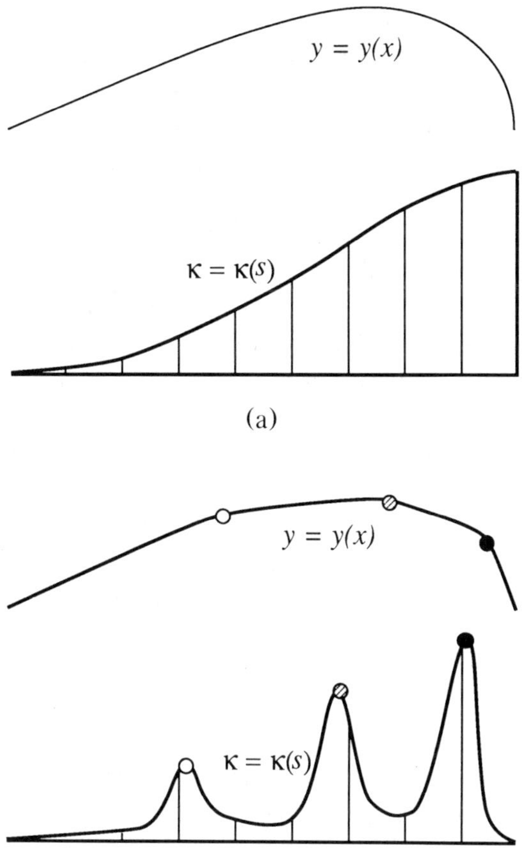

FIG. 1.3. (a) *A curve with monotone curvature.* (b) *The curvature of this curve has several local maxima and minima.*

appearance compared with the one above. This again illustrates the principle of simplest shape; cf. §1.1.2.

We can summarize these remarks as follows: *constraints or aesthetic requirements for a plane curve can be expressed as fairly simple mathematical conditions on the intrinsic equation, i.e., as conditions on the function* $\kappa(s)$.

1.3.2. Monotonicity conditions on the curvature. To develop a computer program that can produce curves as good as a draftsman can draw, we need a "complete set" of aesthetic constraints (and their translation into curvature conditions) used by draftsmen in judging the quality of curves. *Some* of these conditions emerge if we apply the principle of simplest shape to the number of ogees, and of minima and maxima of κ. We list these conditions under (a)–(d), below. Their significance is illustrated in Figs. 1.1 and 1.3.

(a) Sign changes and maxima and minima of the curvature possess obvious visual equivalents. By the principle of simplest shape, the numbers of occurrences of each of these features should be kept as small as possible subject to and consistent with the design intent. This implies that their *numbers* and *locations* should be *determined explicitly*. Some ways to do this are described in (b)–(d).

(b) If data points P_i $(i = 1, \ldots, n)$ for a plane curve are given (points measured from a model or a drawing), then the numbers and locations of sign changes of κ, as well as of the maxima and minima of κ, can be *specified as additional input information* for a computer program intended to compute a smooth, aesthetically pleasing curve passing through the data points. That this can actually be done by a draftsman with minimal mathematical training should be clear from the Introduction, since the same or at least analogous decisions would be made by him on the drawing board.

(c) Using the principle of simplest shape, we can try to write a computer program to replace the draftsman by *computing* numbers and locations of sign changes and maxima and minima of κ. More specifically, given a tolerance ϵ, a program may be written to find the minimal numbers of such features necessary to approximate the data within tolerance ϵ. Simultaneously, the computer program can try to optimize their locations, i.e., it can try to come as close to the data points as possible, subject to the limitation on the number of sign changes, maxima and minima of κ. The justification of this approach is that the draftsman does something quite similar, as described in the Introduction to this chapter.

(d) Approaches (b) and (c) both have their difficulties. For example, a designer sometimes may not be able to specify locations of sign changes of κ with sufficient accuracy. This problem could be managed by the following scheme. First the designer specifies numbers and locations of sign changes, maxima and minima of κ *approximately*—according to (b). This information can then be checked and locations *optimized*—according to (c). The optimization of locations of sign changes, maxima and minima of κ, may be expected to appreciably decrease the deviations of data points from a computed curve. In other situations these locations may not be variable due to design constraints.

Any restriction on the number of maxima and minima of κ is equivalent to a restriction on the number of *separate curve segments* where the curvature κ is monotone (increasing or decreasing). Therefore, we refer to the conditions (a)–(d) on maxima and minima of κ as *monotonicity conditions* on the curvature. Considering curves of class C^3, we could say that monotonicity conditions are sign conditions on $d\kappa/ds$, and we shall do this for convenience, even though our curves are only C^2 or even piecewise C^2. In such cases one can define monotonicity as constancy of the sign of the difference quotient $\Delta\kappa/\Delta s = (\kappa(s_2) - \kappa(s_1))/(s_2 - s_1)$. Accordingly, we can say that conditions (a)–(d) amount to *specifying the signs of κ and $d\kappa/ds$ (or $\Delta\kappa/\Delta s$) at each point of the curve.*

Monotonicity of the curvature is important for a variety of curve design applications [12], [22], [23]. Many curves (most spirals, for example) inherently have monotone curvature, and conditions for curvature monotonicity of some nonspiral curves have been established [14], [24]. However, although monotonicity conditions form a partial set of aesthetic constraints for the fitting of plane curves, they are not a complete set. Additional conditions—on the convexity of κ—are also needed. We discuss these beginning in the next section.

1.4. Aesthetics and Convexity of Curvature

1.4.1. The styling radius. A common practice in engineering is to round off the sharp corners of an object using a circular arc, as shown in Fig. 1.4(a). The resulting curve, sometimes called a *simple radius*, consists of two straight line segments connected by the circular arc. The curvature distribution of this curve is displayed in Fig. 1.4(b). In this case $\kappa(s)$ has two jump discontinuities. Observe, however, that $\kappa(s)$ has no sign changes (the curve has no ogee) and the curve can be divided (at the midpoint of the arc) into two symmetric segments with monotone curvature. This indicates that the curve should be considered rather smooth in light of our earlier remarks. Nonetheless, artists and designers feel that this curve is not *attractive*. To remedy this, when they need to draw a smooth corner linking two relatively flat curves, they use a device known as a *styling radius*.

A styling radius is a template made by first cutting a piece of material such as plexiglass to a simple radius and then "shaving off" a bit where the circular arc joins the straight lines, as shown in Fig. 1.5. This makes the transition between the flat ends and the strongly bent corner more gradual. Artists might say that the simple radius looks "knobby" or "bulgy" as opposed to a styling radius. The styling radius shown in Fig. 1.5 was constructed from two symmetric cubic parametric curves that join at the point of maximum curvature. As with the original curve, the curvature distribution consists of two symmetric parts, one of which is monotone decreasing (shown in Fig. 1.5), and the other, monotone increasing.

In view of this example, where we have a curve with minimally piecewise monotone curvature which is not deemed attractive, we need to discover additional conditions on κ in order to obtain a "complete set" of aesthetic constraints for mathematical curve-fairing.

1.4.2. The need for a smooth curvature distribution. It seems plausible that the attractiveness of the styling radius (Fig. 1.5) versus the original curve (Fig. 1.4(a)) is related to its improved curvature distribution, especially the elimination of the jumps in $\kappa(s)$. For an heuristic argument, one might imagine that the eye, in scanning the curve and sensing the curvature, receives a stimulus related to the size of κ. If κ changes suddenly, so will the stimulus—and this may give rise to an unpleasant sensation.

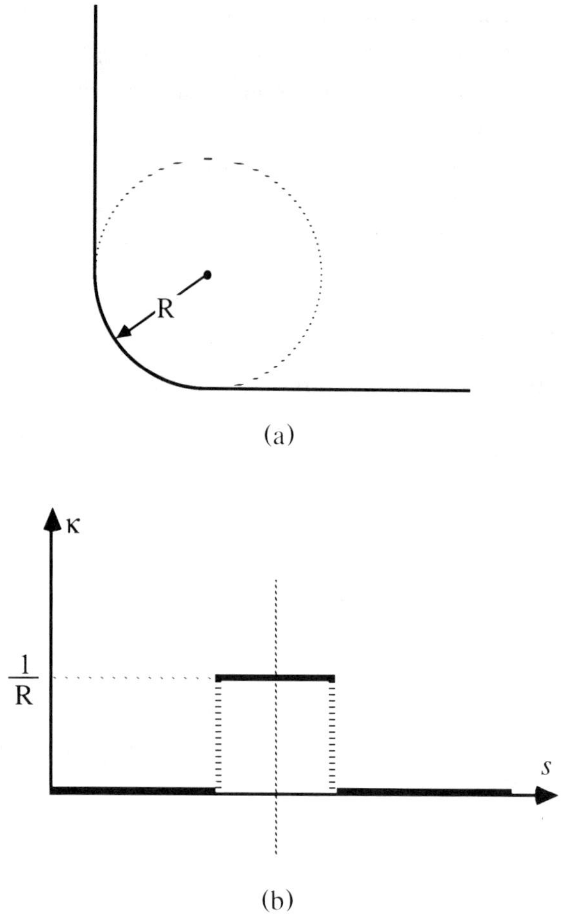

FIG. 1.4. *Simple radius* (a) *constructed by rounding a corner using a circular arc, and its curvature distribution* (b).

A similar situation occurs when a vehicle is driven from a straight road into a curve of constant radius. Passengers experience a sudden (jump) change in the side force (centripetal acceleration). To reduce this unpleasant sensation on high speed roads (and to improve safety), engineers routinely design transition curves (usually Euler spirals) between straight and circular road sections [1]. These replace the discontinuous curvature plot of the plane view of the roadway by one that is continuous (in this case, piecewise linear).

For our purposes, we need to find mathematical conditions which eliminate jumps in the curvature and constrain the shape of the curve to be more attractive, more like the styling radius. As we explain in the sections that follow, our solution to this problem involves the use of *convexity conditions* (and *generalized* convexity conditions) on the curvature. This was suggested to us by an empirical observation: In all of the curvature plots we computed for automobile design lines, a shape like that shown in Fig. 1.6 recurred: *the point or points*

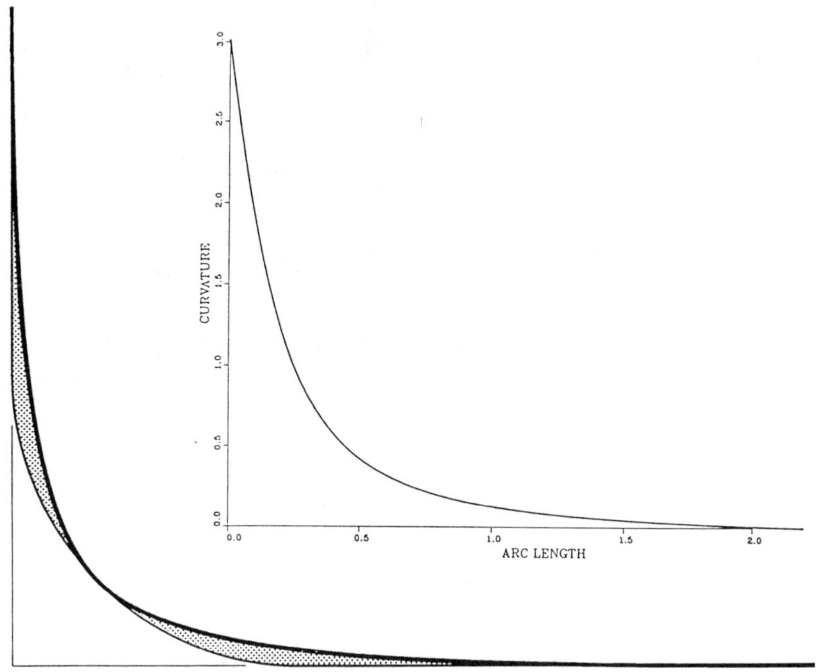

FIG. 1.5. *A styling radius constructed from a simple radius, and the curvature distribution of half of the styling radius.*

of maximum absolute value of the curvature divide(s) the plot of $|\kappa|$ *versus* s *into distinct sections, each of which is convex.*[1] Consequently, at any maximum of $|\kappa|$ the slope $d\kappa/ds$ is discontinuous. This slope discontinuity in $\kappa(s)$ is an advantage in that it unambiguously identifies a unique design feature and its location. As an explanation for these empirical observations we offer that design lines have characteristic shapes. This is reflected in the fact that their intrinsic equation $\kappa = \kappa(s)$ has concentrated high curvature magnitudes on short arcs. Broad curvature maxima with concave $|\kappa(s)|$ simply do not occur.

1.5. Conditions for Convexity of Curvature

We shall now assume that our curve of interest has been partitioned into distinct segments in such a manner that for each the curvature κ and its first derivative $d\kappa/ds$ both have constant signs. In other words, the points where sign changes of curvature (ogees), or maxima and minima of curvature occur have been identified, and we can now focus our attention on any *one* of these segments with monotone curvature of constant sign.

1.5.1. Curvature convexity made precise. For convexity considerations, we are only interested in the properties of the magnitude or absolute value of

[1] When the graph of a function is convex downwards (concave upwards) we shall follow the usual convention and call it simply *convex* [7, p. 246], concave in the opposite case.

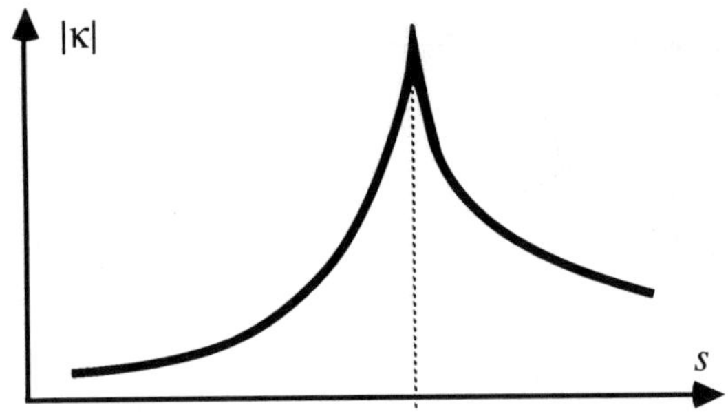

FIG. 1.6. *The point of maximum absolute value of the curvature divides the plot of $|\kappa(s)|$ vs. s into convex sections.*

the curvature $|\kappa|$, as was explained above. Consider a curve segment of class C^4. When referring to *convexity of the curvature* $|\kappa|$ we mean the condition

$$\frac{d^2|\kappa|}{ds^2} \geq 0 \,. \tag{1.10}$$

For curves that are merely piecewise C^2, we define convexity of $|\kappa|$ as monotonicity of the difference quotient

$$\frac{|\kappa(s_3)| - |\kappa(s_2)|}{s_3 - s_2} \geq \frac{|\kappa(s_2)| - |\kappa(s_1)|}{s_2 - s_1} \quad \text{if} \quad s_1 < s_2 < s_3 \,. \tag{1.11}$$

For convenience, we shall refer to convexity of $|\kappa|$ simply as a sign condition on $d^2|\kappa|/ds^2$ as in (1.10). All the statements to be made can be translated to the more general case (1.11).

1.5.2. Intrinsic parameters. In order to study conditions of convexity of curvature more closely, it turns out that we need to understand how these conditions relate to various allowable parametrizations. To be geometrically meaningful, all such conditions must be expressed in terms of the intrinsic equation of a curve. However, only those parameters t that are closely related to the geometry of the curve, i.e., to the intrinsic equation, reflect intrinsic properties of a curve. Such a close relationship is present, for example, if the parametric speed $|ds/dt|$ is a positive continuous function $g(|\kappa|)$ of the curvature κ,

$$\left|\frac{ds}{dt}\right| = g(|\kappa|) > 0 \,. \tag{1.12}$$

In this case we shall say that the allowable parameter t is an *intrinsic parameter*.

Suppose now that t is some intrinsic parameter. Compared with the simple visual interpretations of sign(κ) and sign($d\kappa/ds$) discussed in §1.3, the condition of convexity of the curvature is more difficult to explain and, admittedly, not as well understood. This appears to be related to certain mathematical facts, as follows: If $\kappa = \kappa(t)$ has constant sign on a segment of a curve, then it does so also for any other allowable parameter. Likewise if $\kappa = \kappa(t)$ is a monotone function of t, then it is also monotone as a function of any other allowable parameter. However, $|\kappa|$ *may well be convex as a function of the intrinsic parameter* t *but not for some other intrinsic parameter*. To see this, consider a curve with the intrinsic equation

$$\kappa = s^2 , \qquad 1 \le s \le 2 .$$

Then $|\kappa|$ is a convex function of s. But if we introduce the allowable parameter $t = \kappa^2$, we get

$$|\kappa| = \sqrt{t} , \qquad 1 \le t \le 16 ,$$

and $\sqrt{t}$ is not a convex function of t.

This example shows that we can expect to obtain convexity conditions of different geometric significance (and hence different aesthetic meaning) by considering convexity conditions of the form

$$\frac{d^2}{dt^2}|\kappa| \ge 0 \tag{1.13}$$

for various intrinsic parameters t.

1.5.3. Intrinsic convexity conditions. If t is an intrinsic parameter, we consider *intrinsic convexity conditions* of the form (1.13), and show that such conditions are equivalent to a condition of the form

$$\frac{d^2}{ds^2} f(|\kappa|) \ge 0 , \tag{1.14}$$

where $f(|\kappa|)$ is some twice differentiable function of $|\kappa|$. Both conditions (1.13) and (1.14) may be called *intrinsic convexity conditions* on $|\kappa|$. Carrying out the indicated differentiations and using (1.12) leads to the inequalities

(1.15)
$$\frac{d^2|\kappa|}{dt^2} = \frac{d}{ds}\Big(\frac{d|\kappa|}{ds}\frac{ds}{dt}\Big)\frac{ds}{dt} = \frac{d}{ds}\Big(\frac{d|\kappa|}{ds}\Big|\frac{ds}{dt}\Big|\Big)\Big|\frac{ds}{dt}\Big| = \frac{d}{ds}\Big(\frac{d|\kappa|}{ds} g(|\kappa|)\Big)\cdot g(|\kappa|) \ge 0$$

and

$$\frac{d^2}{ds^2} f(|\kappa|) = \frac{d}{ds}\Big(\frac{d|\kappa|}{ds} f'(|\kappa|)\Big) \ge 0 . \tag{1.16}$$

Evidently conditions (1.15) and (1.16) are *equivalent*, provided it is known that

$$f'(|\kappa|) = g(|\kappa|) = \Big|\frac{ds}{dt}\Big| > 0 . \tag{1.17}$$

Hence, we see that the condition of *convexity of* $|\kappa|$ *with respect to an intrinsic parameter* t *is equivalent to the condition of convexity of a function* $f(|\kappa|)$ *with respect to arc length* s, the definition of t and its relation to $f(|\kappa|)$ being contained in (1.17). Henceforth we shall assume that $f(|\kappa|)$ and $g(|\kappa|)$ satisfy (1.17), and therefore that (1.15) and (1.16) are equivalent.

1.5.4. Example. As we might expect from (1.5), the tangent angle θ turns out to be an intrinsic parameter. To see this we write, using (1.3),

$$(\log|\kappa|)' = \frac{1}{|\kappa|} = \left|\frac{ds}{d\theta}\right| > 0 \qquad (\kappa \neq 0)\ . \tag{1.18}$$

Clearly the requirement (1.12) for an intrinsic parameter is satisfied if $g(|\kappa|) = 1/|\kappa|$. Moreover, (1.18) is of the form (1.17) with $f(|\kappa|) = \log|\kappa|$. Applying the above results, we then find that the following two convexity conditions are equivalent:

$$\frac{d^2|\kappa|}{d\theta^2} \geq 0 \tag{1.19}$$

and

$$\frac{d^2}{ds^2}\log|\kappa| \geq 0\ . \tag{1.20}$$

In words, *the condition of convexity of* κ *with respect to* θ *is equivalent to the condition of logarithmic convexity of* κ *with respect to* s.

1.5.5. Another form of the convexity condition. Convexity condition (1.14) can be written in still another form, by carrying out the differentiation in (1.16) and using $f'(|\kappa|) > 0$, namely

$$\frac{d^2}{ds^2}|\kappa| \geq -\left(\frac{d|\kappa|}{ds}\right)^2 \frac{f''(|\kappa|)}{f'(|\kappa|)}\ . \tag{1.21}$$

Thus an intrinsic convexity condition on $|\kappa|$ amounts to imposing a lower bound on $(d^2|\kappa|)/ds^2$. From this it can be seen that such conditions have varying *strength* depending on the size of the lower bound.

1.5.6. Alpha convexity. We propose in this chapter that a safe and workable method of ensuring aesthetically pleasing curves may be obtained by imposing on the curves a sufficiently strong form of curvature convexity condition. Ideally a designer should have a *family* of such conditions from which to choose. In this section we define such a family, and in the remainder of the chapter we explore the effectiveness of this approach.

We choose a family of conditions, referred to as *α-convexity* conditions, for which the strength depends on a single parameter α in a simple way. In (1.14) let

$$f(|\kappa|) = f_\alpha(|\kappa|) = \begin{cases} |\kappa|^\alpha\,, & \alpha > 0; \\ \log|\kappa|, & \alpha = 0; \\ -|\kappa|^\alpha\,, & \alpha < 0. \end{cases} \tag{1.22}$$

When equality is required in (1.14) instead of the inequality we obtain the intrinsic equation $d^2 f_\alpha(|\kappa|)/ds^2 = 0$. This includes some well-known families of curves [27] which are thus seen to be *α-convex*:

- $\alpha = 1$: *Euler spirals*, also called *Cornu spirals* or *clothoids*. $f(|\kappa|) = f_1|\kappa| = |\kappa|$ is linear in s, $\kappa = as + b$.
- $\alpha = -1$: *logarithmic* or *equiangular spirals*.
- $\alpha = -2$: *involutes of circles*. Involutes of circle have interesting and possibly useful mathematical properties. The radius of curvature R is a linear function of the tangent angle θ, i.e., $R = a\theta + b$. This can be integrated exactly giving a curve with linear parameters if θ is used as the allowable intrinsic parameter.

With the function f defined by (1.22), formula (1.21) now reads

$$\frac{d^2|\kappa|}{ds^2} \geq (1-\alpha)\frac{\left(\frac{d|\kappa|}{ds}\right)^2}{|\kappa|}\,. \tag{1.23}$$

We considered the case $\alpha = 0$ in §1.5.4. Evidently the strength of (1.23) increases as α decreases and vice versa.

1.6. Alphaconvexity and Aesthetics

By imposing condition (1.23) on the curvature, jumps are indeed eliminated because of a classical result that a convex function on an open interval must be continuous. In spite of this true mathematical fact, tests have shown that the condition $(d^2|\kappa|)/ds^2 \geq 0$ still permits curves that may be unsatisfactory for automobile exteriors, and in fact subject to objections similar to those made against the curve in Fig. 1.4(a) that gave rise to the styling radius. In an effort to elucidate this situation, we conducted further experiments which seemed to indicate that replacing the condition

$$\frac{d^2|\kappa|}{ds^2} \geq 0 \tag{1.24}$$

(α-convexity with $\alpha = 1$) by the condition

$$\frac{d^2\log|\kappa|}{ds^2} \geq 0 \tag{1.25}$$

(α-convexity with $\alpha = 0$) produced perfectly acceptable curves.

In trying to explain why the condition (1.25) of logarithmic convexity of κ is sufficient and the condition (1.24) of ordinary convexity is not, we first point out that (1.25) is *stronger* than (1.24). This can be seen from (1.23), where $\alpha = 0$ gives (1.25) and $\alpha = 1$ gives (1.24). Furthermore, a curvature plot obtained when (1.24) is used in our optimizer tends to look like the one shown in Fig. 1.7(a), approximately piecewise linear.[2] For the same curve, plots of $\log|\kappa| = f_0(|\kappa|)$ and $-|R| = -(1/|\kappa|) = f_{-1}(|\kappa|)$ appear as in Figs. 1.7(b) and 1.7(c). The change in appearance is a consequence, as well as an illustration, of the fact that α-convexity becomes stronger as α decreases. The point is that even though (1.24) eliminates jumps in the curvature, the plots of functions of κ corresponding to stronger convexity conditions may approach the shape of step functions. If (large) jumps or sudden changes in κ are undesirable, as suggested by the styling radius, then it may well be the case that jumps in functions of κ such as $f_\alpha(|\kappa|)$ are likewise undesirable. This is consistent with the idea of a visual stimulus depending on κ. We do not know *how* such a stimulus might depend on κ; it might well be proportional to a function $f(|\kappa|)$ like $\log|\kappa|$ or $|R|$. If this were the case, it would again be plausible that an unpleasant sensation is connected with scanning a curve that has characteristics like those shown in Fig. 1.7.[3]

Experiments with the approach to curve smoothing outlined here have been successful. So far, α-convexity with $\alpha = 0$ was used to full satisfaction. However, with $\alpha = 1$, the convexity condition is (probably) too weak, and α-convexity with $\alpha = -1$ appears to be too strong.

1.7. Computing Circular Splines with Curvature Constraints

Here we summarize the computational details of one possible algorithm that has been implemented and that has given satisfactory results.

1.7.1. Discretization of curvature constraints. Numerical procedures for curve fitting subject to aesthetic constraints can be implemented in several ways, but in any such method it is necessary to *discretize* those inherently *continuous* differential inequalities which represent the aesthetic constraints.

[2] The piecewise linear curve in Fig. 1.7(a) is due to the principle of complementary slackness from linear and nonlinear optimization theory. Our computer program uses a customized optimizer to obtain a best fit. The details are described in §1.7. The constraints are discretized as in (1.30), including inequalities $\Delta(\Delta\kappa_i/\Delta t_i) \geq 0$. A count of the variables (including slacks) versus constraints (equations) then reveals that equality must hold in most of the preceding inequalities, except for roughly as many strict inequalities as there are data points that the faired curve is able to match exactly.

[3] The equivalence of (1.15) and (1.16) suggests considering a varying scanning speed ds/dt depending on κ. One can actually observe an adaptation of this kind, for example, by scanning Fig. 1.4. The scanning speed is slower for larger $|\kappa|$, perhaps with a relation like (1.18). In this case, the unpleasant sensation might be due to sudden changes in the scanning speed and/or to the fact that κ or $f(\kappa)$ is no longer convex as a function of t, but approaches a step function. All these explanations are mathematically equivalent.

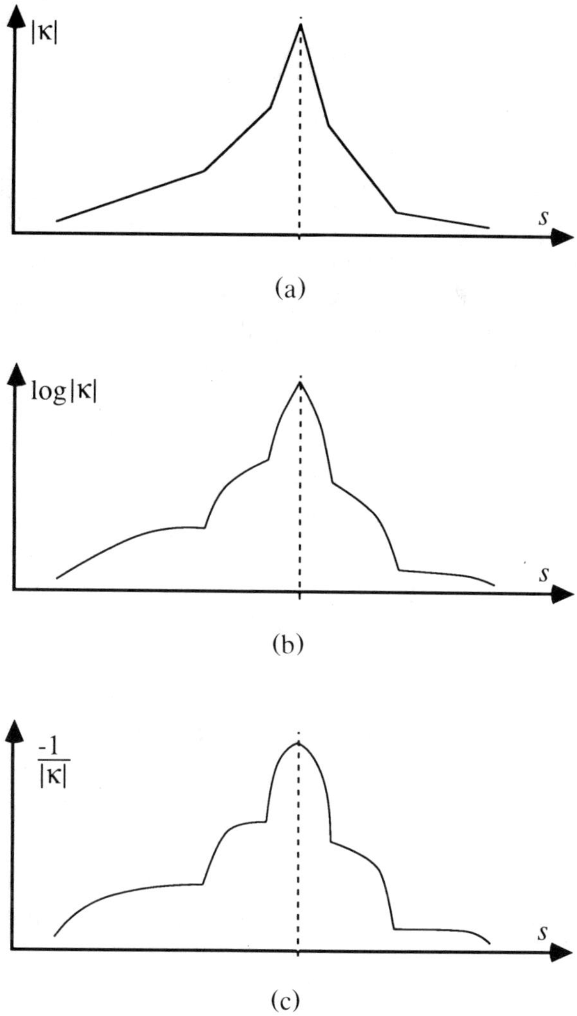

FIG. 1.7. *Although $|\kappa| = |\kappa(s)|$ is far from being a step function, this may not be true for some functions $f(|\kappa|)$, e.g., $f(|\kappa|) = -1/|\kappa|$.*

To simplify, the following discussion is restricted to the situation of a curve consisting of a single segment with constant signs of the curvature, its first and second derivatives with respect to some allowable parameter t,

$$\kappa > 0, \qquad \frac{d\kappa}{dt} \geq 0, \qquad \frac{d^2\kappa}{dt^2} \geq 0. \tag{1.26}$$

At this point we settle on a definite choice of intrinsic parameter, the tangent angle $t = \theta$. This occurs for $f(\kappa) = f_\alpha(\kappa)$, with $\alpha = 0$; cf. §1.5.4. This means that we require *logarithmic convexity* of κ with respect to arc length, a fairly strong grade of convexity condition, yet not excessive. According to our experiments this condition appears to be suitable for general purpose automotive design.

There are several options for carrying out the required discretization. One way is the approximation by some type of spline function for which it is easy to enforce the aesthetic constraints. Discretization by means of *circular splines*, i.e., C^1-curves made up of circular arcs, is attractive due to the inherent simplicity of geometry and algebra connected with circles. Of course, the circular arcs of the spline must be sufficiently short so that the jumps in curvature between arcs are too small to be noticeable (because, after all, convexity of $|\kappa|$ was introduced to eliminate jumps in the first place). This is discussed further below. A circular spline with short arcs and monotone curvature satisfying a discretized form of convexity, that we discuss next, would be indistinguishable to a working tolerance from a curve with genuinely log-convex curvature. Finally, we note that the procedure described below could also be used, with the appropriate modifications, for computing various kinds of splines other than circular splines, for instance parabolic or cubic Fowler–Wilson splines [13], subject to α-convexity for any real α.

That a circular spline with sufficiently short circular arcs can be an acceptable discretization of a curve with monotone and logarithmically convex curvature is fairly obvious. This approximation is justified because of the need to ultimately produce control points for N/C drafting or milling. Given raw data from a clay model or drawing, the N/C control points must be selected from an aesthetically acceptable curve, and they must be placed sufficiently close to each other along the curve so that straight line interpolation between adjacent points produces a polygonal approximation that is indistinguishable from the smooth curve to within a working tolerance ϵ. Of course, today some N/C equipment can move a tool along circular arcs, in addition to straight line segments. Now imagine replacing the linear segments between control points by the arcs of a circular spline. It seems plausible, and our experience confirms, that a C^1 circular spline could be made to "hug" much closer to the ideal curve than the line segments of the N/C polygon and hence that *a circular spline with knots at the N/C control points can provide an acceptable discretization* for aesthetic curvature constraints (1.26) to within a working tolerance from the data.

With raw production data the point spacing usually would not be sufficiently close. In this case extra points may be filled in. Such preliminary fill-in is allowed to be fairly crude because the final computed spline will satisfy the aesthetic curvature constraints, even if the raw filled points are not within tolerance. A familiar formula expresses the chord length L of a circle of curvature κ given the chord height ϵ, $L = \sqrt{|\kappa|^{-1}(8\epsilon - 4\epsilon^2\,|\kappa|)}$. Given the curvature of an arc of a circular spline this formula guarantees tolerance ϵ for deviation from the curve by limiting the chord length L (one neglects the ϵ^2-term because, it seems, that curvatures $\kappa \gg 1$ tend not to occur along design lines).

1.7.2. Basic facts of circular splines. Given data points (x_i, y_i), $i = 1, \ldots, n$, we abbreviate $\Delta x_i = x_{i+1} - x_i$, $\Delta y_i = y_{i+1} - y_i$, and chord lengths

$$L_i = \sqrt{(\Delta x_i)^2 + (\Delta y_i)^2}, \quad M_i = \sqrt{(\Delta x_i + \Delta x_{i-1})^2 + (\Delta y_i + \Delta y_{i-1})^2}\ .$$

Define the vectors $\mathbf{u}_i = (\Delta x_i, \Delta y_i)$, and let ω_i denote the angle between $\mathbf{u}_{i-1}$ and $\mathbf{u}_i$ with the positive sense of ω_i shown in Fig. 1.8. Consider a C^1-curve, such as a circular or other spline, that interpolates the given points. For this curve, there arise angles ϕ_i and ψ_{i-1}, respectively, between the vectors $\mathbf{u}_i$ and $\mathbf{u}_{i-1}$ and the tangent line to the curve at the point (x_i, y_i). If the positive sense of ϕ_i and that of ψ_{i-1} are as shown in Fig. 1.8, we observe that

$$\omega_i = \psi_{i-1} - \phi_i.$$

We may define *numerical curvatures* $\tilde{\kappa}_i$, $i = 2, \ldots, n-1$, i.e., the curvature of the oriented circle passing through three successive points (x_j, y_j), $j = i-1, i, i+1$. From elementary geometry we have

$$\tilde{\kappa}_i = \frac{2}{M_i} \sin \omega_i. \tag{1.27}$$

Hence

$$\tilde{\kappa}_i = \frac{2}{M_i} (\cos \phi_i \ \sin \psi_{i-1} - \cos \psi_{i-1} \ \sin \phi_i). \tag{1.28}$$

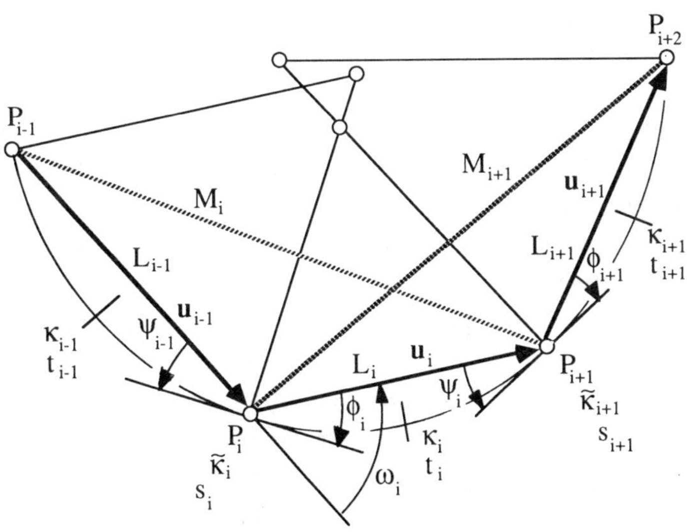

FIG. 1.8. *A C^1 continuous circular spline.*

This is the *tangency condition* that must be satisfied by the angles ϕ_i and ψ_{i-1}, in order for the segments of the curve to combine into a C^1-curve as was stipulated.

For circular splines we may simplify (1.28). Denote by κ_i the curvature of the ith circular segment of the spline. From the properties of circles we have $\psi_i = -\phi_i$ and $\sin \phi_i = -\frac{1}{2} L_i \kappa_i$. Therefore (1.28) takes on the form

$$L_{i-1} \cos \phi_i \, \kappa_{i-1} + L_i \cos \phi_{i-1} \, \kappa_i = M_i \tilde{\kappa}_i. \tag{1.29}$$

These tangency conditions constitute a set of nonlinear equations that must be solved subject to discretized aesthetic constraints

$$\kappa_i > 0, \quad \Delta \kappa_i \geq 0, \quad \Delta \frac{\Delta \kappa_i}{\Delta t_i} \geq 0, \tag{1.30}$$

where we associate the parameter value t_i and the spline curvature κ_i with the midpoint of the ith circular arc of the spline (due to the need of harmonizing the Δt_i with the quantities $\Delta \kappa_i$ which relate to two adjacent circular arcs). It is convenient to associate the arc length s_i with the ith point so that $\Delta s_i \approx L_i$. Discretizing (1.18), we obtain the relationship $\Delta t \approx \kappa \Delta s$. Using this, the log-convexity inequality in (1.30) may be discretized over intervals from the midpoint of one spline segment to the midpoint of the next by letting

$$\Delta \kappa_i = \kappa_{i+1} - \kappa_i \quad \text{and} \quad \Delta t_i = \tilde{\kappa}_{i+1} \frac{\Delta s_i + \Delta s_{i+1}}{2}. \tag{1.31}$$

Use of the numerical curvature $\tilde{\kappa}_i$ in the expression for Δt_i is justified in §1.7.4.

1.7.3. Computational procedure. The game plan is to linearize the entire problem and then to use the simplex algorithm of linear programming to solve the discretized approximation problem.[4] The nonlinearity of the problem requires that we carry out an iterative procedure, analogous to Newton's method, at least in principle. In our experience the first iteration step has almost always given satisfactory answers, but an iterative repetition of the first step is easy to implement. We describe some of the details of the first iteration briefly, under the following headings.

Fill-in, optimal spacing. Often, the spacing of the raw data points (picked from a clay model) is far from optimal for reasons of practicality. Therefore, points must be filled in to achieve optimal point spacing. Filled points optimally spaced along connecting straight lines or circles may be estimated in a fairly reliable manner, provided that the raw data points are spaced reasonably far apart. Again, it is important to remember there is no need to have the filled points improve the definition of the curve implied by the raw data points,

---

[4] For an application of linear programming to the fairing of ships' lines, see [2].

since this will be taken care of by enforcing the aesthetic constraints. The chord height formula can be stated in the form $8\epsilon = \kappa L^2$ (neglecting the ϵ^2-term), however, in terms of numerical curvatures we interpret this in the form $32\epsilon = \tilde{\kappa}_{i+1}(\Delta s_i + \Delta s_{i+1})^2$ because $\tilde{\kappa}_{i+1}$ relates to two successive arcs. With the simplification $L = \Delta s$, justified because of small angles ϕ_i due to the optimal spacing, we substitute from the preceding formula in formula (1.31),

$$\Delta t_i \approx \frac{16\epsilon}{\Delta s_i + \Delta s_{i+1}} \approx \frac{16\epsilon}{L_i + L_{i+1}}. \tag{1.32}$$

The actual computation uses the numerical curvatures $\tilde{\kappa}_i$ in a more indirect way to guarantee that segments are not too long. All points, original (i.e., raw) or filled, are numbered consecutively (x_i, y_i), $i = 1, \ldots, n$. The original data, which are the only ones to contain any information regarding the faired curve, are known as $(x_{\text{inj}(i)}, y_{\text{inj}(i)})$, $i = 1, \ldots, n^{\text{orig.}}$, with $\text{inj}(i)$ the entries of a suitable integer array that are computed at the time of fill-in.

Approximate tangent vectors and normal vectors, displacements. Approximate tangent vectors $\mathbf{t}_i$, and orthogonal to these, approximate normal vectors $\mathbf{n}_i$ can also be estimated from the raw data. As shown in Fig. 1.9, the input points, raw or filled, are to be displaced by an amount δ_i in the direction of the normal vectors $\mathbf{n}_i$ and the desired N/C control points computed in the form

$$(x_i^*, y_i^*) = (x_i, y_i) + \delta_i \mathbf{n}_i.$$

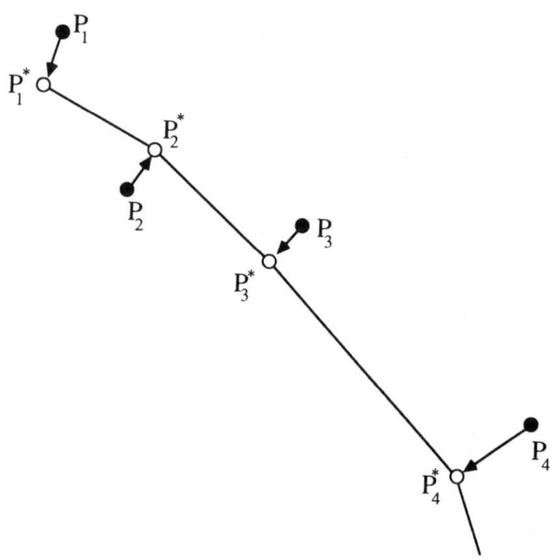

FIG. 1.9. *Input and output points.*

Objective function. We seek to minimize the sum of displacements of the circular spline from the "nonfilled" points, i.e., the objective function

$$z = \sum_{i=1}^{n^{\text{orig.}}} |\delta_{\text{inj}(i)}|.$$

This objective function can be treated in terms of linear programming. There are certain advantages to minimizing the sum versus the maximum of displacements, in that the sum allows some of the displacements to come out somewhat larger than average.

Tangency condition. Here, due to the optimal spacing, the angles ϕ_i are so small that we can use the approximation $\cos\phi_i \approx 1$. Hence (1.29) becomes

$$L_{i-1}\kappa_{i-1} + L_i\kappa_i = M_i\tilde{\kappa}_i, \qquad M_i = L_{i-1} + L_i. \tag{1.33}$$

Linearization of numerical curvatures and aesthetic constraints. In agreement with (1.31) and (1.32) log convexity of curvatures can be linearized in the form

$$(L_{i-1} + L_i)\Delta\kappa_{i-1} \leq (L_i + L_{i+1})\Delta\kappa_i. \tag{1.34}$$

To apply the simplex method of linear programming, we must linearize in (1.33) the complicated nonlinear dependence of numerical curvatures $\tilde{\kappa}_i$ on displacements δ_i. This can be done by a first degree Taylor formula. The results of a lengthy but otherwise elementary and enjoyable calculation by means of vector calculus is the expansion

$$\tilde{\kappa}_i^* \approx \tilde{\kappa}_i + T_{i1}\delta_{i-1} + T_{i2}\delta_i + T_{i3}\delta_{i+1},$$

with

$$\begin{aligned}
T_{i1} &= \frac{2}{L_{i-1}^2 M_i^2}[\mathbf{u}_{i-1}\cdot(\mathbf{u}_{i-1}+\mathbf{u}_i)][\mathbf{t}_i^- \cdot \mathbf{t}_{i-1}]\ ,\\
T_{i2} &= -\frac{2}{L_{i-1}^2 L_i^2}[\mathbf{u}_{i-1}\cdot \mathbf{u}_i][\mathbf{t}_i^0 \cdot \mathbf{t}_{i-1}]\ ,\\
T_{i3} &= \frac{2}{L_i^2 M_i^2}[\mathbf{u}_i\cdot(\mathbf{u}_{i-1}+\mathbf{u}_i)][\mathbf{t}_i^+ \cdot \mathbf{t}_{i-1}]\ .
\end{aligned}$$

The $\mathbf{t}_i$ are the approximate tangent vectors introduced earlier, and the vectors $\mathbf{t}_i^-$, $\mathbf{t}_i^0$, $\mathbf{t}_i^+$ are tangent vectors to the circle that is used to define the numerical curvature $\tilde{\kappa}_i$ at the three defining points. The latter are given by the formulas

$$\begin{aligned}
\mathbf{t}_i^- &= \frac{1}{L_i}\left[\frac{M_i}{L_{i-1}}\mathbf{u}_{i-1} - \frac{L_{i-1}}{M_i}(\mathbf{u}_{i-1}+\mathbf{u}_i)\right]\ ,\\
\mathbf{t}_i^0 &= \frac{1}{M_i}\left[\frac{L_i}{L_{i-1}}\mathbf{u}_{i-1} + \frac{L_{i-1}}{L_i}\mathbf{u}_i\right]\ ,\\
\mathbf{t}_i^+ &= \frac{1}{L_{i-1}}\left[\frac{M_i}{L_i}\mathbf{u}_i - \frac{L_i}{M_i}(\mathbf{u}_{i-1}+\mathbf{u}_i)\right]\ .
\end{aligned}$$

1.7.4. An alternative, purely discrete approximation. In this section we establish a theorem for the linearized version (1.33) of the tangency (C^1)-condition that also holds for the nonlinear version, provided the angles are sufficiently small. Using this result it turns out that if the points are optimally spaced, we can completely avoid using the curvatures κ_i of the circular spline. In this case, (1.32) should be changed to the simpler form

$$\Delta t_i \approx \frac{8\epsilon}{\Delta s_i} \approx \frac{8\epsilon}{L_i} \,. \tag{1.35}$$

Instead of requiring the *spline* curvatures κ_i to satisfy (1.34), we now require log convexity of *numerical* curvatures $\tilde{\kappa}_i$. Due to (1.35), this condition is best linearized in the form

$$L_{i-1}\Delta\tilde{\kappa}_{i-1} \leq L_i\Delta\tilde{\kappa}_i. \tag{1.36}$$

The result on which this new solution method is based is contained in the following.

THEOREM 1.7.1. *Given optimally spaced points* (x_i, y_i), $i = 1, \ldots, n$, *with positive, nondecreasing, and approximately log-convex numerical curvatures* $\tilde{\kappa}_i$, $i = 2, \ldots, n-1$, *i.e., satisfying* (1.36), *for* $i = 2, \ldots, n-2$, *we can solve the equations* $L_{i-1}\kappa_{i-1} + L_i\kappa_i = M_i\tilde{\kappa}_i$, $M_i = L_{i-1} + L_i$, $i = 2, \ldots, n-1$, *for the curvatures* κ_i *of a circular spline interpolating the given points in such a way that* $0 < \kappa_1 \leq \tilde{\kappa}_2 \leq \kappa_2 \leq \tilde{\kappa}_3 \leq \ldots \leq \tilde{\kappa}_{n-2} \leq \kappa_{n-2} \leq \tilde{\kappa}_{n-1} \leq \kappa_{n-1}$.

Proof. First, for $i = 1$ choose $\kappa_1 > 0$ such that the condition $\tilde{\kappa}_2 - (L_2/L_1)(\tilde{\kappa}_3 - \tilde{\kappa}_2) \leq \kappa_1 \leq \tilde{\kappa}_2$ is guaranteed, the lower bound on κ_1 being a consistent requirement. Now, for $i \geq 2$ and by induction let $\tilde{\kappa}_{i-1} \leq \kappa_{i-1} \leq \tilde{\kappa}_i$. Due to $L_{i-1}\kappa_{i-1} + L_i\kappa_i = M_i\tilde{\kappa}_i$ we may then solve for

$$\kappa_i = \tilde{\kappa}_i + \frac{L_{i-1}}{L_i}(\tilde{\kappa}_i - \kappa_{i-1}). \tag{1.37}$$

Hence (1.37) implies $\tilde{\kappa}_i \leq \kappa_i$, advancing the induction in part (the left-hand side). Next, rewriting the convexity condition $L_{i-1}\Delta\tilde{\kappa}_{i-1} \leq L_i\Delta\tilde{\kappa}_i$, we find that

$$\tilde{\kappa}_i - \frac{L_i}{L_{i-1}}(\tilde{\kappa}_{i+1} - \tilde{\kappa}_i) \leq \tilde{\kappa}_{i-1} \leq \kappa_{i-1}.$$

The condition $\tilde{\kappa}_i - L_i(\tilde{\kappa}_{i+1} - \tilde{\kappa}_i)/(L_{i-1}) \leq \kappa_{i-1}$ (which for $i = 2$ is given by the consistent lower bound for κ_1) implies

$$L_{i-1}(\tilde{\kappa}_i - \kappa_{i-1}) \leq L_i(\tilde{\kappa}_{i+1} - \tilde{\kappa}_i),$$

hence, by (1.37),

$$\kappa_i = \tilde{\kappa}_i + \frac{L_{i-1}}{L_i}(\tilde{\kappa}_i - \kappa_{i-1}) \leq \tilde{\kappa}_{i+1}.$$

This completes the induction (the right-hand side).

Due to this result, we can avoid computing explicit mathematical representations for spline curves, from which we can pick the N/C control points, because Theorem 1.7.1 guarantees that a satisfactory circular spline exists that interpolates the control points. Moreover, optimal spacing (cf. above, §1.7.3) guarantees that polygonal line interpolation is visually indistinguishable from the curve. This results in a considerable reduction in the number of variables that need to be carried in the simplex algorithm and thus amounts to a savings in execution time.

1.8. Smoothing and Aesthetics

Here we summarize how our approach to curve smoothing relates to other aspects of the subject.

Spline smoothing has a long history in the automobile industry. An early reference is de Boor's work on bicubic spline interpolation [5].

Curvature monotonicity and convexity may be viewed as nonlinear forms of generalized convexity conditions. The general subject of approximation under higher order generalized convexity was studied by one of us [8]–[10].

We emphasize that our approach concerns only *limited* aspects of the aesthetics of curves and surfaces. We consider only plane sections of automobile body surfaces, and we view these sections in isolation, each being a *single line* (curve). Moreover, we consider only those aspects of the single line that concern "smoothness" or "fairness," meaning "freedom from undesirable wiggles"—with some appropriate definition of "wiggle." This concept appears to include global as well as local properties of a curve, but there might be global properties of curves other than fairness that are relevant to aesthetics.

Our concept of "smoothness" does not seem to be too different from the one used in the mathematical discipline of *smoothing*, in which aesthetic considerations clearly play a role [4], [18], [26]. The principle of simplest shape appears to be generally applicable. We feel that this is the reason for the success of smoothing with cubic splines with a minimal number of optimal knots [5], [6]. In cubic spline smoothing algorithms one can *measure* the "simplicity of shape" by counting the number of cubic segments used to represent the curve. In this sense, this approach is an application of the principle of simplest shape in a way quite analogous to that of counting the number of sign changes, as well as maxima, and minima of κ having been discussed earlier.

On the other hand, in geometric modeling of exterior automotive design lines aesthetics is *all* that matters, while in mathematical smoothing, aesthetic considerations only enter in when a "complete theory" of the processes producing the data to be smoothed is not available. If a "complete theory" is available, mathematical smoothing reduces to parameter estimation. (Part of the "complete theory" is usually the assumption that errors are due to random noise, normally distributed and superimposed on the "basic" process.) Lacking such a "complete theory," mathematical smoothing may attempt to simulate manual smoothing, which can be done by fitting a curve through the "cloud" of data by means of French curves. It is here that aesthetics enters into mathematical smoothing. Our experience suggests that for automobile lines a "complete theory" of curve fairness can be provided by the principle of simplest shape and an analysis of curvature properties like the ones we have studied here.

References

[1] K. G. Baass, *The use of clothoid templates in highway design*, Transportation Forum, 1 (1984), pp. 47–52.

[2] S. A. Berger and W. C. Webster, *An application of linear programming to the fairing of ship's lines*, in Recent Advances in Mathematical Programming, R. L. Graves and P. Wolfe, eds., Mac Graw-Hill, New York, 1963.

[3] G. D. Birkhoff, *Aesthetic Measure*, Harvard University Press, Cambridge, MA, 1933.

[4] M. T. L. Bizley, *A measure of smoothness and a new principle of graduation*, J. Inst. Actuaries, 84 (1958), pp. 125–144.

[5] C. R. de Boor, *Bicubic spline interpolation*, J. Math. Phys., 41 (1962), pp. 212–218.

[6] C. R. de Boor and J. R. Rice, *Least-squares approximation by cubic splines. I: Fixed knots*, and II: *Variable knots*, reports CSD TR 20 and CSD TR 21, Computer Science Department, Purdue University, West Lafayette, IN, 1968.

[7] I. N. Bronshtein and K. A. Semendyayev, *Handbook of Mathematics*, K. A. Hirsch, editor of English translation, Verlag Harri Deutsch, Van Nostrand Reinhold Company, New York, 1985.

[8] H. G. Burchard, *Interpolation and approximation by generalized convex functions*, Ph.D. Thesis, Purdue University, West Lafayette, IN, 1968.

[9] ———, *Extremal positive splines, with applications*, in Approximation Theory, G. G. Lorentz, ed., Academic Press, New York (1973), pp. 291–294.

[10] ———, *Extremal positive splines with applications to interpolation and approximation by generalized convex functions*, Bull. Amer. Math. Soc., 79 (1973) pp. 959–963.

[11] G. Farin, *Curves and Surfaces for Computer Aided Geometric Design*, Academic Press, Boston, 1988.

[12] G. Farin and N. Sapidis, *Curvature and the fairness of curves and surfaces*, IEEE Computer Graphics and Applications, 9, 2 (1989) pp. 52–57.

[13] A. H. Fowler and C. W. Wilson, *Cubic spline, a curve fitting routine*, report Y-1400, Oak Ridge National Laboratory, Oak Ridge, TN, 1963.

[14] W. H. Frey and D. A. Field, *Designing Bézier conic segments with monotone curvature*, General Motors Research Publication GMR-7485, GM Research Laboratories, Warren, MI, 1991.

[15] P. J. Grillo, *Form, Function, and Design*, Dover, New York, 1975.

[16] G. Howard, *Evolution of Motorcar Shapes and Design*, International Association for Vehicle Design, Interscience Enterprises, Ltd., Geneva, Switzerland, 1985.

[17] J. I. Jones, *A system for designing and approximating aesthetically smooth curves with interactive graphic controls*, Ph.D. Thesis, University of Detroit, Detroit, MI, 1970.

[18] P. Lancaster and K. Salkauskas, *Curve and Surface Fitting: an Introduction*, Academic Press, London, 1986.

[19] M. M. Lipschutz, *Theory and Problems of Differential Geometry*, Schaum's Outline Series, McGraw-Hill, New York, 1969.

[20] E. Mehlum, *Non-linear splines*, in Computer Aided Geometric Design, R. E. Barnhill and R. F. Riesenfeld, eds., Academic Press, Boston, MA, 1974.

[21] H. Nelms, *Thinking with a Pencil*, Barnes & Noble Division, Harper and Row, New York, 1984.

[22] J. Roulier, T. Rando, and B. Piper, *Fairness and monotone curvature*, in Approximation Theory and Functional Analysis, C. K. Chui, ed., Academic Press, Boston, MA, 1990.

[23] N. Sapidis and G. Farin, *Automatic fairing algorithm for B-spline curves*, Comput. Aided Design, 22, 2 (1990), pp. 121–129.

[24] N. S. Sapidis and W. H. Frey, *Controlling the curvature of a quadratic Bézier curve*, Comput. Aided Geom. Design, 9 (1992), pp. 85–91.

[25] D. J. Struik, *Differential Geometry*, Constable and Co., Ltd., London, 1950.

[26] E. T. Whittaker and G. G. Robinson, *The Calculus of Observations*, fourth ed., Blackie and Son Ltd, London, 1944.

[27] R. C. Yates, *Curves and their Properties*, reprint by the National Council of Teachers of Mathematics, Washington, D.C., of the original 1952 edition by J. W. Edwards, Ann Arbor, MI, 1974.

CHAPTER

2

Curvature Integration through Constrained Optimization

Alan K. Jones

2.1. Introduction

To an aerodynamicist, smoothness of a curve is defined in terms of the qualitative features of its curvature profile. Some of these features, such as the lack of isolated sharp peaks or extraneous zero crossings, are straightforward. Some of them are not, and can be lumped into the requirement that the curvature plot "looks right" to an experienced engineer. But an airfoil curve must possess other properties as well in order to be useful. It must meet spanwise defining member curves, at least at the leading and trailing edges, and possibly at intermediate locations. It must have prescribed tangent vectors, at least at the leading and trailing edges. Finally, but not trivially, it must be expressible in the mathematical basis used by the CAD (computer-aided design) system at hand. Curvature profiles which can be obtained from curves satisfying these background constraints will be called *feasible profiles.*

If the desired qualitative features of the curvature distribution can be characterized adequately by additional constraints or penalty terms, then it is possible to regard the curve design as a single constrained optimization problem, and solve it accordingly. The observations of §2.4 may be of some use in this regard. Unfortunately, such optimization problems do not always have solutions, and one is faced with exploring tradeoffs. Even more commonly, the smoothness criteria are imprecise. An engineer may decide that a curvature plot "should not be so rough in this region," or "should have more of the curvature concentrated over there." Thus, there is room for a man-in-the-loop mode in which the space of feasible profiles is manually explored by an experienced engineer. This also happens to coincide with the habits of thought and work of many such users. Constrained optimization is an appropriate technique in this mode as well, allowing the user to, in effect, sketch a portion of the target curvature profile freehand, and then project the modified profile into the space of feasible profiles.

This chapter discusses the current state of curvature modification procedures, and suggests that they can be significantly improved by a combination of spline technology and modern software for constrained optimization. It is organized as follows. Sections 2.2 and 2.3 discuss both the problem and the currently available solutions. Section 2.4 provides the key technical insight, namely that the curvature of a polynomial spline curve is almost a ratio of splines, and a great deal of control can be exerted over the curvature distribution by manipulating these auxiliary splines. Sections 2.5 and 2.6 describe the application of this principle, and §2.7 presents typical experimental results.

2.2. The Problem

The problem is to find a smooth, feasible curvature profile and its associated airfoil curve. The *curvature profile* of a planar parametric curve, denoted by c, is the signed curvature expressed as a function of the curve parameter. For a given curve design problem, a *feasible profile* will be one that corresponds to a curve satisfying the following requirements.

(1) *Mathematical constraints.* The curve must be chosen from a specified function space. In this study, we will assume it is a parametric (nonrational) polynomial spline. Furthermore, because the airfoil must be embedded in a wing surface, typically represented as a parametric tensor product polynomial spline, we will also assume that the polynomial degree and knot set have been fixed a priori.

(2) *Geometrical constraints.* In this study, we will consider only linear equality constraints. The curve must pass through prescribed points and/or possess prescribed tangent vectors. Again, because the curve is regarded as embedded in a tensor product spline surface, these constraints will be imposed at specified parameter values. From the user's point of view, the tangent conditions to be fixed will be G^1, not C^1. That is, the ratios of parametric derivatives are significant, not the individual components. However, the two become equivalent in cases where the velocity magnitudes can be held constant.

(3) *Smoothness constraints.* Smoothness is an inherently ill-defined term. The following classes of criteria, by no means exhaustive, are among those that subjectively determine smoothness of a curvature profile. It should be emphasized that these criteria are not just aesthetic judgments on the part of the designer, as is often the case, for example, in automobile body design. They do correspond to objective aerodynamic properties of the airfoil.

(a) *Inequality constraints.* Keep the profile positive (negative), monotone, or convex over certain parameter intervals.

(b) *High frequency features.* Prevent high, narrow peaks, and general "roughness" associated with high-frequency components in a signal.

(c) *Low frequency features.* Prescribe the bulk distribution of curvature.

Smoothness criteria of the inequality constraint type can be directly incorporated into a constrained optimization procedure. It is likely that criteria of the second type, which corresponds intuitively to band limiting the Fourier transform of the curvature profile, can be incorporated also, although that effort is beyond the scope of this chapter. Criteria of the third type, however, seem better adapted to interactive graphical input.

2.3. Current State of the Art

Users typically envision the design process as the following.

(1) Construct a curve by some means, and examine its curvature profile, regarded typically as a string of plot points.

(2) Modify the profile interactively to make it smooth. Move individual points or groups of points, apply smoothing filters, etc. Call the result the *target profile*.

(3) Iterate the following steps as needed.

(a) Apply heuristics to modify the target profile to allow the geometric constraints to be satisfied.

(b) Integrate the modified target profile numerically and obtain a dense set of values (and derivatives) along the resulting *ideal curve*.

(c) Best-fit the ideal curve with a curve satisfying the mathematical constraints.

(d) Check that the geometric constraints and smoothness criteria are satisfied.

The best-fitting procedure in the above cannot simply minimize least-squares deviation from the ideal curve locations at a dense set of plot points. In general, this gives no hope of reproducing the target profile. At a minimum, the fitting procedure must attempt to reproduce first- and second-order derivative data of the ideal curve. Better still, the objective function should measure directly deviations from the modified target curvature profile. Furthermore, the deviation measure chosen must give information at *every* point in the interval. It is not enough just to sample at a dense set of discrete points.

We mention in passing an approach that has been adopted in the past to finesse this issue. This is to approximate the modified target profile up to tolerance by a low degree spline, and then to define the corresponding curve procedurally as a solution to a system of differential equations which guarantees that its curvature distribution is exactly that spline. See, for example, [1] or §1.1.2.4 of [4]. In the special case of piecewise constant c, the resulting curve is piecewise circular, but cannot of course have curvature continuity. Curvature profiles which are polynomial splines of nonzero degree do not integrate to give spline curves. For linear c, the solution is the so-called Cornu spiral. Generating points along such a curve requires numerical evaluation of Fresnel integrals, of the form

$$\begin{aligned} C(s) &= \int_0^s \cos(\pi\sigma^2/2)d\sigma, \\ S(s) &= \int_0^s \sin(\pi\sigma^2/2)d\sigma. \end{aligned} \tag{2.1}$$

See [3] for a numerical evaluation of these. A system to construct and evaluate curves which are piecewise Cornu spirals is described in [8]–[13]. A related but fundamentally different approach is taken in [2], where C^1 piecewise smooth curves are constructed to minimize the L_2 norm of curvature subject to various interpolation constraints. We summarize both approaches by emphasizing that the computation involved is nontrivial, and that the resulting curves do not satisfy our mathematical constraints.

2.4. Viewing Curvature Profiles as (Almost) Rational Splines

Let $\hat{c}$ be a target curvature profile, and c the one actually obtained. Some mechanism must be found to bound an appropriate function norm of $c-\hat{c}$. Now, if c and $\hat{c}$ were in fact splines, then it would be enough to express them with a common knot set and degree and measure the distance between corresponding coefficients c_i and $\hat{c}_i$. For example, an upper bound on

$$\|c - \hat{c}\|_\infty = \sup |c(t) - \hat{c}(t)|$$

is given by $\max |c_i - \hat{c}_i|$. This would not be a suitable objective function for a constrained optimization problem because the max function destroys continuity. A similar but more useful norm is

$$\|c - \hat{c}\|^2_{\ell^2,\mathbf{t}} = \sum_{i=1}^{N} |c_i - \hat{c}_i|^2 (t_{i+n+1} - t_i), \tag{2.2}$$

which was shown in [6] and [7] to be equivalent to the usual L_2 norm $\|c-\hat{c}\|_2^2 = \int |c(t) - \hat{c}(t)|^2 dt$. Here n is the degree of the splines and $\mathbf{t} = \{t_1, \ldots, t_{N+n+1}\}$ is the knot set.

Unfortunately, as we have seen, c is not a spline in general if the underlying curve is. However, c is completely determined by two auxiliary functions which are splines, and hence can be controlled by such means. We will now see just how much mileage can be gained out of that simple observation.

Write a typical point on the curve as $(x(s), y(s))$, where x and y are splines, and let the curvature at parameter value s be given by $c(s)$.

Recall that

$$c(s) = \frac{p(s)}{q(s)^{3/2}}, \tag{2.3}$$

where

$$\begin{aligned} p &= \dot{x}\ddot{y} - \dot{y}\ddot{x}, \\ q &= (\dot{x})^2 + (\dot{y})^2, \end{aligned} \tag{2.4}$$

where dots represent differentiation with respect to the curve parameter.

If x and y are polynomial splines, then p and q are also splines, which we call, respectively, the *curvature numerator* and the *curvature denominator*. These are the auxiliary splines that will form the basis of our calculations. If x and y are of degree n and class C^k, then q is of degree $2n-2$ and class C^{k-1}, while p is of class C^{k-2}, and formally of degree $2n-3$. It is well known, and easy to see by expanding (x, y) locally in a power basis, that p is in fact of degree $2n-4$. Here, and throughout this paper, we adopt the convention that C^j actually means $C^{\max(-1,j)}$. A curve is of class C^{-1} if it is piecewise smooth, with only jump discontinuities.

A well-known corollary of the above calculation is the following.

PROPOSITION 2.4.1. *The curvature of a nonsingular parametric spline curve is nonnegative over the parametric interval* $[a, b]$ *if the curvature numerator spline* p *is nonnegative there.*

Here nonsingular means that the first parametric derivative never vanishes. Repeated differentiation of the curvature formula yields two similar results.

PROPOSITION 2.4.2. *The curvature profile of a nonsingular* C^2 *parametric spline curve is monotone increasing over the parametric interval* $[a, b]$ *if the spline* $r = 2\dot{p}q - 3p\dot{q}$ *is nonnegative there.*

PROPOSITION 2.4.3. *The curvature profile of a nonsingular* C^3 *parametric spline curve is convex up over the parametric interval* $[a, b]$ *if the spline* $4\ddot{p}q^2 - 12\dot{p}q\dot{q} + 15p\dot{q}^2 - 6pq\ddot{q}$ *is nonnegative there.*

The differentiability requirements on the plane curve can be relaxed with due care, but the result is to add additional constraints. For example, we have the following.

PROPOSITION 2.4.4. *The curvature profile of a nonsingular parametric spline curve is monotone increasing over the parametric interval* $[a, b]$ *if the spline* $2\dot{p}q - 3p\dot{q}$ *is nonnegative wherever it exists in that interval, and if at each knot* t *of the* p *spline, the value jump* $\Delta p(t) \equiv p(t_+) - p(t_-) \geq 0$.

In practice, we force nonnegativity of splines like p or r over an entire interval $[a, b]$ by requiring that the spline coefficients corresponding to the B-splines whose supports intersect $[a, b]$ are nonnegative, and noting that the B-splines themselves are nonnegative. This is a sufficient but not necessary condition, and can occasionally be too strong in the sense that the solutions simply fail to exist. However, it is straightforward to code, and mathematically tractable from the point of view of modern optimization codes. Since the splines p and q are quadratic expressions in the coefficients of the curve (x, y), these propositions describe algebraic inequality constraints of degrees 2, 4, and 6, respectively, on the spline coefficients of (x, y). Constraints such as those on Δp are quadratic inequality constraints of a particularly simple form, involving only differences of coefficients of p.

Proposition 2.4.1 gives a test for convexity of a spline curve which is much simpler to code than the tests proposed in Theorem 3.7 of [5], which were based on purely geometric criteria for the convexity of the curve's spline polygon. It has not yet been shown that these two tests are logically equivalent in every case. However, brute force computation verifies the equivalence for a Bezier cubic curve, and we conjecture that it is true in general. In any case, it is advisable to express p using its actual degree of $2n-4$, where x and y have degree n, rather than the naive estimate $2n-3$. Artificial degree raising of the numerator spline does not produce a "tighter" bound, but simply adds redundant constraints.

Propositions 2.4.2 and 2.4.4 appear to be a new observation, and have applications in automating the elimination of "spikes" and "wiggles" from curvature profiles. We will refer to

$$r \equiv 2\dot{p}q - 3p\dot{q} \tag{2.5}$$

as the *curvature slope numerator*. It is a spline of class C^{k-3} and degree $4n-7$. No additional cancellation of high-order terms occurs in general.

Proposition 2.4.3, however, is of doubtful practical value, since most users do not need to micro-manage their curvature profiles to this extent. Note also that this condition ensures convexity only when the curvature is plotted against the curve parameter. A curvature versus arclength plot, for example, could still fail to be convex.

2.5. An Optimization-Based Algorithm

In light of the above, the algorithm of §2.2 can be recast as follows.

(1) Define mathematical constraints, i.e., spline degree and knot set.

(2) Define geometrical constraints as linear equality constraints on the spline coefficients.

(3) Define an initial guess curve satisfying the mathematical constraints and as many of the geometrical constraints as can easily be done.

(4) Iterate as needed.

(a) Define smoothness inequality constraints, i.e., regions where curvature must be positive (negative), or monotone increasing (decreasing).

(b) Define qualitative curvature constraints, i.e., regions in which the curvature should qualitatively resemble a user-provided *target profile*, expressed as a pair $(\hat{p}, \hat{q})$ of target curvature numerator and denominator splines.

(c) Find a curve satisfying the mathematical, geometrical, and smoothness inequality constraints whose associated p and q splines are closest (in some sense) to $\hat{p}$ and $\hat{q}$.

(d) Check the smoothness properties of the resulting curve.

Techniques for obtaining $\hat{p}$ and $\hat{q}$ through interactive modification of p and q profiles are discussed in the next section. Note that, if no qualitative smoothness constraints are imposed, we can simply set $\hat{p}$ and $\hat{q}$ to be those of the initial guess curve.

All of the serious computation involved is hidden in step 4(c), where it can be handled by off-the-shelf software for solving constrained optimization problems. Thus, all three stages of the original algorithm, integration, approximation, and imposition of constraints, can be combined into a single operation and solved automatically. The fitting tolerance can be measured in the curvature domain, where it belongs, and the geometrical constraints are imposed on the physical curve, where they belong.

The next question is, how do we measure closeness of (p, q) to $(\hat{p}, \hat{q})$? The most straightforward answer is just to best-fit $\hat{p}$ and $\hat{q}$ separately, as follows.

Problem 1. Minimize the objective functional

$$P_1(x; y) = \|p\|^2_{\ell^2,\mathbf{t}} + \|q\|^2_{\ell^2,\mathbf{u}}$$

subject to the linear equality constraints on $x, y, \dot{x}, \dot{y}$ that give interpolation of the specified locations and tangents, and the nonlinear inequality constraints that determine the sign of curvature and/or its slope. The optimization variables are just the spline coefficients of the functions x and y. Here $\mathbf{t}$ and $\mathbf{u}$ are the knot vectors appropriate to the functions p and q, respectively.

This is a quadratic programming problem with linear equality constraints, and quadratic and quartic inequality constraints. If necessary, the Hessians can be written out explicitly, and there is additional structure that we may be able to exploit. The functions $p_i(x; y)$ and $q_j(x; y)$ are block (anti)symmetric, and each block is banded because of the local support of B-splines. Note that eliminating unnecessary inequality constraints will simplify this stage considerably.

In practice, the unweighted least-squares problem described above does not work well for at least two reasons. First, it ignores the question of relative scaling between the norms of p and q. Second, it is inappropriate to give equal weight to errors in all of the coefficients because of the very large dynamic range typical of curvature distributions around an airfoil. The qualitative shape of the target distribution is better preserved if instead relative errors of the form $(p_i - \hat{p}_i)/\hat{p}_i$ are controlled. This will discourage the optimizer from, say, introducing ripples into previously flat regions of the curvature profile while removing a notch from a peak.

Once this idea is adopted, an estimate of the relative scaling of p and q norms is suggested by the calculation

$$\Delta c = \Delta(pq^{-3/2}) \approx \Delta p q^{-3/2} - \frac{3}{2} p q^{-5/2} \Delta q, \tag{2.6}$$

which we can write as

$$\frac{\Delta c}{c} = \frac{\Delta p}{p} - \frac{3}{2}\frac{\Delta q}{q}. \tag{2.7}$$

Thus, the contribution to the relative error in c of relative changes in q is $3/2$ that of relative changes in p, which motivates the following.

Problem 2. Minimize the objective functional

$$P_2(x;y) = \sum\left(\frac{p_i - \hat{p}_i}{\hat{p}_i}\right)^2 (t_{i+2n-3} - t_i) + \frac{9}{4}\sum\left(\frac{q_j - \hat{q}_j}{\hat{q}_j}\right)^2 (u_{j+2n-1} - u_j),$$

subject to the constraints of Problem 1.

So far the p and q splines have been handled independently, with their own natural orders and knot sets, and only weakly coupled by the definition of the objective function in Problem 2. Better results may come from writing the two splines in a common form, as a C^{k-2} spline of degree $2n-2$, over a knot set v. The objective functional can then be enhanced with the cross terms omitted from Problem 2, yielding the next problem.

Problem 3. Minimize the objective functional

$$P_3(x;y) = \sum\left(\frac{p_k - \hat{p}_k}{\hat{p}_k} - \frac{3}{2}\frac{q_k - \hat{q}_k}{\hat{q}_k}\right)^2 (v_{k+2n-1} - v_k)$$

subject to the constraints of Problem 1.

2.6. Implementation Details

The implementation approach described in this paper was chosen specifically to work with an existing CAD system in which airfoil curves typically have the following properties.

(1) They are stored as parametric polynomial splines, usually nonrational.

(2) They are densely defined, with tens or hundreds of spline coefficients per curve. However, it is usually not desirable to add additional knots.

(3) They are approximately arc length parametrized; variations in parametric velocity are negligible compared to variations in curvature.

Where qualitative smoothness constraints are required, the target splines $(\hat{p}, \hat{q})$ are constructed by interactive modification of the splines $(\bar{p}, \bar{q})$ associated with the initial guess curve. This takes place in two phases.

In the first phase, the denominator $\bar{q}$ is modified to remove behaviors that are identified with undesired features of the curvature profile. In practice, any sort of variation at all in $\bar{q}$ usually turns out to be undesirable, so we typically try to impose $q \approx$ constant. The best feasible approximation to this can be achieved by minimizing in the ℓ_2 norm the coefficient vector of the spline $\dot{x}\ddot{x} + \dot{y}\ddot{y}$. However, in the host CAD system, $\bar{q}$ is typically almost constant. See, for example, Fig. 2.4. Hence, in most cases, it is reasonable to take $\hat{q} = \bar{q}$, and the q modification step is omitted.

In the second phase, the user adjusts the spline coefficients $\hat{p}_k$ interactively. Ideally, visual feedback would be available, consisting of the spline polygon of $\hat{p}$ superimposed on the plot of $\hat{p}/\bar{q}$. However, with well-behaved $\hat{q}$, this has not proved necessary.

A consequence of the very large knot arrays characteristic of the host system is that the spline polygon for the curvature numerator p has a large number of vertices. Thus it is impractical to manipulate it vertex-by-vertex. Probably the right way to deal with this problem is to use [6] to approximate p by a spline with a much sparser knot set, hence there are much fewer vertices on its spline polygon, then manipulate these sparse vertices individually to get a qualitatively good curvature profile, then perform knot addition to find the corresponding polygon with respect to the full knot set. The method actually implemented in the test code is cruder in concept but similar in effect for the test curves we have seen. The spline polygon of p is itself regarded as a piecewise linear curve, with the coefficient value p_i plotted at a parameter value given by the average of knots t_i through t_{i+2n-3}. The user is then allowed to replace individual p_i values, or to specify linear variation of the p_i (with respect to the curve parameter) over a particular range of i.

The objective function described in Problem 2 has been found the most satisfactory in practice. Note that this is only possible because our concerns are typically for regions around the leading edge of an airfoil, where $\hat{p}$ is bounded away from zero, so the relative error norm makes sense. We note in passing that the spline representations for p, q, r, etc., were actually generated, not by formal differentiation and multiplication of the B-spline series, but rather by oversampling and least-squares fitting. That is, pick the points s_i so that they uniquely determine the splines p and q, given their known degrees and knot sets. Then the difference between the norms $\sum(p(s_i) - \hat{p}(s_i))^2$ and $\sum(p_k - \hat{p}_k)^2$ is just the linear transformation that takes the data $\{p(s_i)\}$ to the spline coefficients $\{p_k\}$. Similarly, this is also true for q. This procedure was adopted primarily to speed up code development for proof of concept purposes. The resulting code may execute slightly slower, but there is little or no loss in accuracy.

Two classes of geometric linear equality constraints have been implemented. First, discrete point locations or parametric first derivatives can be "frozen" at values read off the input curve. Second, a vertical or horizontal tangent can be specified at any point, without fixing the magnitude of the parametric derivative. In addition, one type of nonlinear inequality constraint is supported. Curvature monotonicity constraints as prescribed by Proposition 2.4.4 may be specified in the following manner. The user enters a parameter value t_{peak} where the curvature peak is to be located. The region of the curve to be modified is then assumed to have monotonically increasing curvature for $t < t_{\text{peak}}$, and monotonically decreasing curvature for $t > t_{\text{peak}}$.

No attempt has been made to implement the optimization step efficiently by exploiting its special structure as described in §2.5. For proof of concept, it has been coded in a very naive fashion, with gradients and Hessians evaluated by the optimization package using finite differences. Performance in this mode is not impressive. A large case such as the first example in the next section can take tens of seconds on a typical engineering workstation.

2.7. Examples

Two test cases will be reported, both of which are C^1 cubic splines with airfoil-like geometry. Case 1 is taken from a curve defined by 203 spline coefficients. This density is typical of the host CAD system, and, again typically, the parametric speed q is nearly constant. Case 2 is at the opposite extreme. This is a very sparsely defined curve with only 14 coefficients, where the curvature irregularities are in fact dominated by the variation in q.

Figure 2.1 shows an analytically defined symmetrical airfoil. Case 1 is a curve fit to the upper half of this shape, and extended by reflection symmetry. Figure 2.2 shows the curvature behavior for the this curve, with the lower surface trailing edge at parameter value -1, the leading edge at parameter value 0 and the upper surface trailing edge at parameter value $+1$. The "sawtooth" curvature at the leading edge is a serious defect.

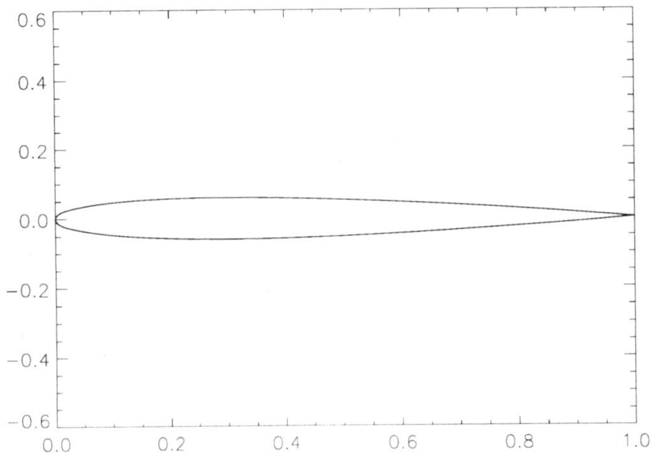

FIG. 2.1. *An analytically defined airfoil.*

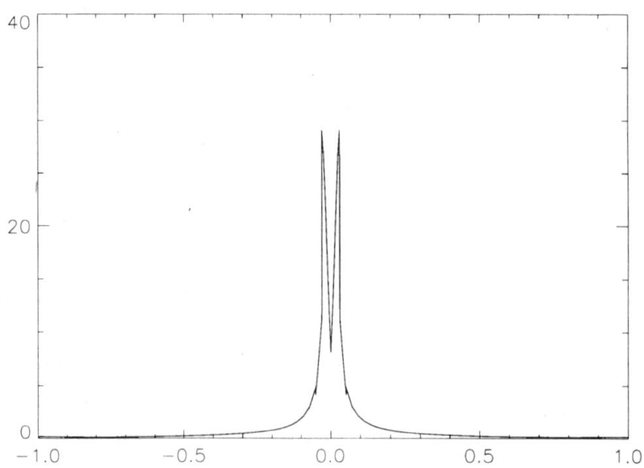

FIG. 2.2. *Curvature plot of the airfoil.*

The solid trace in Fig. 2.3 is a blowup of the curvature plot in the leading edge area. (From now on, only the upper surface will be plotted.) The solid trace in Fig. 2.4 shows q, which is nearly constant.

Thus, the solid trace in Fig. 2.5, which is the p function plotted over this interval, closely mimics the form of the curvature plot. The dotted curve in Fig. 2.5 shows the target p distribution after interactive manipulation to force $p(0) = 50$ and linear variation of p between parameter values 0 and 0.03. The target q distribution is unchanged from the input curve. The dotted trace in Fig. 2.3 shows the curvature distribution implied by the target p and q. The optimization procedure was run with linear equality constraints forcing the leading edge position to be fixed and the leading edge tangent to be vertical, together with a nonlinear inequality constraint to force the curvature to be monotone decreasing over the top half of the airfoil. Nothing outside the

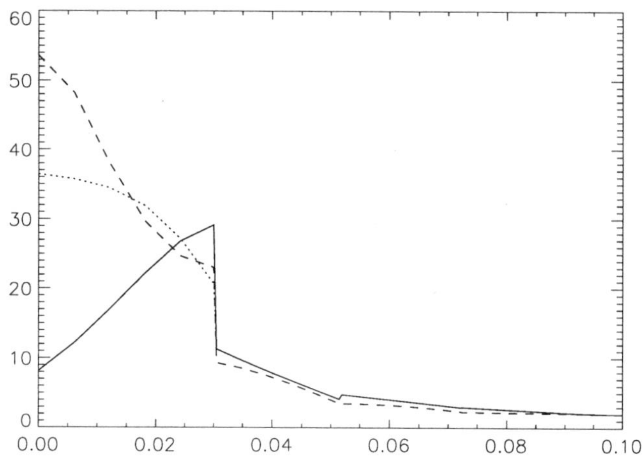

FIG. 2.3. *Curvature behavior near the leading edge.*

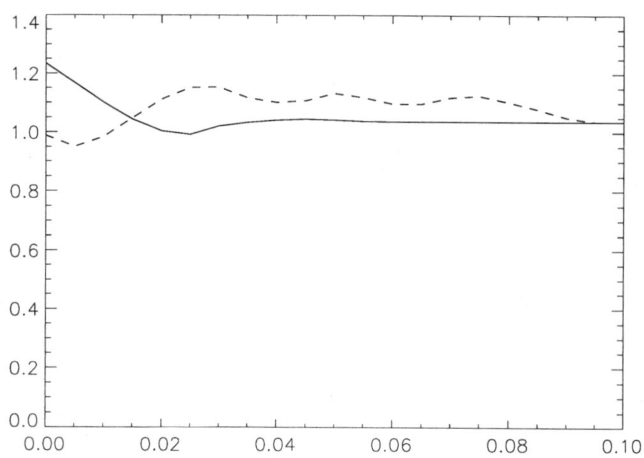

FIG. 2.4. *q distributions for case* 1.

parameter interval $[0, 0.1]$ was altered. This was an optimization problem with 16 variables and 36 constraints, and took almost 6 minutes of CPU time on a Sun SPARCstation IPX. The p, q, and curvature distributions of the output curve are the dashed traces in Figs. 2.3–2.5.

The solid and dashed traces in Fig. 2.6 show the r distributions of the input and output curves. It will be seen that $r \geq 0$ everywhere in the output curve, as required.

Finally, the solid and dashed traces in Fig. 2.7 show the input and output curves, respectively, in the leading edge region. Note that the input curve did not satisfy the symmetry condition that the leading edge tangent be vertical. The output curve does satisfy this constraint, and has eliminated the "saw-tooth" curvature behavior as well. On the other hand, the output curvature

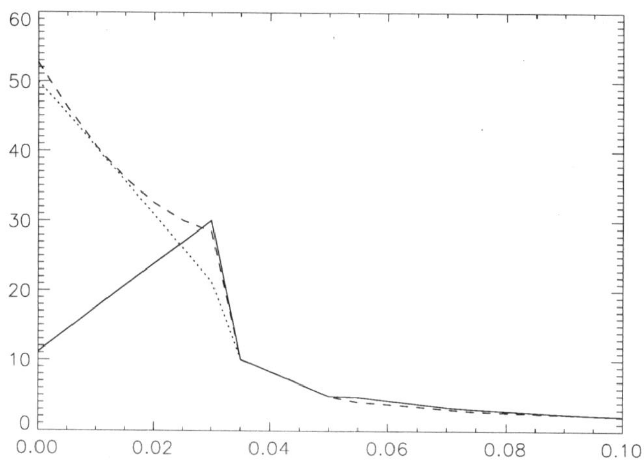

FIG. 2.5. *p distributions for case* 1.

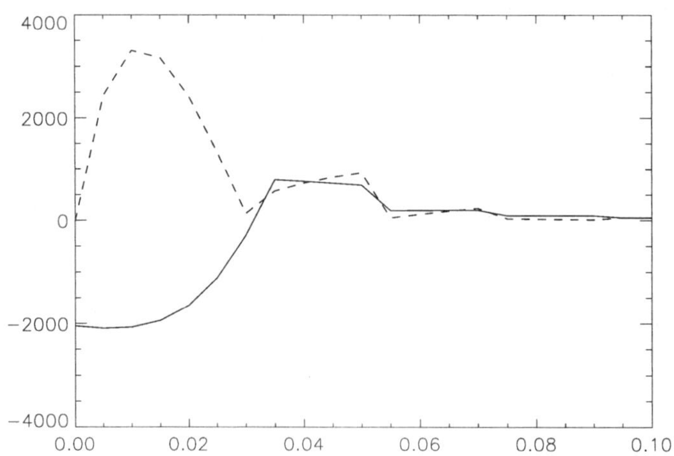

FIG. 2.6. *r distributions for case* 1.

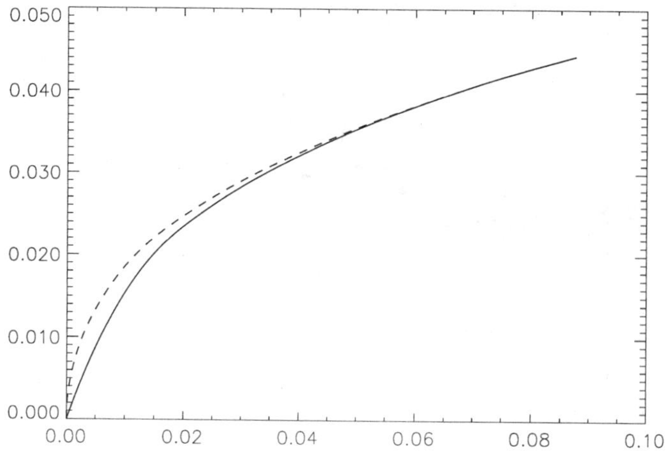

FIG. 2.7. *Input and output curves for case* 1.

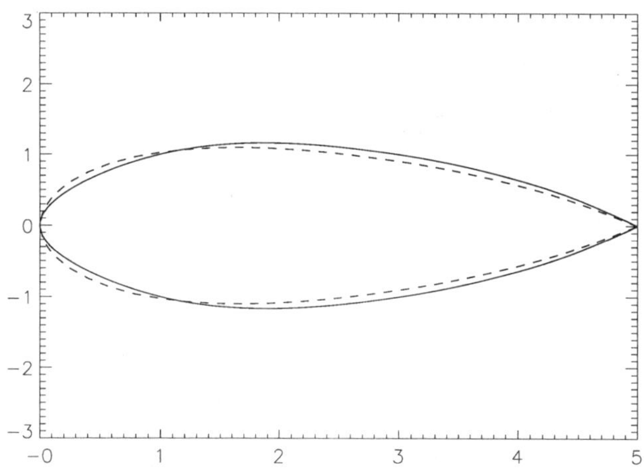

FIG. 2.8. *Input curve for case* 2.

plot is not particularly close to the target plot shown in Fig. 2.3. This demonstrates that the freehand sketching mode is useful mainly for influencing the bulk distribution and gross properties of the curvature distribution, but cannot be relied upon to prescribe fine details of a plot. The mathematical constraint that the underlying curve be contained in a fixed spline space simply does not allow that sort of fine control.

The solid trace in Fig. 2.8 shows the input curve for Case 2. The solid trace in Fig. 2.9 is the q profile for this curve. It displays large, short-period oscillations in the region of high curvature, which are typical of sparsely defined curves in the host CAD system. Comparison of the p profile in Fig. 2.10 with the curvature plot of Fig. 2.11 reveals that these oscillations have greatly exaggerated the depth of the "dips" in curvature that flank the leading edge

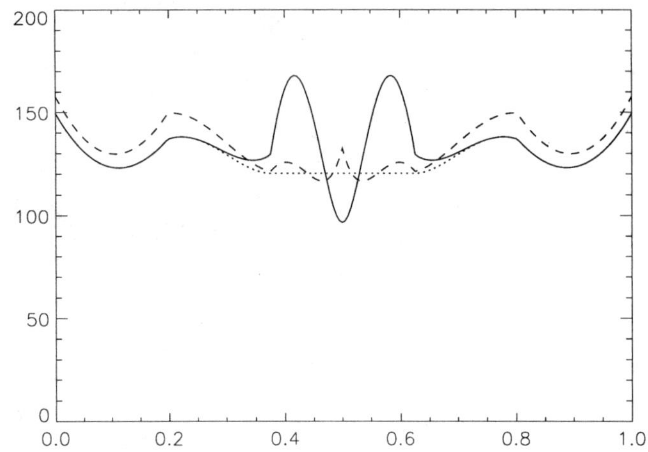

FIG. 2.9. *q distributions for case* 2.

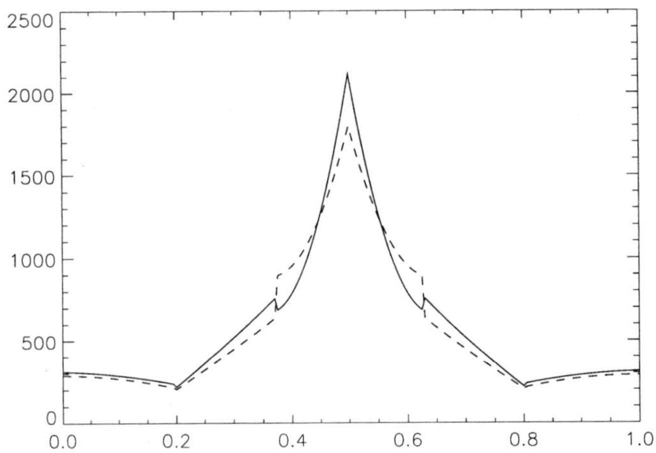

FIG. 2.10. *p distributions for case* 2.

peak. In this case, the main payoff will come from forcing better behavior in the q profile. The dotted trace in Fig. 2.9 shows the modified target q. The optimization procedure was run with linear equality constraints to freeze the positions at the leading and trailing edges, and to hold a vertical tangent at the leading edge. This was an optimization problem with 28 variables and 6 constraints, and ran in 9 seconds of CPU time, almost 40 times faster than Case 1. As before, dashed traces show the results of the optimization. Finally, in Fig. 2.12, we see the spline coefficient polygons before and after modification. It will be seen that after optimization the curvature "dips" are filled in, and the central peak is significantly blunted. Whether this is a satisfactory improvement, or a further iteration is required, would depend on the user's engineering judgment.

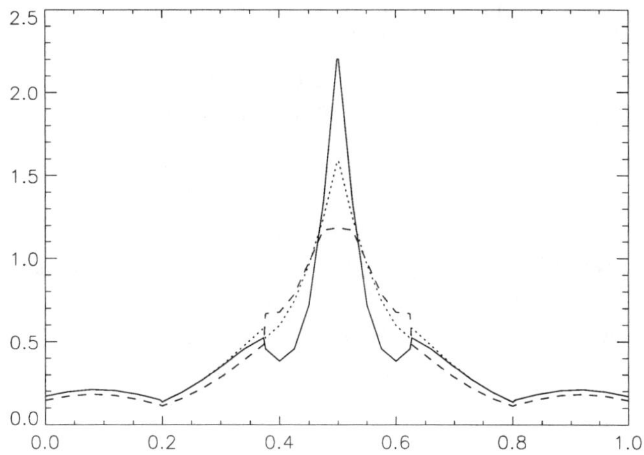

FIG. 2.11. *Curvature plots for case 2.*

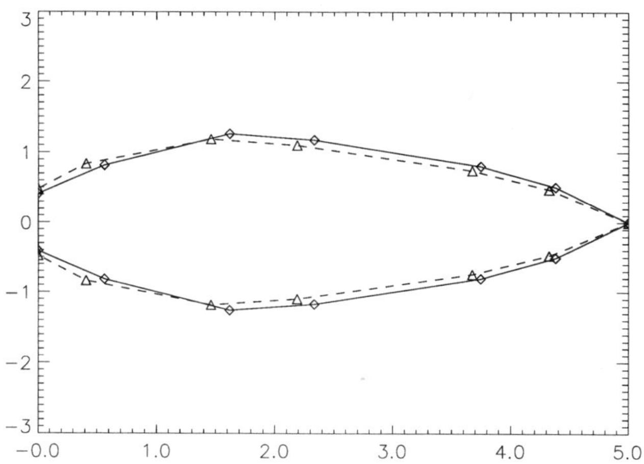

FIG. 2.12. *Spline polygons for case 2.*

References

[1] J. A. Adams, *The intrinsic method for curve definition*, CAD, 7 (1975), pp. 243–249.

[2] G. Brunnett, *Properties of Minimal-Energy Splines*, in Curve and Surface Design, H. Hagen, ed., Society for Industrial and Applied Mathematics, Philadelphia, PA, 1992.

[3] W. J. Cody, *Chebyshev approximations for the Fresnel integrals*, Math Comp., 11 (1988), pp. 450–453.

[4] I. D. Faux and M. J. Pratt, *Computational Geometry for Design and Manufacture*, Ellis Horwood, Chichester, UK, 1979.

[5] D. R. Ferguson, P. D. Frank, and A. K. Jones, *Surface shape control using constrained optimization on the B-spline representation*, Comput. Aided Geom. Design, 5 (1988), pp. 87–103.

[6] T. Lyche and K. Morken, *Knot removal for parametric B-spline curves and surfaces*, Comput. Aided Geom. Design, 4 (1987), pp. 217–230.

[7] ———, *A data reduction strategy for splines with applications to the approximation of functions and data*, IMA J. Numer. Anal., 8 (1988) pp. 185–208.

[8] A. W. Nutbourne, P. M. McClellan, and R. M. L. Kensit, *Curvature profiles for plane curves*, CAD, 4 (1972), pp. 176–184.

[9] ———, *A cubic spline package part 2—the mathematics*, CAD, 5 (1973), pp. 7–13.

[10] T. K. Pal and A. W. Nutbourne, *Two-dimensional curve synthesis using linear curvature elements*, CAD, 9 (1977), pp. 121–134.

[11] ———, *Intrinsic spline curve with local control*, CAD, 10 (1978), pp. 19–29.

[12] ———, *Mean tangent rotational angles and curvature integration*, CAD, 10 (1978), pp. 30–34.

[13] A. Schechter, *Synthesis of 2d curves by blending PL curvature profiles*, CAD, 10 (1978), pp. 8–18.

CHAPTER

3

Automatic Fairing of Point Sets

Matthias Eck and Rainer Jaspert

3.1. Introduction

It is a frequent problem that appears in the everyday life of a CAD system: working on measured data points that came, for example, from digitizing processes. Therefore, such a system must have the capability to handle possible deviations or data errors in order to avoid problems during later calculations.

The most common way of solving this task in the case of planar data is to determine an approximating spline curve which minimizes a certain fairness functional (cf. [3], [7], and [9]). In [13] a remarkable iterative and local procedure is described that makes use of the geometry of the curve.

At first an initial cubic B-spline curve $\mathbf{X}$ is determined which interpolates the planar data points $\mathbf{P}_i = \mathbf{X}(t_i)$. Then with help of the *local fairness criterion*[1] $z_i = |\dot{\kappa}(t_i^-) - \dot{\kappa}(t_i^+)|$, and the *global fairness criterion* $\xi = \sum_i z_i$, the actual fairing procedure works as follows.

1. Find the largest z_j.
2. Fair the curve in the area of $\mathbf{X}(t_j)$ by a local procedure called "knot removal–knot reinsertion."
3. Determine the new criteria $\hat{z}_i$ and $\hat{\xi}$.
4. If $\hat{\xi} < \xi$ then goto step 1.
   else undo last iteration and exit.

As illustrated in [13], this automatic fairing procedure works well.

In contrast to such curve-oriented algorithms, the heuristic method given in [12] modifies the data points nonautomatically by improving their first (or second) divided differences plot.

Therefore, the aim of this chapter is to deduce an automatic and local fairing process according to [13] for given planar or spatial point sets which is based on direct manipulations of the data.

[1] $\dot{\kappa}(t_i^-)$ (respectively, $\dot{\kappa}(t_i^+)$) denotes the left-handed (respectively, right-handed) first derivative of the curvature with respect to arc length, at the point $\mathbf{X}(t_i)$.

The idea is to interpret the point set as an open or closed polygon. Then it results that *difference geometry* is advantageous in formulating suitable fairness criteria as the polygons are a simple model of this theory. For this reason the basic ideas of difference geometry are gone over before the proper planar and spatial discrete fairing algorithms are described.

An ineffective and nonautomatic proposal of such a planar algorithm was worked out in [16] and therefore is not considered in this chapter.

3.2. Basic Ideas of Difference Geometry

The development of the nearly 60-year-old theory of difference geometry is outlined in the textbook of Sauer [14] which provides a simple and readable introduction.

The *polygons* $\mathcal{P}_N^d := (\mathbf{P}_0, \mathbf{P}_1, \ldots, \mathbf{P}_N) = (\mathbf{P}_i)_{i=0}^N$ are the simplest models of difference geometry. They represent the polygonal connection of the $N+1$ vertices or points $\mathbf{P}_i \in I\!R^d$.

A polygon is called *closed* if, first, the additional condition $\mathbf{P}_0 = \mathbf{P}_N$ is fulfilled and, second, each vertex is an inner vertex. Otherwise it is called *open* and possesses the inner vertices $\mathbf{P}_1, \ldots, \mathbf{P}_{N-1}$.

If we interpret such a polygon $\mathcal{P}_N^d$ as a *discrete curve* in $I\!R^d$, then usual differential geometry of planar ($d = 2$) or spatial ($d = 3$) curves is contained in difference geometry as a limit. This is described in the following paragraphs.

Assume a regular parametric curve $\mathbf{X}(t)$, $t \in [a, b]$, to be given in $I\!R^d$. Then we associate a polygon $\mathcal{P}_N^d$ with the curve $\mathbf{X}$ by the $N+1$ vertices $\mathbf{P}_i := \mathbf{X}(t_i)$. The sequence $\tau = (t_i)_{i=0}^N$ of parameter values has to increase strictly monotone fulfilling $t_0 = a$ and $t_N = b$.

Now, the length s^N of this polygon $\mathcal{P}_N^d$ is simply the sum of the edge lengths

$$s^N = \sum_{i=1}^{N} L_i \qquad \text{with} \qquad L_i = \|\mathbf{L}_i\| \qquad \text{and} \qquad \mathbf{L}_i = \mathbf{P}_i - \mathbf{P}_{i-1}.$$

During the limiting process $N \to \infty$ (meaning a repeated refinement of the parameter sequence τ and therefore of the polygon $\mathcal{P}_N^d$), the polygon length s^N tends (together with $\max_i\{L_i\} \to 0$) to the following limiting value $s(a, b)$:

$$s(a, b) = \lim_{N\to\infty} s^N = \int_a^b \|\mathbf{X}'(u)\| du,$$

which is the well-known formula for the *arc length* of the curve in the interval $[a, b]$.

The above has been frequently used in CAGD (computer-aided geometric design) to approximate the arc length parametrization $s = s(a, t)$. For example, if suitable parameter values $(s_i)_{i=0}^N$ for a given point set $(\mathbf{P}_i)_{i=0}^N$ have to be determined (in order to start interpolation or approximation processes), then mainly the so-called *chord length* parametrization is chosen (cf. [4] and [8]):

$$s_0 = 0, \qquad s_j = \sum_{i=1}^{j} L_i \qquad (j = 1, \ldots, N).$$

Moreover, if we assign to every inner vertex of a polygon $\mathcal{P}_N^3$ a *discrete Frenet frame* by defining a tangent vector, a main normal vector, and a binormal vector in a suitable way, then it is also possible to formulate *discrete Frenet–Serret formulas* by moving the discrete Frenet frame [14]. The limiting process $N \to \infty$ then leads again to the well-known Frenet frame and Frenet–Serret formulas of the curve theory in differential geometry of $\mathbb{R}^3$ (cf. [4]). Similar arguments are valid for plane polygons and plane differential geometry.

For the following context it is only necessary to define the *discrete curvature* and the *discrete torsion* in an inner vertex of a polygon, which are the coefficients of these discrete Frenet–Serret formulas.

First, the discrete curvature K_i is defined simply as the inverse of the radius R_i of the circle passing through the three points $\mathbf{P}_{i-1}$, $\mathbf{P}_i$ and $\mathbf{P}_{i+1}$, as illustrated in Fig. 3.1.

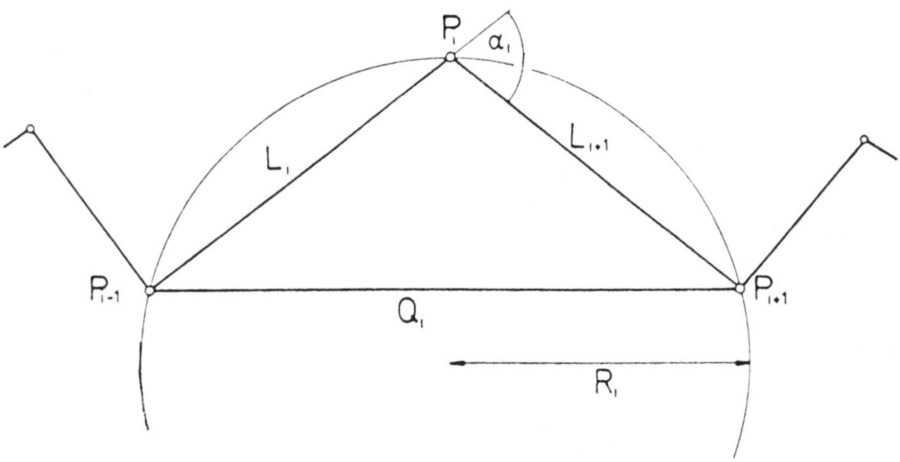

FIG. 3.1. *Discrete curvature K_i of a polygon.*

To express K_i in formulas, we have to distinguish the planar polygons $\mathcal{P}_N^2$ and the spatial polygons $\mathcal{P}_N^3$ because of the spatial problem of orientation. The same distinction is used in differential geometry where, in the case of space curves, the curvature is defined to be always positive.

So we get in the planar case

$$K_i = \frac{2\triangle_i}{L_i L_{i+1} Q_i} = \operatorname{sgn}(\triangle_i)\frac{2\sin(\alpha_i)}{Q_i} \tag{3.1}$$

with

$$Q_i = \|\mathbf{Q}_i\|, \qquad \mathbf{Q}_i = \mathbf{P}_{i+1} - \mathbf{P}_{i-1},$$

$$\triangle_i = \det(\mathbf{L}_i, \mathbf{L}_{i+1}),$$

$$\cos(\alpha_i) = \frac{\mathbf{L}_i \cdot \mathbf{L}_{i+1}}{L_i L_{i+1}}, \qquad 0 \le \alpha_i \le \pi$$

and the sign function sgn( ) as usual. Finally, in the spatial case,

$$K_i = \frac{2\sin(\alpha_i)}{Q_i}. \tag{3.2}$$

Second, the discrete torsion $\hat{T}_i$ is defined in the space with help of the angle β_i between *the osculating planes* through $\mathbf{P}_{i-2}$, $\mathbf{P}_{i-1}$, $\mathbf{P}_i$ and $\mathbf{P}_{i-1}$, $\mathbf{P}_i$, $\mathbf{P}_{i+1}$ (see Fig. 3.2):

$$\hat{T}_i = \frac{\overline{\triangle}_i}{V_{i-1}V_i} = \operatorname{sgn}(\overline{\triangle}_i)\frac{\sin(\beta_i)}{L_i}, \tag{3.3}$$

with

$$\cos(\beta_i) = \mathbf{N}_{i-1} \cdot \mathbf{N}_i, \qquad 0 \le \beta_i \le \pi,$$

$$\mathbf{N}_i = \frac{\mathbf{V}_i}{V_i}, \quad \mathbf{V}_i = \mathbf{L}_i \times \mathbf{L}_{i+1} \neq 0, \quad V_i = \|\mathbf{V}_i\|,$$

$$\overline{\triangle}_i = \det(\mathbf{L}_{i-1}, \mathbf{L}_i, \mathbf{L}_{i+1}).$$

If $V_{i-1} = 0$ or $V_i = 0$ we define $\hat{T}_i$ to be zero as no deviation of the polygon $\mathbf{P}_{i-2}$, $\mathbf{P}_{i-1}$, $\mathbf{P}_i$, $\mathbf{P}_{i+1}$ from a plane occurs.

In formula (3.3) there is the obvious handicap that no unique assignment of $\hat{T}_i$ to a vertex of the polygon is possible because of the nonsymmetric definition. To avoid this, we symmetrize the discrete torsion by

$$T_i = \frac{L_i\hat{T}_{i+1} + L_{i+1}\hat{T}_i}{L_i + L_{i+1}}. \tag{3.4}$$

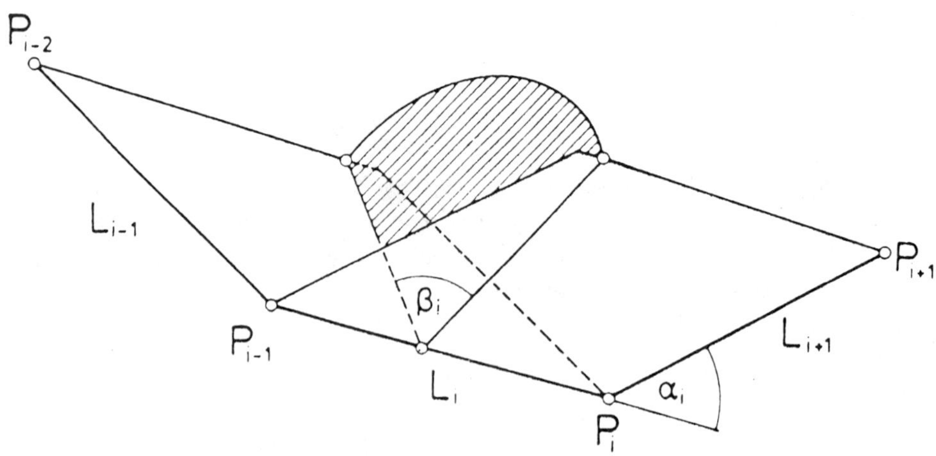

FIG. 3.2. *Discrete torsion $\hat{T}_i$ of a polygon.*

This equation can be derived by linear interpolation of $\hat{T}_i$ and $\hat{T}_{i+1}$ according to the parameter values $t_i - \frac{L_i}{2}$ and $t_i + \frac{L_{i+1}}{2}$. This choice of the parametrization may seem artificial but it represents the fact that, e.g., $\hat{T}_i$ corresponds to P_{i-1} as well as to P_i. The resulting linear polynomial $T(t)$ is

$$T(t) = \hat{T}_i + \frac{2}{L_i + L_{i+1}}\left(\hat{T}_{i+1} - \hat{T}_i\right)\left(t - t_i + \frac{L_i}{2}\right).$$

So it is obvious that $T_i = T(t_i)$. Furthermore we can use $T(t)$ to determine a (later needed) approximation of the first derivative of T_i in the inner vertex $\mathbf{P}_i$:

$$T_i' = T'(t_i) = \frac{2}{L_i + L_{i+1}}\left(\hat{T}_{i+1} - \hat{T}_i\right). \tag{3.5}$$

If we proceed in the same way to determine the first and second derivatives of the discrete curvature K_i with help of the quadratic interpolating polynomial $K(t)$ of K_{i-1}, K_i and K_{i+1} according to the chord length parametrization, we get

$$K_i' = K'(t_i) = \frac{1}{L_i + L_{i+1}}\left(L_i\frac{K_{i+1} - K_i}{L_{i+1}} + L_{i+1}\frac{K_i - K_{i-1}}{L_i}\right), \tag{3.6}$$

$$K_i'' = K''(t_i) = \frac{2}{L_i + L_{i+1}}\left(\frac{K_{i+1} - K_i}{L_{i+1}} - \frac{K_i - K_{i-1}}{L_i}\right). \tag{3.7}$$

Remarks.

- Vertices $\mathbf{P}_i$ with $K_i = 0$ are called *inflection points.*
- The straight polygons (respectively, plane polygons) are characterized by $\alpha_i = 0$ (respectively, $\beta_i = 0$) or equivalent $K_i = 0$ (respectively, $T_i = 0$) for all i.
- L_i, K_i, and T_i are euclidean invariants, i.e., they are not changed by a rigid motion. On the other hand, this is not true for scaling. Therefore, before fairing, we scale the data in all directions by a scaling factor $\lambda = N/s^N$, to compare data sets of different magnitude (up to general affine transformations). After the fairing process they are rescaled by λ^{-1}.
- A polygon $\mathcal{P}_N^3$ is (except for rigid body motions) defined uniquely by all edge lengths L_i, all curvatures K_i, and all torsions T_i. This polygon can be constructed stepwise. We assume the first three points $\mathbf{P}_0$, $\mathbf{P}_1$, and $\mathbf{P}_2$ to be given so that L_1, L_2, and K_1 are fulfilled. The position of these three points in space is not unique, but with help of a rigid body motion any other position is possible. Then the next point $\mathbf{P}_3$ is uniquely evaluated by help of the length L_3 and the two angles α_2 and β_2, which are implicitly given by K_2 and T_2. Repeating this procedure for $\mathbf{P}_4$ and so on, we finally obtain the whole polygon.

- More accurate values of K_i', K_i'' (respectively, T_i') could be achieved by interpolating polynomials of degree higher than two (respectively, one). This would increase the globality of the algorithm since more values of K_j and T_j are incorporated in the evaluations of the above derivatives. Additionally, higher-order polynomial approximations may introduce unwanted oscillations; thus we are not going to consider this idea below.
- For the evaluation of T_1 and T_{N-1} in the case of open polygons, the two points $\mathbf{P}_{-1}$ and $\mathbf{P}_{N+1}$ are necessary and these can, e.g., be defined as lying on the line through $\mathbf{P}_0$ and $\mathbf{P}_1$, respectively, $\mathbf{P}_{N-1}$ and $\mathbf{P}_N$.

3.3. Fairness Criteria

It is a major point in every existing fairing algorithm to find suitable fairness criteria, as most "definitions" of fairness are subjective and mathematically not unique (cf. [4] and [13]). Therefore, we are interested only in practical concepts which are useful for the problem considered here.

In most of the literature, a curve is declared to be fair if it minimizes the energy functional $I = \int \kappa^2(s)ds$ (where s is the arc length) which possesses a physical meaning (cf. [10] and [11]). A fair curve in such a sense has to fulfill the following two differential equations:

$$\kappa'' + \frac{1}{2}\kappa^3 - \kappa\tau^2 = 0, \tag{3.8}$$

$$\kappa'\tau + \frac{1}{2}\kappa\tau' = 0. \tag{3.9}$$

These equations were developed by Radon (cf. [1] and [15]) by determing the first variation of I. Now to get practical and suitable fairness criteria in the case of fairing of polygons (or discrete curves), we proceeded as follows:

First, we took a sufficiently large number of points belonging to a smooth curve (e.g., a cubic space curve) in order to get a smooth polygon. Then we perturbed a lot of points of this point set slightly. Afterwards, we examined all the curve invariants together with their derivatives which appear in equations (3.8) and (3.9) of Radon using the discrete curvatures and discrete torsions as introduced in §3.2.

After a systematic study of these results, we concluded that the values of K_i'' and T_i' are the most significant characteristics of a perturbed vertex $\mathbf{P}_i$. This means that fairing with respect to K_i'' and T_i' has the best fairing influence on the curvature and torsion plots of the polygon.

Therefore, we will use the following fairness criteria similarly to [13]:

local fairness: $z_i = (K_i'')^2 + (T_i')^2$ in the spatial case,
$z_i = (K_i'')^2$ in the planar case.

global fairness: $\xi = \sum_i z_i$.

Remarks.

- Here the discrete values are squared to avoid numerical problems in later calculations.
- In the spatial case both values of the local fairness criterion are weighted in a nearly equal manner because of the prescaling by factor λ.
- Please note that in the planar case the used fairness criteria are very close to the ones used in [13] which are described in §3.1. Using (3.7) we can approximately rewrite the second derivative of the discrete curvature as $K_i'' \approx \frac{1}{L}(K_{i+1}' - K_i')$ by the assumption of nearly equal point distances ($L_i \approx L$ for all i).

After completing the derivation of the fairness indicators, we separately describe the actual fairing processes for the planar and the spatial polygons in the following two sections. This is necessary because some peculiarities occur in the spatial case.

3.4. Planar Fairing Algorithm

The planar (as well as the spatial) fairing process will be organized iteratively by fairing only one point or inner vertex (denoted in the following by $\mathbf{P}_j$) of the given polygon $\mathcal{P}_N^2$ in each iteration.

In order to restrict the possible new positions $\mathbf{P}_j^\star$ of the point $\mathbf{P}_j$ to the neighbourhood of $\mathbf{P}_j$, we allow the new point $\mathbf{P}_j^\star$ to lie on the straight line $\mathbf{G}$ defined below. That represents a heavy constraint for the fairing process, yet it produces a number of advantages (discussed in the remarks later on). This line $\mathbf{G}$ is defined by

$$\mathbf{G}(\mu) = \mathbf{P}_j + \mu \mathbf{X}_j \qquad \text{with} \qquad \mathbf{X}_j = \frac{\alpha \mathbf{P}_{j+1} + \beta \mathbf{P}_{j-1}}{\alpha + \beta} - \mathbf{P}_j,$$

whereby $\mathbf{P}_j^\star = \mathbf{G}(\overline{\mu})$, as illustrated in Fig. 3.3.

Remarks.

- The definition of $\mathbf{G}(\mu)$ makes sense only at points $\mathbf{P}_j$ with nonzero curvature ($K_j \neq 0$).
- The positive numbers α and β can be chosen arbitrarily. For example, each of the two choices $\alpha = 1$ and $\beta = 1$ or $\alpha = |\mathbf{L}_j \cdot \mathbf{Q}_j|$ and $\beta = |\mathbf{L}_{j+1} \cdot \mathbf{Q}_j|$ has the advantage that the final polygon has a nearly uniform structure (meaning that no unwanted gaps occur). Next we will mainly use the second choice.
- Fairing at $\mathbf{P}_j$ is a local procedure since only the curvatures $K_{j-1}, \ldots, K_{j+1}$ and $K_{j-2}'', \ldots, K_{j+2}''$ are changed. The calculation of these values is numerically stable as only divisions by the lengths L_i and Q_i are necessary, which are always nonzero.
- Replacing $\mathbf{P}_j$ by $\mathbf{P}_j^\star = \mathbf{G}(\overline{\mu})$ in the expression of K_j'' (see (3.7)) gives a term which, unfortunately, depends on $\overline{\mu}$ in a highly nonlinear manner. This means that $\overline{\mu}$ cannot be calculated explicitly in order to minimize the local fairness criterion $z_j = (K_j'')^2$. This nonlinear effect is illustrated

in Fig. 3.4. Here the graph of the local criterion z_j is shown depending on $\overline{\mu}$ for the case of only five perturbed points where $\mathbf{P}_j$ is the midpoint and has the largest deviation.

For the proper fairing algorithm it is necessary to limit the variable $\overline{\mu}$ in an interval. To determine such an admissible range $[\overline{\mu}_{\min}, \overline{\mu}_{\max}]$ we proceed in two steps.

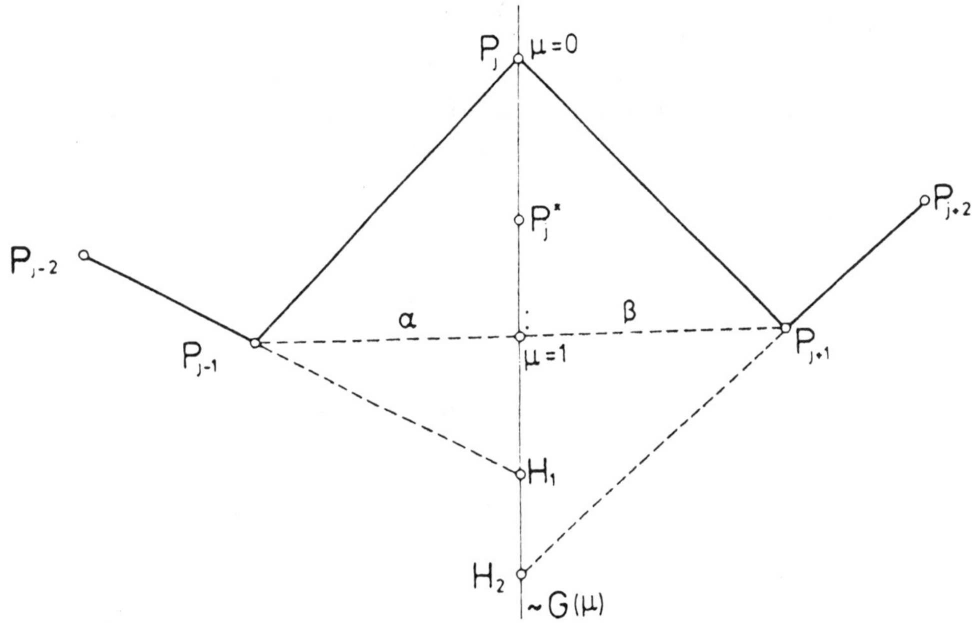

FIG. 3.3. *Changing of the point* $\mathbf{P}_j$ *on the line* $\mathbf{G}$.

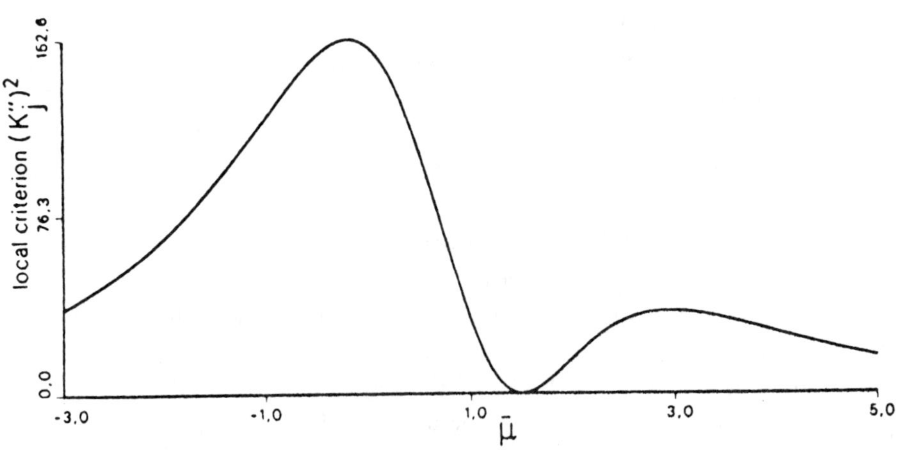

FIG. 3.4. *The local fairness criterion depending on* $\overline{\mu}$.

First, we determine the intersection points $\mathbf{H}_1 = \mathbf{G}(\mu_1)$, respectively, $\mathbf{H}_2 = \mathbf{G}(\mu_2)$ of the lines through $\mathbf{P}_{j-2}$ and $\mathbf{P}_{j-1}$ respectively, $\mathbf{P}_{j+1}$ and $\mathbf{P}_{j+2}$ with the line $\mathbf{G}$ (see Fig. 3.3). Then we set

$$\mu_{\min} = \min\{0, \mu_1, \mu_2\} \qquad \text{and} \qquad \mu_{\max} = \max\{1, \mu_1, \mu_2\}.$$

Second, we incorporate a given distance tolerance $\epsilon > 0$ by determining the two real solutions μ_3 and μ_4 (with $\mu_3 \leq \mu_4$) of the quadratic equation $\|\mathbf{P}_j - \mathbf{P}_j^\star\| = \epsilon$. Then we calculate the boundaries of the final interval:

$$\overline{\mu}_{\min} = \max\{\mu_{\min}, \mu_3\} \qquad \text{and} \qquad \overline{\mu}_{\max} = \min\{\mu_{\max}, \mu_4\}.$$

Remarks.
- For practical applications ϵ should not exceed $\frac{1}{8}$ (recall the prescaling). In general the magnitude of ϵ depends on the degree of deviation in the given data.
- If $\mathbf{H}_1$ (or similarly $\mathbf{H}_2$) does not exist because of parallelism, we use

$$\mu_{\min} = \min\{0, \mu_2\} \qquad \text{and} \qquad \mu_{\max} = \max\{1, \mu_2\}.$$

- Using $\mu_{\min}$ and $\mu_{\max}$ preserves local convexity (or similarly concavity) in the data: if $K_{j-1} > 0$, $K_j > 0$ and $K_{j+1} > 0$, then also $K_{j-1}^\star > 0$, $K_j^\star > 0$ and $K_{j+1}^\star > 0$ after changing of $\mathbf{P}_j$ to $\mathbf{P}_j^\star$.

All of the derived ideas from above are integrated in the following planar fairing algorithm (cf. [2]):

Discrete Curvature Method

0. Determine the initial ξ^0.
1. Find the largest z_j with $K_j \neq 0$.
2. Determine $\overline{\mu} \in [\overline{\mu}_{\min}, \overline{\mu}_{\max}]$ by nonlinear optimization so that $z_j^\star$ is minimized with the constraint $\xi \geq \xi^\star$.
3. Replace the changed values.
4. If convergence has been achieved exit else goto step 1.

Remarks.
- It is of further advantage to restrict the maximal number of possible modifications of each point to, e.g., five. This prevents the algorithm from significantly altering the polygon at a particular point.
- For the nonlinear optimization process in each step it is necessary to use more than one initial value to treat the problem of local minima (cf. [5]). The proper optimization is carried out by the routine DNCONF of the IMSL library (used for the spatial algorithm, too).
- If sharp corners are wanted in any area of the polygon, then an angle criterion has to be used before executing the algorithm in order to detect such regions and leave them unchanged.

To illustrate the working mode of the planar fairing method we consider an academic example. We take 31 equidistant points of the strophoid

$$\mathbf{X}(t) = \frac{t^2-1}{t^2+1}\begin{pmatrix} t \\ 1 \end{pmatrix}, \qquad t \in [-2,2],$$

and perturb any 24 inner vertices of this polygon with help of random numbers (Fig. 3.5). The amount of perturbation is nearly 30% in relation to the scaling factor λ. In Fig. 3.6, we show the discrete curvature (of the 29 inner vertices) as a function of the chord length of the polygon. The deviations are obvious as all plots are compared with the exact ones (dashed lines).

The initial global criterion ξ^0 has nearly the value 284, whereas the faired polygon, Fig. 3.7, has $\xi \approx 0.1$. This result was obtained after 120 iterations. The polygon is smooth, although the original polygon cannot be reconstructed in detail because of the relatively large perturbation.

Also the graph of the discrete curvature is smoother now and has no unwanted changes in the sign (Fig. 3.8). This is very important for further calculations with the polygon. Two further examples are presented and discussed in [2].

3.5. Spatial Fairing Algorithm

The main structure of the spatial algorithm is similar to the one of the planar algorithm described in the previous section.

Again, the first step is to define a possible new position (denoted by $\mathbf{P}_j^\star$) of a point $\mathbf{P}_j$ chosen for fairing in any iteration step. Here it would be too restrictive to limit the point $\mathbf{P}_j^\star$ on any particular line (as done in the planar case). Thus, we allow this new point to be anywhere on the plane

$$\mathbf{E}(\mu,\nu) = \mathbf{P}_j + \mu\mathbf{X}_j + \nu\mathbf{N}_j,$$

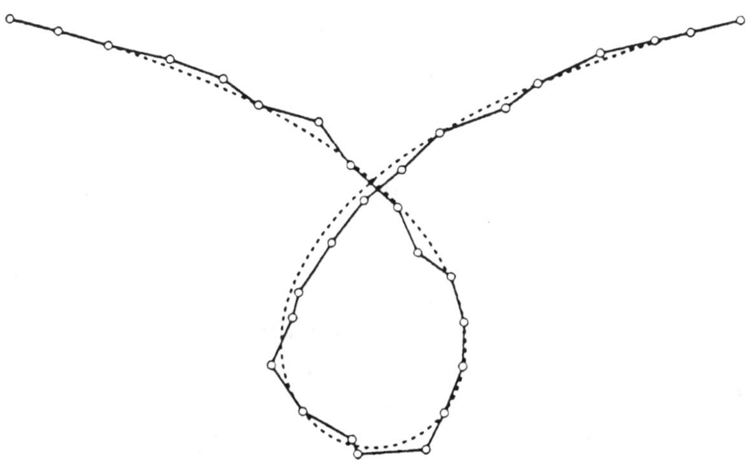

FIG. 3.5. *31 perturbed points of a strophoid.*

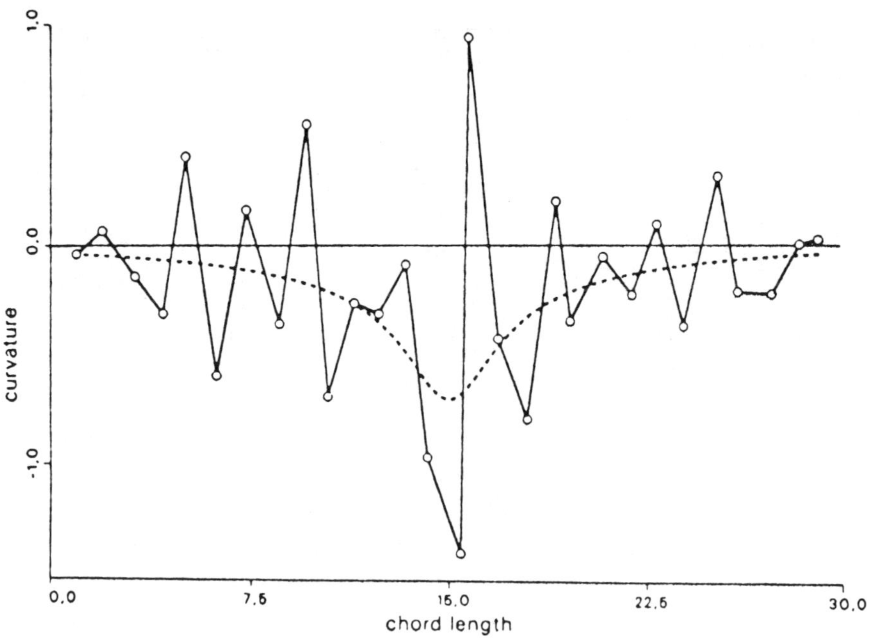

FIG. 3.6. *Curvature plot of Fig. 3.5.*

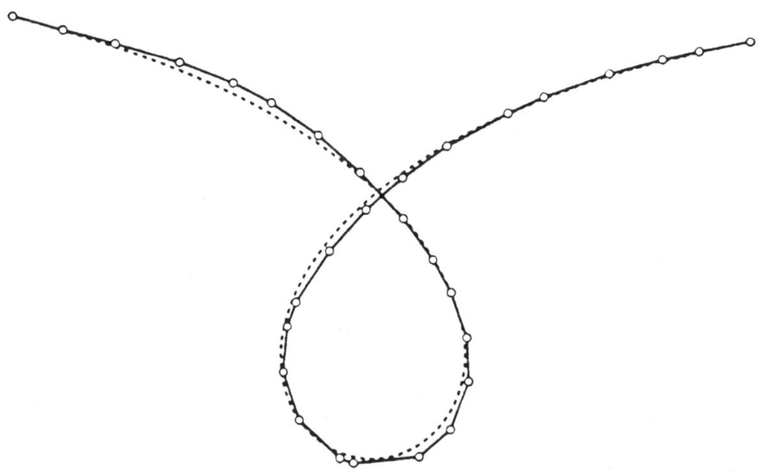

FIG. 3.7. 31 *faired points of a strophoid.*

where the normal vector $\mathbf{N}_j$ is given by (3). Then we consider two fixed parameters $\overline{\mu}$ and $\overline{\nu}$ to determine the new position by $\mathbf{P}_j^\star = \mathbf{E}(\overline{\mu}, \overline{\nu})$.

Remarks.

- This choice of $\mathbf{P}_j^\star$ ensures a nearly uniform structure of the resulting smoothed polygon as in the planar case.
- Fairing at $\mathbf{P}_j$ is a local procedure as only $K_{j-1}, \ldots, K_{j+1}$, $K''_{j-2}, \ldots$, K''_{j+2}, $T_{j-2}, \ldots, T_{j+2}$, and $T'_{j-2}, \ldots, T'_{j+2}$ are changed. But in contrast

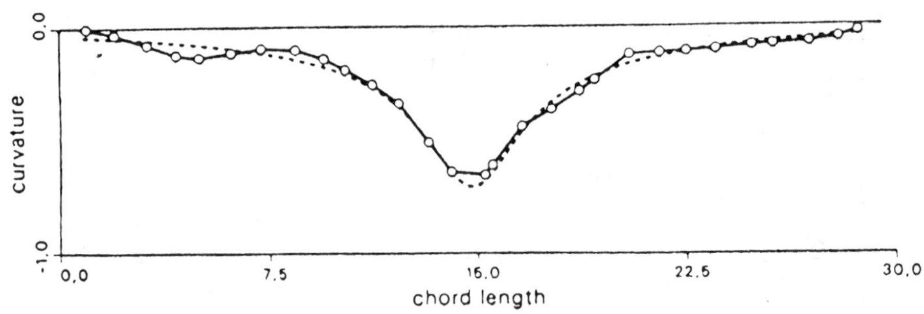

FIG. 3.8. *Curvature plot of Fig. 3.7.*

to the discrete curvatures (compare with the planar case) the numerical computations of the discrete torsion values are problematic. The reason for this can be seen in (3) where we divide by the length V_j in order to obtain the discrete normal $\mathbf{N}_j$. Otherwise the fact that V_j becomes very small during any step of the optimization process generally cannot be avoided. Therefore, a modified definition of the normal $\mathbf{N}_j$ is used to make the numerical calculations more stable:

$$\mathbf{N}_j = \frac{\tilde{\mathbf{V}}_j}{\|\tilde{\mathbf{V}}_j\|} \qquad \text{with} \qquad \tilde{\mathbf{V}}_j = \mathbf{V}_{j-1} + 50\mathbf{V}_j + \mathbf{V}_{j+1}.$$

So the normal is given by five points and depends only on $\mathbf{V}_{j-1}$ and $\mathbf{V}_{j+1}$ if $\mathbf{V}_j$ is vanishing. Here, the factor 50 is chosen arbitrarily. Numerical tests have shown that this factor influences the result only slightly if it is chosen much larger than 10.

- Again, the local criterion $z_j = (K_j'')^2 + (T_j')^2$ depends nonlinearly on $\overline{\mu}$ and $\overline{\nu}$.

An a priori determination of admissible ranges for the variables $\overline{\mu}$ and $\overline{\nu}$ (as described in the planar case) is not possible in the case of spatial fairing, so we use the distance tolerance condition $\|\mathbf{P}_j - \mathbf{P}_j^\star\| < \epsilon$ directly as a constraint for the optimization in the following algorithm:

<u>*Discrete Curvature–Torsion Method*</u>

0. Determine the initial ξ^0.
1. Find the largest z_j with $K_j \neq 0$.
2. Determine $\overline{\mu}$ and $\overline{\nu}$ by nonlinear optimization so that $z_j^\star$ is minimized with the constraints $\xi \geq \xi^\star$ and $\|\mathbf{P}_j - \mathbf{P}_j^\star\| < \epsilon$.
3. Replace the changed values.
4. If convergence has been achieved exit else goto step 1.

So far the algorithm is mainly like the planar one. However, problems occur in the spatial case if the data set is locally nearly planar. (Remember that the discrete curvature has no sign!) So it can happen that points are changed into the opposite direction than expected in order to minimize the local criterion z_j.

To avoid this misbehaviour of the algorithm, it is advantageous to define local signs for the curvatures K_{j-1}, K_j and K_{j+1}. These signs are used exclusively for the calculation of the quantity K_j'' due to equation (3.7). For this purpose, the five points $\mathbf{P}_{j-2}, \ldots, \mathbf{P}_{j+2}$ are projected onto the plane defined by the normal $\mathbf{N}_j$ and the point $\mathbf{P}_j$. Then we determine the local signs of K_{j-1}, K_j and K_{j+1} relative to this local plane with help of the five projected points in the same way as in the planar situation. Thus, the number of sign changes can be 0, 1, or 2 as pointed out in detail in [6].

These additional modifications provide good results, as illustrated in the following example of 42 points of the screwing line:

$$\mathbf{X}(t) = \begin{pmatrix} \sin(2\pi t) \\ \cos(2\pi t) \\ 32t(t-\frac{3}{4})^2 - 1 \end{pmatrix}, \qquad t \in [0,1].$$

In Fig. 3.9 two different projections of the polygon are illustrated. It can be seen that all 40 inner points are slightly perturbed in all three dimensions with help of random numbers. In Fig. 3.10 the associated diagrams of the discrete curvatures and torsions (of the inner vertices) are shown in comparison to the original (not perturbed) ones.

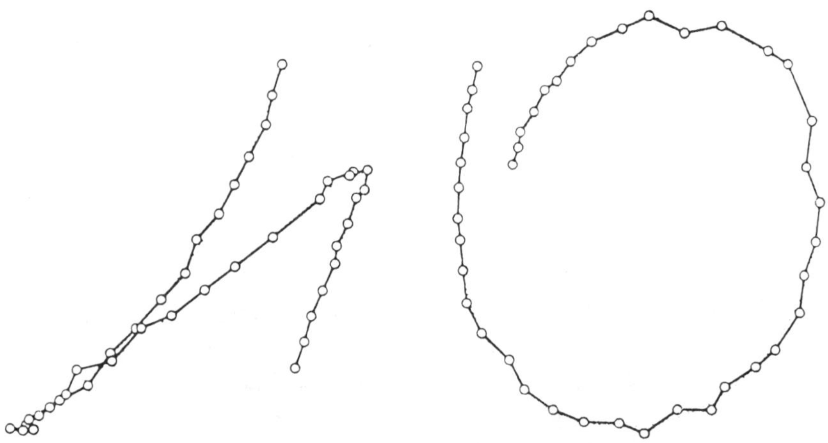

FIG. 3.9. *42 perturbed points of a screwing line (two different projections).*

The result of the spatial fairing algorithm (Fig. 3.11) is obtained after nearly 200 iterations using a slightly modified version of the algorithm as follows: (a) the algorithm is allowed to operate on a specific point at most once (this produces a more *regular* fairing of the points), and (b) during the first two iterations of the algorithm, the local fairness criterion used is $z_j = 5(K_j'')^2 + (T_j')^2$, so that at first fairing is focused on curvature.

The graphs of the discrete curvature as well as the discrete torsion are smoother after fairing, although the torsion plot has more changes in the sign

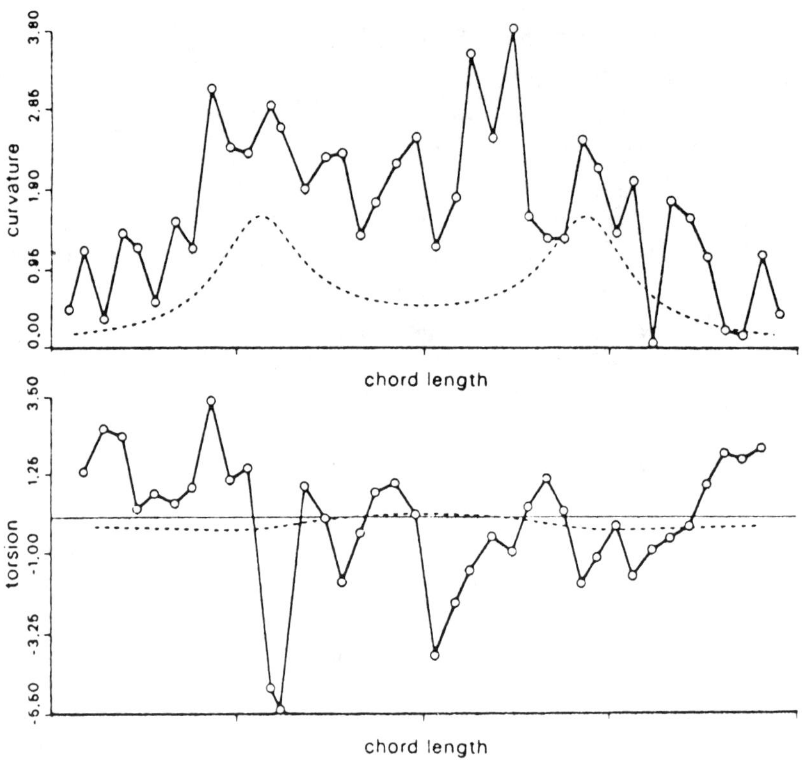

FIG. 3.10. *Curvature and torsion plots of Fig.* 3.9.

than the one of the original polygon (Fig. 3.12). The perturbation of the polygon was too large for exactly reconstructing the torsion plot.

3.6. Concluding Remarks

The methods described in this chapter aim at solving the problem of fairing planar and spatial point sets. The fairing is accomplished by using some geometrical invariants from difference geometry like discrete curvature and discrete torsion. The planar and the spatial fairing algorithms are different in some way as the theory distinguishes between the two- and three-dimensional case.

Both algorithms are working on principle by iteratively changing only one point in each iteration step in order to minimize a local criterion (we have also tested the changing of two points in each step without obtaining good results). Although the problem is inherently nonlinear, the proposed fairing techniques are simple yet effective, as the discussed examples demonstrate.

Furthermore it is guaranteed that in every iteration, (a) the *faired* point is within a given distance tolerance from the original point, and (b) a global fairness criterion is also decreased (this is important for ensuring that the final point set is, as a whole, fairer than the initial one). Finally, we note that the

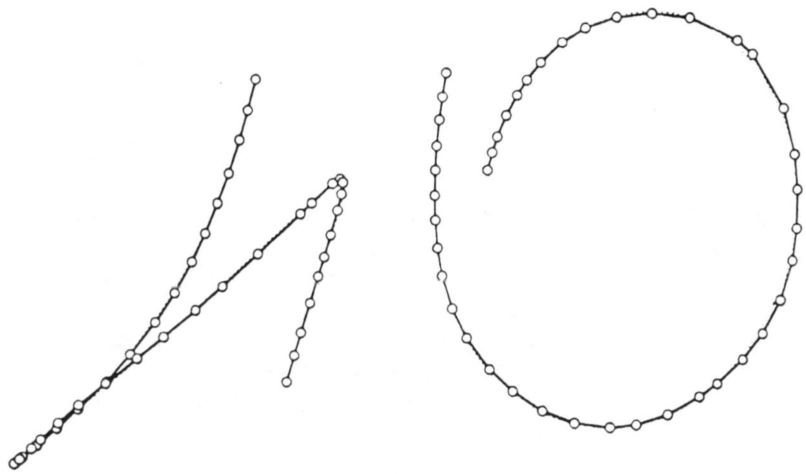

FIG. 3.11. 42 *faired points of a screwing line (two different projections).*

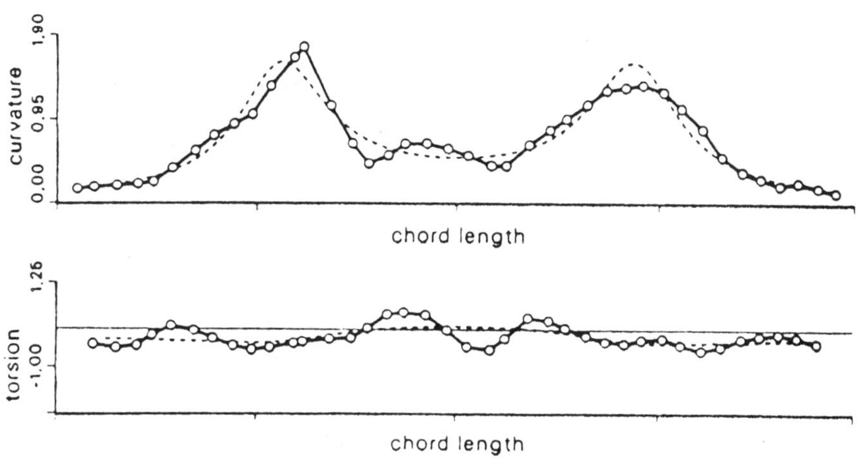

FIG. 3.12. *Curvature and torsion plots of Fig.* 3.11.

planar algorithm guarantees *shape preservation* by restricting modification of points.

Regarding stability, we have observed that the planar algorithm performs satisfactorily even with large deviations in the point set (assuming that the points are not too irregular—i.e., the point set includes no gaps). On the other hand, the spatial algorithm does have some stability problems due to some insufficiencies of the theory of difference geometry in three dimensions (see §3.2). Nevertheless, the proposed algorithm gives good results for data which are not *too rough* (see Figs. 3.9–3.12).

Acknowledgments

The authors thank the Editor, Nickolas S. Sapidis, for many helpful hints and comments.

References

[1] W. Blaschke, *Vorlesungen ueber Differentialgeometrie* I. Third ed., Chelsea, New York, 1967.

[2] M. Eck, *Geometrische Verfahren zur dreidimensionalen Osteotomieplanung.* Ph.D thesis, Tech. Hochschule Darmstadt, Germany, 1991.

[3] G. Farin, G. Rein, N. Sapidis, and A. J. Worsey, *Fairing cubic B-spline curves.* Comput. Aided Geom. Design, 4 (1987), pp. 91–103.

[4] G. Farin, *Curves and Surfaces for Computer Aided Geometric Design. A Practical Guide.* Academic Press, New York, 1988.

[5] R. Fletcher, *Practical Methods of Optimization.* John Wiley, New York, 1990.

[6] T. N. T. Goodman, *Inflection on curves in two and three dimensions.* Comput. Aided Geom. Design, 8 (1991), pp. 37–50.

[7] M. Hosaka, *Theory of curves and surface synthesis and their smooth fitting.* Information Processing in Japan, 9 (1969), pp. 60–68.

[8] J. Hoschek and D. Lasser, *Grundlagen der geometrischen Datenverarbeitung.* Teubner, Leipzig, Germany, 1989.

[9] J. A. P. Kjellander, *Smoothing of cubic parametric splines.* Comput. Aided Design, 15 (1983), pp. 288–293.

[10] E. T. Y. Lee, *Energy, fairness and a counterexample.* Comput. Aided Design, 22 (1990), pp. 37–40.

[11] H. Meier and H. Nowacki, *Interpolating curves with gradual changes in curvature.* Comput. Aided Geom. Design, 4 (1987), pp. 297–306.

[12] W. Renz, *Interactive smoothing of digitized point data.* Comput. Aided Design, 14 (1982), pp. 267–269.

[13] N. Sapidis and G. Farin, *Automatic fairing algorithm for B-spline curves.* Comput. Aided Design, 22 (1990), pp. 121–129.

[14] R. Sauer, *Differenzengeometrie.* Springer-Verlag, New York, 1970.

[15] G. Schulze, *Elastische Wege und Nichtlineare Splines im CAGD.* Ph.D thesis, University of Kaiserslautern, Germany, 1990.

[16] B. Su and D. Liu, *Computational Geometry. Curve and Surface Modelling,* Academic Press, New York, 1989.

CHAPTER

4

Tight String Method to Fair Piecewise Linear Curves

Mark Feldman

4.1. Introduction

By fairing of a piecewise linear curve (PLC), or a pointset, we mean adjusting its vertices within given tolerance bands while minimizing variation in discrete curvature.

In computer-aided design (CAD), fairing of a PLC plays an important role. A PLC usually emerges as a set of sampled points of a three-dimensional object, or as a sequence of measured parameters of a physical process.

For example, a clay model of an automobile body is produced by a designer at a certain stage of the design process. The model is then digitized along the feature lines, and PLCs thus obtained are used to create a computer model of the automobile body.

Another general example of a PLC is a control polygon of a B-spline curve.

B-spline curves possess a number of properties that make them indispensable for CAD applications. The following three items are the key properties that allow the shape of a B-spline curve to be improved by fairing its control polygon:

- B-spline curves have the variation diminishing property.
- B-spline curves have the local control property.
- B-spline curves preserve the maximum perturbation distance.

The last property means that if the control polygon of a B-spline curve is perturbed, the distance between the perturbed and the original B-spline curve is not greater than the maximum of the perturbation distances of the control points. This property simply follows from the convex hull property.

To the best of this author's knowledge, there are few publications and reports on fairing of PLCs (see [2], [4], and [6]).

In this chapter, we introduce an approach to a fairing procedure that has been developed and implemented at Camax Systems, Inc.(see [8]) which meets the following requirements:

- Fair nonuniformly spaced pointsets with large deflections.

- Create a faired pointset to conform to given tolerances.
- Create a faired pointset with a minimal number of inflection points.
- Make the curvature of the faired pointset vary evenly.

The fairing procedure is implemented in two stages:

(1) Rough fairing.

(2) Fine fairing.

The mechanical analogy to the rough fairing process would be the transformation of the shape of a string put loosely through the tolerance gates (the original shape) and then tightened by pulling its ends (the tight string shape).

We introduce a finite algorithm that recursively creates a rough-faired shape from the extreme points of convex sets.

The rough fairing stage creates the shape that satisfies all the above requirements except for the last, i.e., this shape does not have even curvature variation.

The fine fairing stage makes this necessary improvement by using the tight string shape as an initial vector for the optimization procedure, and the number of the inflection points of the tight string shape as well as the tolerance gates as constraints.

Using the length of the curvature polygon as a measure of even curvature variation, we find the fine-faired shape as a solution of the constrained optimization problem.

4.2. Extreme Points of Convex Sets

In this section, we recall some definitions and facts from the theory of convex sets (see [1] and [7] for more details).

Let X be a linear vector space, and let K be a subset of X.

DEFINITION 4.2.1. *K is called convex if for every k_1, $k_2 \in K$, and $0 < a < 1$, $ak_1 + (1-a)k_2 \in K$.*

DEFINITION 4.2.2. *The convex hull of K is a minimal convex set that contains K. We denote the convex hull of K by $\mathcal{CH}(K)$.*

DEFINITION 4.2.3. *A nonempty subset $A \subset K$ is called an extreme subset of K, if for every k_1, $k_2 \in K$, and $0 < a < 1$, $ak_1 + (1-a)k_2 \in A$ implies k_1, $k_2 \in A$.*

DEFINITION 4.2.4. *An extreme subset of K which consists of a single point is called an extreme point.*

Example. If K is a cube in R^3, vertices, edges, and facets are the extreme subsets of K, and vertices are the extreme points of K.

DEFINITION 4.2.5. *Let f be a real function whose domain is a subset K of R^m. The set $\{(k,a) | k \in K, a \in R^1, a \geq f(k)\}$ is called the epigraph of f and is denoted by* epi(f). *We define f to be a convex function on K if* epi(f) *is convex as a subset of R^{m+1}. A concave function on K is a function whose negative is convex.*

FACT 4.2.1. *$\mathcal{CH}(K)$ can be constructed by taking convex combinations of the points of K, i.e., $\mathcal{CH}(K) = \{\sum a_i k_i, k_i \in K, a_i > 0, \sum a_i = 1\}$.*

FACT 4.2.2. *If e is an extreme point of $\mathcal{CH}(K)$, then $e \in K$.*

FACT 4.2.3. *Let $K = \{P_i\}$, $(i = 1, \ldots, n)$ be a finite set of points in R^m, and E be the set of extreme points of $\mathcal{CH}(K)$. Then E is not empty, and E is a subset of K.*

FACT 4.2.4. *Let $\mathcal{P}, \mathcal{Q}$ be two closed polygons without self-intersections in R^2, and let $\mathcal{P}$ be within $\mathcal{Q}$. If the region bounded by $\mathcal{P}$ is convex, the length of $\mathcal{P}$ is not greater than the length of $\mathcal{Q}$.*

FACT 4.2.5. *Let f be a real function whose domain $K \subset R^m$ is convex. Then f is convex on K if and only if $f((1-a)k_1 + ak_2) \leq (1-a)f(k_1) + af(k_2)$, $0 < a < 1$, for every k_1 and k_2 in K.*

Fact 4.2.3 follows from Fact 4.2.1, Fact 4.2.2, and the following theorem.

THEOREM 4.2.1 (*Krein-Milman*, 1940). *If K is a convex compact subset in a locally convex linear topological space and E is the set of extreme points of K, then K is equal to the closure of the convex hull of E.*

4.3. Rough Fairing Process

In this section, we introduce an algorithm to find the shape of a tight string that passes through the tolerance gates.

Let $(L_i, X_i)\,(i = 0, \ldots, n)$ be a two-dimensional pointset, such that the first coordinate L is strictly monotonic.

We define a tolerance gate $G_i\,(i = 0, \ldots, n)$ as a segment $[(L_i, X_i + LoTol_i), (L_i, X_i + UpTol_i)]$, where the upper tolerance $UpTol_i$ is greater than or equal to the lower tolerance $LoTol_i$, and $LoTol_0 = UpTol_0$, $LoTol_n = UpTol_n$.

For each position i, $0 < i < n$, we will find a solution value $\tilde{X}_i$ (or a solution point $\tilde{P}_i = (L_i, \tilde{X}_i)$) such that (1) $\tilde{P}_i$ belongs to the tolerance gate G_i; (2) points $\tilde{P}_i(i = 0, \ldots, n)$ form a tight string shape.

We will call the pointset $\mathcal{P} = \{\tilde{P}_i\}$ $(i = 0, \ldots, n)$ a solution pointset, and the polygon through these points a solution polygon.

We will call the pointset $\mathcal{Q} = \{Q_i\}$ $(i = 0, \ldots, n)$ a feasible pointset if Q_i belongs to the tolerance gate $G_i(i = 0, \ldots, n)$. The polygon through points Q_i will be called a feasible polygon.

For any two-dimensional pointset $Q_i(i = 0, \ldots, n)$, a segment $[Q_j, Q_{j+1}]$ $(0 < j < n-1)$ is called an inflection segment if the two shapes Q_{j-1}, Q_j, Q_{j+1}, and Q_j, Q_{j+1}, Q_{j+2} are such that one is concave, and one is convex.

We will say that (i_1, i_2) $(0 \leq i_1 < n, 0 < i_2 \leq n)$ is an unsolved interval at some stage of recursion, if we have the following:

(1) positions i_1 and i_2 are already solved;

(2) $i_1 + 1 < i_2$;

(3) all positions between i_1 and i_2 are not solved yet.

We denote by $SL(P_1, P_2)$ a straight line through points P_1, P_2.

If $P(x, t_1)$ and $Q(x, t_2)$ are two points in the plane, we will say that Q is located above (below) P, if t_2 is greater (less) than t_1.

Tight String Algorithm

1. Initialization

Solve the first and the last position by setting $\tilde{X_0} = X_0 + LoTol_0$, $\tilde{X}_n = X_n + LoTol_n$.

2. Perform recursive step until all positions are solved

For each unsolved interval (i_1, i_2)

- create pointset
  $S_{la} = S_{la}(i_1, i_2) = \{A_i = (L_i, X + LoTol_i) : i_1 < i < i_2$ and A_i is located above the $SL(\tilde{P}_{i_1}, \tilde{P}_{i_2})\}$, where la stands for *lower above*
- create pointset
  $S_{ub} = S_{ub}(i_1, i_2) = \{B_i = (L_i, X_i + UpTol_i) : i_1 < i < i_2$ and B_i is located below the $SL(\tilde{P}_{i_1}, \tilde{P}_{i_2})\}$, where ub stands for *upper below*
- create pointset
  $S = S(i_1, i_2) = S_{la} \cup S_{ab} \cup \{\tilde{P}_{i_1}\} \cup \{\tilde{P}_{i_2}\}$
- *if*
  S includes only solution points $\tilde{P}_{i_1}$ and $\tilde{P}_{i_2}$, i.e., S_{la} and S_{ub} are both empty sets, then all the positions between i_1 and i_2 are solved.
  For each i, $i_1 < i < i_2$, set $\tilde{X}_i$ equal to the second coordinate of the point that belongs to the $SL(\tilde{P}_{i_1}, \tilde{P}_{i_2})$ and has the first coordinate equal to L_i.
  In other words, position all the solution points points $\tilde{P}_i$, $i_1 < i < i_2$, on the $SL(\tilde{P}_{i_1}, \tilde{P}_{i_2})$.
  We call these positions (points) indifferent because they are not active in forming the shape of the tight string.
- *else*
  find the convex hull $\mathcal{CH}(S)$ of the pointset S
  for each point $A_i \in S_{la}$ $(B_i \in S_{ub})$:
  if A_i (B_i) is an extreme point of $\mathcal{CH}(S)$, position i is solved. Set $\tilde{P}_i$ equal to $A_i(B_i)$.

Now we will discuss some important properties of the tight string algorithm and the solution polygon.

PROPERTY 4.3.1. *The tight string algorithm is finite, i.e., all the positions will be solved after a finite number of steps.*

The proof follows simply from Fact 4.2.3.

PROPERTY 4.3.2. *Let (i_1, i_2) be an unsolved interval at some step of the recursion. Then every solution point $\tilde{P}_i$, $i_1 < i < i_2$, belongs to the $\mathcal{CH}(S(i_1, i_2))$.*

Proof. It is sufficient to show that if (j_1, j_2), $i_1 \leq j_1 < j_2 \leq i_2$, is an unsolved interval and $\tilde{P}_{j_1}$, $\tilde{P}_{j_2}$ belong to the $\mathcal{CH}(S(i_1, i_2))$, then every solution $\tilde{P}_i$, $j_1 < i < j_2$, belongs to the $\mathcal{CH}(S(i_1, i_2))$ (see Fig. 4.2).

If i is an indifferent position, the proof follows simply from the fact that $\tilde{P}_i$ belongs to the $SL(\tilde{P}_{j_1}, \tilde{P}_{j_2})$.

Let us assume that $\tilde{P}_i$ belongs to $S_{la}(j_1, j_2)$ (the case of S_{ub} is analogous). Now we have to consider the following three cases:

(a) $\tilde{P}_i$ is located above the $SL(\tilde{P}_{i_1}, \tilde{P}_{i_2})$.

In this case, $\tilde{P}_i$ belongs to $S_{la}(i_1, i_2)$. Therefore, $\tilde{P}_i$ belongs to $S(i_1, i_2)$ and to the $\mathcal{CH}(S(i_1, i_2)$.

(b) $\tilde{P}_i$ is located on the $SL(\tilde{P}_{i_1}, \tilde{P}_{i_2})$. This case is trivial.

(c) $\tilde{P}_i$ is located below the $SL(\tilde{P}_{i_1}, \tilde{P}_{i_2})$.

Because of the assumption above, $\tilde{P}_i$ is located above the $SL(\tilde{P}_{j_1}, \tilde{P}_{j_2})$. Therefore $\tilde{P}_i$ belongs to the $\mathcal{CH}(\tilde{P}_{i_1}, \tilde{P}_{i_2}, \tilde{P}_{j_1}, \tilde{P}_{j_2})$, that is included in $\mathcal{CH}(S(i_1, i_2))$. □

PROPERTY 4.3.3. *Let three sequential solution points $\tilde{P}_i$, $\tilde{P}_{i+1}$, $\tilde{P}_{i+2}$ form a concave (convex) shape. If $\tilde{P}_{i+1}$ is not an indifferent point, it is a lower (upper) point of the tolerance gate G_{i+1}.*

Proof. Let $\tilde{P}_{i+1}$ be located above the $SL(\tilde{P}_i, \tilde{P}_{i+2})$, i.e., 3 points form a concave shape (the convex case is analogous). It is sufficient to show that $\tilde{P}_{i+1}$ is not an upper point of the tolerance gate.

Let us assume, on the contrary, that $\tilde{P}_{i+1}$ is an upper point of G_{i+1}. Suppose that a solution $\tilde{P}_{i+1}$ was found when some unsolved interval (i_1, i_2), $i_1 \leq i, i+2 \leq i_2$, was solved. Thus $\tilde{P}_{i+1}$ is located below the $SL(\tilde{P}_{i_1}, \tilde{P}_{i_2})$ because $\tilde{P}_{i+1}$ can only belong to $S_{ub}(i_1, i_2)$.

Now, we have $\tilde{P}_{i+1}$ located above the $SL(\tilde{P}_i,\ \tilde{P}_{i+2})$ and below the $SL(\tilde{P}_{i_1},\ \tilde{P}_{i_2})$. Therefore $\tilde{P}_{i+1}$ belongs to the $\mathcal{CH}(\tilde{P}_{i_1}, \tilde{P}_{i_2}, \tilde{P}_i, \tilde{P}_{i+2})$, so it can be represented as a convex combination of these 4 points, and is not equal to any of them. As it follows from Property 4.3.2, $\tilde{P}_i$ and $\tilde{P}_{i+2}$ belong to the $\mathcal{CH}(S(i_1, i_2))$. Therefore, $\tilde{P}_i$ is not an extreme point of $\mathcal{CH}(S(i_1, i_2))$. This contradiction completes the proof.

THEOREM 4.3.1. *From all feasible polygons, the solution polygon has a minimal number of inflection segments, and minimal length.*

Proof. Let $\mathcal{P} = \{\tilde{P}_i\}$ $(i = 0, \ldots, n)$ be a solution polygon. Without loss of generality we can assume that $\mathcal{P}$ does not have indifferent points.

A. First we will prove that $\mathcal{P}$ has a minimal number of inflection segments. In order to simplify the proof, let us assume first that $\mathcal{P}$ has one inflection segment, namely $[\tilde{P}_{i_0}, \tilde{P}_{i_0+1}]$. Without loss of generality, we can assume that shape $\tilde{P}_0, \ldots, \tilde{P}_{i_0}, \tilde{P}_{i_0+1}$ is concave, and the shape $\tilde{P}_{i_0}, \tilde{P}_{i_0+1}, \ldots, \tilde{P}_n$ is convex.

As it follows from Property 4.3.3, $\tilde{P}_{i_0}$ is the lower point of G_{i_0}, and $\tilde{P}_{i_0+1}$ is the upper point of G_{i_0+1}.

Let $\mathcal{Q} = \{Q_i\}$ $(i = 0, \ldots, n)$ be a feasible polygon. As it follows from the conclusion above, Q_{i_0} is not below $\tilde{P}_{i_0}$, and Q_{i_0+1} is not above $\tilde{P}_{i_0+1}$.

Therefore segments $[Q_{i_0}, Q_{i_0+1}]$ and $[\tilde{P}_{i_0}, \tilde{P}_{i_0+1}]$ have at least one common point, namely T.

Let us assume that T is an end point of the segment $[\tilde{P}_{i_0}, \tilde{P}_{i_0+1}]$; for definiteness, let T coincide with $\tilde{P}_{i_0}$. As far as Q_{i_0+1} is not located above $\tilde{P}_{i_0+1}$, it is clear that Q_{i_0} is located above the $SL[Q_0, Q_{i_0+1}]$. This shows that the portion of the polygon $\mathcal{Q}$ between positions 0 and $i_0 + 1$ is not convex. Analogously, the portion of the polygon $\mathcal{Q}$ between positions i_0 and n is not concave. This proves that the polygon $\mathcal{Q}$ has at least one inflection segment.

Now assume that T is not the end point of the segment $[\tilde{P}_{i_0}, \tilde{P}_{i_0+1}]$. As far as $\tilde{P}_{i_0}$ is located above the $SL[\tilde{P}_0, T]$, it is clear that Q_{i_0} is located above the same line, and the portion of the polygon $\mathcal{Q}$ that begins at Q_0 and ends at T is not convex. Analogously, the portion of the polygon $\mathcal{Q}$ that begins at T and ends at Q_n is not concave, and therefore $\mathcal{Q}$ has at least one inflection segment.

Now let us consider the general case. As has been already shown, the solution polygon $\mathcal{P}$ and the feasible polygon $\mathcal{Q}$ intersect each other within each inflection segment of the polygon $\mathcal{P}$.

Let m be the number of the inflection segments of $\mathcal{P}$, and let T_j, $(j = 1, \ldots, m)$ be the sequence of the intersection points within the inflection segments. We add to this sequence two end points, namely $T_0 = \tilde{P}_0$ and $T_{m+1} = \tilde{P}_n$. Now for each $j = 1, \ldots, m$, we can consider portions of polygons $\mathcal{P}$ and $\mathcal{Q}$ from the point T_{j-1} to the point T_{j+1}. These portions coincide at the ends, and the solution polygon has only one inflection segment between T_{j-1} and T_{j+1}. Therefore, we can apply the reasoning above to show that the feasible polygon $\mathcal{Q}$ has at least one inflection segment between T_{j-1} and T_{j+1}. This completes the proof of §**A**.

B. Now we will prove that polygon $\mathcal{P}$ has minimal length. Let us first assume that $\mathcal{P}$ has no inflection segments. Without loss of generality, we can assume that $\mathcal{P}$ is concave.

As it follows from Property 4.3.3, every point Q_i is positioned not below the point $\tilde{P}_i (i = 0, \ldots, n)$. As far as $Q_0 = \tilde{P}_0$ and $Q_n = \tilde{P}_n$, we can conclude that the closed polygon $\tilde{P}_0, \ldots, \tilde{P}_n, \tilde{P}_0$ is within the closed polygon $\tilde{Q}_0, \ldots, \tilde{Q}_n, \tilde{Q}_0$.

From the definition of a convex function, it is easy to see that polygon $\tilde{P}_0, \ldots, \tilde{P}_n, \tilde{P}_0$ bounds a convex region. Therefore we can apply Fact 4.2.4 to prove that the length of $\mathcal{P}$ is not greater than the length of $\mathcal{Q}$.

To prove the general case, we use the same approach as in §**A**. That is, for each $j = 0, \ldots, m$, consider portions of $\mathcal{P}$ and $\mathcal{Q}$ that begin at T_j and end at T_{j+1}, and apply to them the reasoning above. This completes the proof of the theorem.

The process of rough fairing is illustrated in Figs. 4.1 and 4.2. We conclude this section with a complexity analysis of the tight string algorithm.

PROPOSITION 4.3.1. *The tight string algorithm can be performed in $O(N^2)$ time, where N is the number of points in a two-dimensional pointset. A concave (convex) tight string shape can be constructed in $O(N)$ time.*

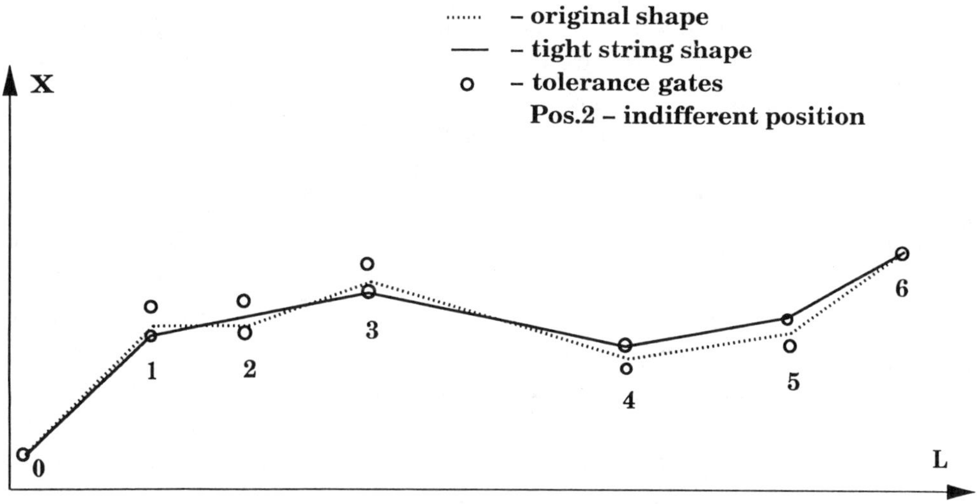

FIG. 4.1. *Shape transformation after rough fairing.*

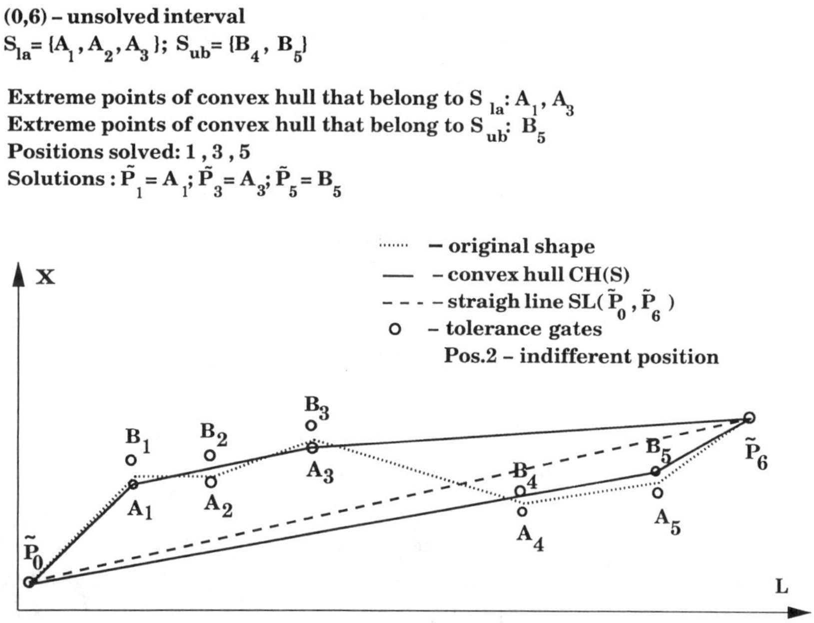

FIG. 4.2. *Rough fairing: main step of recursion.*

Proof. As is shown in [5], the convex hull of a N-vertex polygon without self-intersections can be constructed in $O(N)$ time. Taking into account that all other employed computational procedures are linear in N, we can conclude that each recursive step can be performed in $O(N)$ time. As far as the number of unsolved positions decreases at least by one at each step of recursion, $O(N^2)$ gives us a worst-case performance time.

Assume now that the tight string shape is concave (convex). It is easy to see that all the positions that are not solved at the first step of recursion are the indifferent positions. Therefore, it will take no more than $O(N)$ time to position each unsolved point onto the appropriate straight line. □

4.4. Fine Fairing Process

The stage of rough fairing builds the shape that satisfies all the aforementioned requirements except for the last, i.e., that the shape has an even curvature variation. We will further consider the length of the curvature polygon as a measure for even curvature variation.

The stage of fine fairing minimizes the length of the curvature polygon by using the vector $\{\tilde{X}_i\}$ $(i = 0, \ldots, n)$ as an initial vector for the optimization procedure, and the number of the inflection segments of the tight string shape as well as the tolerance gates as constraints.

Let C_i denote the discrete curvature at point P_i that is a signed reciprocal of the radius of a circle built through points P_{i-1}, P_i, P_{i+1} (the sign is set to minus if points P_{i-1}, P_i, P_{i+1} form a concave shape, and is set to plus otherwise).

Let $LCRVPOL(P_i)(i = 0, \ldots, n)$ denote the length of the curvature polygon that is the polygon built through points $(L_i, C_i)(i = 0, \ldots, n)$.

We find the fine-faired shape as a solution of the following optimization problem.

Optimization problem

Find $\tilde{\tilde{X}}_i(i = 0, \ldots, n)$ *that minimize* $LCRVPOL((L_i, \tilde{\tilde{X}}_i))$ *subject to the following constraints:*

(A) $\tilde{\tilde{X}}_i$ *belongs to* $G_i(i = 0, \ldots, n)$*;*

(B) *the number of sign changes in the sequence* $\{\tilde{\tilde{C}}_i\}$ *is not greater than the number of sign changes in the sequence* $\{\tilde{C}_i\}$ $(i = 0, \ldots, n)$*.*

This is a multivariate constraint optimization problem with a simple bound constraint (A) and a nonsmooth constraint (B).

Although in theory constraint (B) seems to be redundant, it is sometimes necessary in practical implementation, especially in the case of almost flat shapes. Methods for solving the optimization problem depend on whether only constraint (A) or both constraints (A) and (B) are chosen (see [3] for possible methods to solve this optimization problem).

The process of fine fairing is illustrated in Fig. 4.3.

Remark 4.4.1. So far, we have discussed fairing of two-dimensional pointsets. If a three-dimensional pointset $(X_i, Y_i, Z_i)(i = 0, \ldots, n)$ has to be faired, practical and sound results can be achieved by fairing each of the three two-dimensional pointsets (L_i, X_i), (L_i, Y_i), and (L_i, Z_i) $(i = 0, \ldots, n)$, where L_i is the cumulative chordal length of the original pointset.

An example of such a approach is illustrated in Figs. 4.4–4.7. The initial set of 69 data points was obtained by digitizing along the feature line of a clay

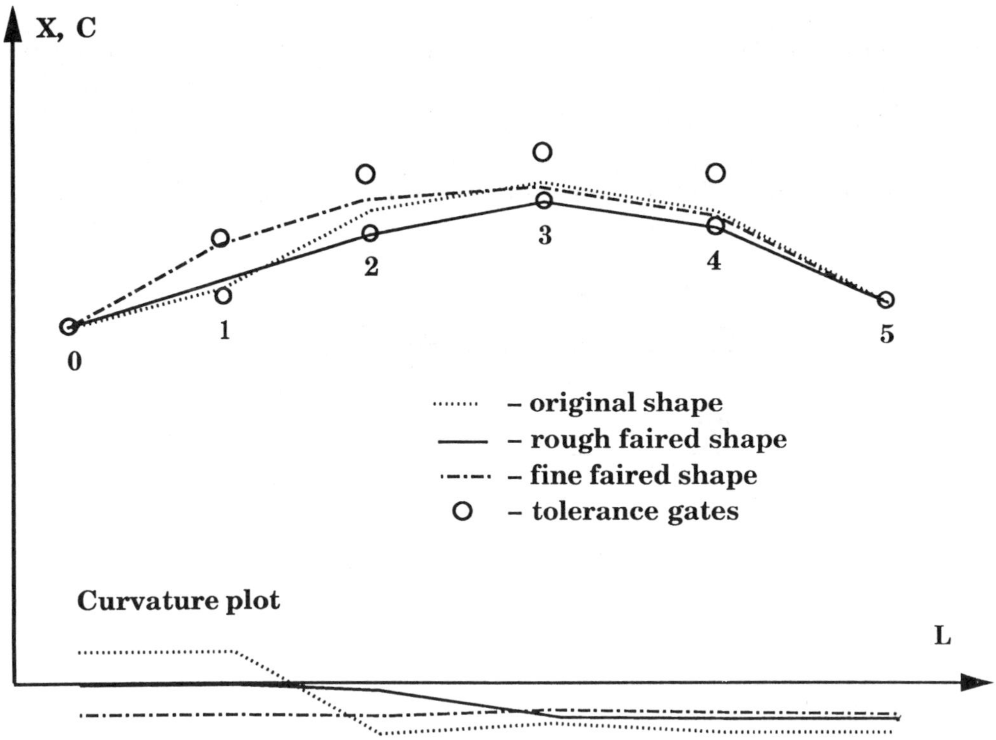

FIG. 4.3. *Fine fairing.*

model of a car hood (see Fig. 4.4). Because of the large deflections of data points, input tolerances varied from 0.5 to 2 mm. After the stage of rough fairing the initial pointset was reduced to 13 points that captured the shape.

The fine fairing procedure was applied to each of the three two-dimensional pointsets (L_i, X_i), (L_i, Y_i), and (L_i, Z_i). As a result of fairing, all inflection points were eliminated, and the length of the curvature polygon has decreased by 1.9%, 62.0%, and 10.8%, respectively. Discrete curvature plot for the projections of the initial and faired pointset onto coordinate planes is given in Fig. 4.5.

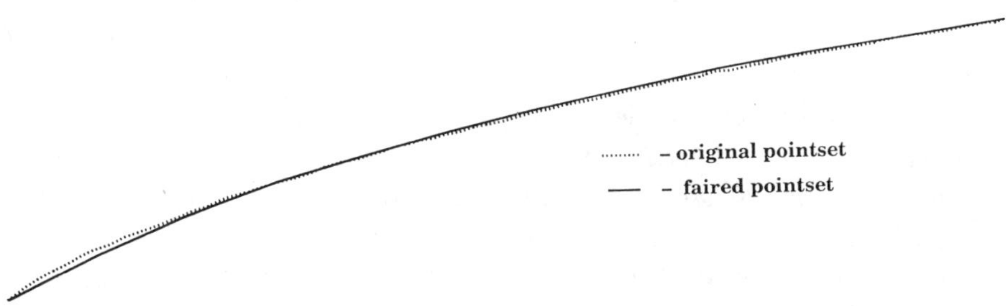

FIG. 4.4. *Fairing of digitized data points.*

Curvature plot for the projections of the interpolating spline and spline itself is given in Figs. 4.6 and 4.7. As a result of fairing, all inflection points of the interpolating spline were eliminated, and the length of the curvature plot for the projections of the interpolating spline has decreased by 2.8%, 70.8%, and 11.9%, respectively. It is worth mentioning shape similarity between the discrete curvature plot of a pointset and the curvature plot of the interpolating spline.

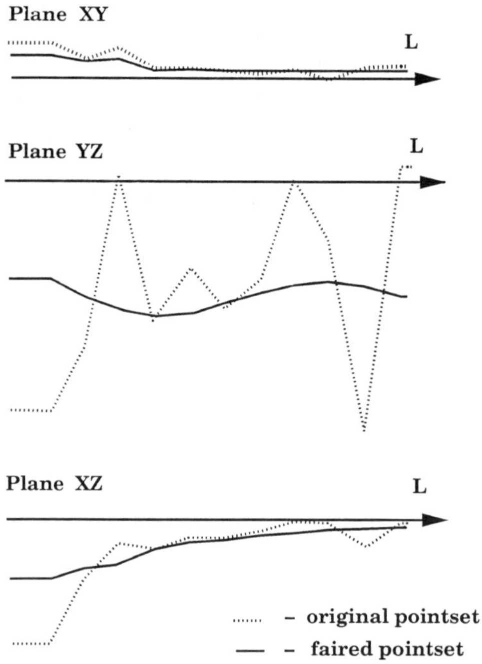

FIG. 4.5. *Discrete curvature plot.*

4.5. Conclusions

A tight string method for rough fairing of piecewise linear curves has been demonstrated. The results obtained indicate that extreme points of convex sets are a useful tool for creating fair shapes.

As a by-product, the proposed method for rough fairing solves another geometric problem: *it finds the shortest path between two points in the plane that passes through the system of gates.*

The results obtained also demonstrate that length of the curvature polygon is a robust measure of fairness at the stage of fine fairing, and that the fine-faired shape can be created as a solution of an optimization problem that uses the rough-faired shape as an initial vector.

It is of evident interest to understand how the tight string method can be generalized to higher dimensions and more complicated *gates*, and if the extreme points of convex sets can still be used as a tool for finding *tight shapes.*

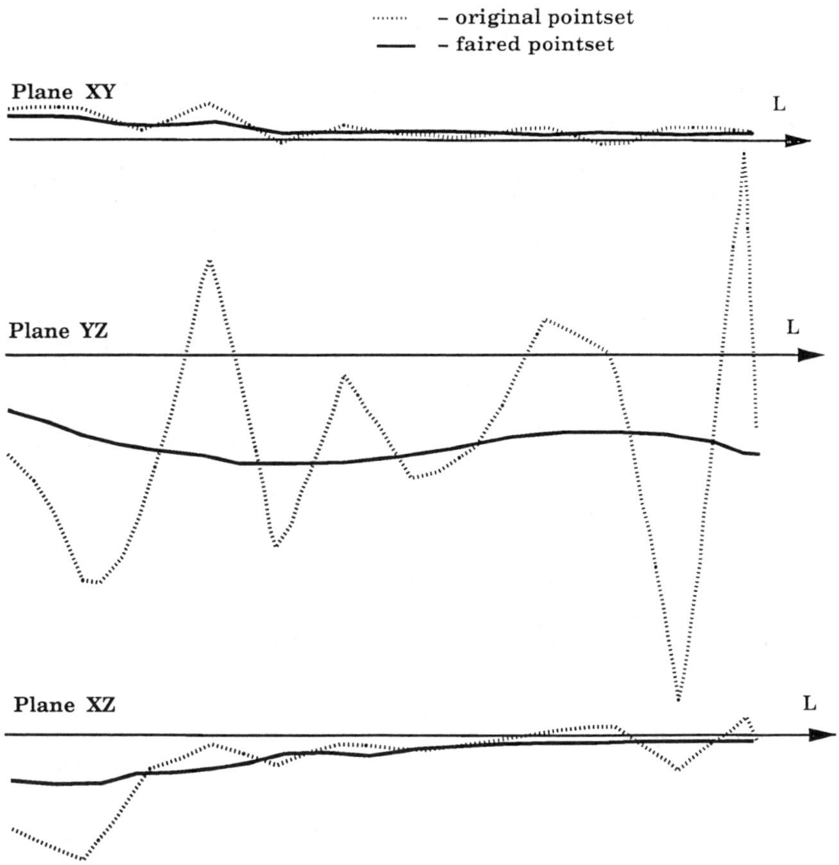

FIG. 4.6. *Curvature plot of the interpolating spline.*

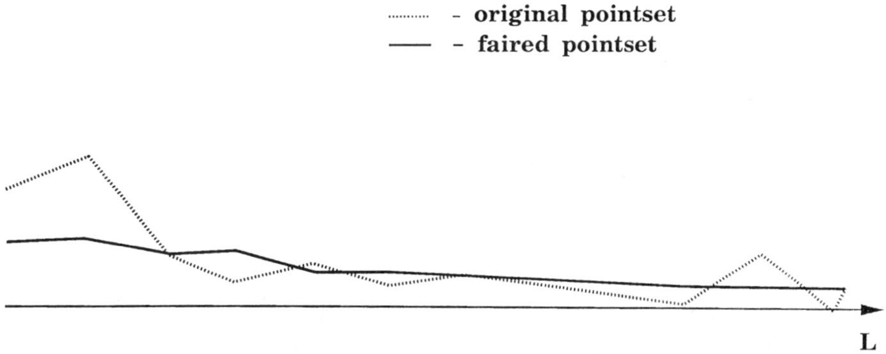

FIG. 4.7. *Three-dimensional curvature plot of the interpolating spline.*

One of the important questions is how to build a *tight net* from a rectangular grid of points in a three-dimensional space.

4.6. Acknowledgments

I would like to thank Camax Systems, Inc. for allowing me to publish these results, and my colleagues at Camax for help in preparation of the chapter. I am also grateful to Professor E. Fabes for a fruitful discussion, and to the referees and the editor for their valuable comments.

References

[1] N. Dunford and J. T. Schwartz, *Linear Operators, Part* 1: *General Theory*, Interscience Publishers, New York, London, 1958.

[2] M. Feldman, *Optimized Fairing of Digitized Curves, Second SIAM Conference on Geometric Design*, Tempe, Arizona, 1991, Society for Industrial and Applied Mathematics, Philadelphia, PA, extended abstracts, p. 8.

[3] P. Gill, W. Murray, and M. H. Wright, *Practical Optimization*, Academic Press, London, 1981.

[4] I. O'Neil, *Convexity of Data with Errors*, presentation at Second SIAM Conference on Geometric Design, Tempe, AZ, 1991.

[5] F. P. Preparata and M. Shamos, *Computational Geometry*, Springer-Verlag, New York, 1985.

[6] W. Renz, *Interactive Smoothing of Digitized Point Data*, Comput. Aided Des., 14 (1982), pp. 267–269.

[7] R. T. Rockafellar, *Convex Analysis*, Princeton University Press, Princeton, NJ, 1970.

[8] $CAMAND^{TM} CAD/CAM$ *Software*, Reference Manual, Release 5.0, Minneapolis, MN, 1992.

PART

2

Designing Fair Surfaces

CHAPTER

5

Measures of Fairness for Curves and Surfaces

John Roulier and Thomas Rando

5.1. Introduction

The concepts of fairness and faired curves play an important but elusive role in the field of computer-aided geometric design (CAGD). For curves, CAGD must provide tools to solve two related problems: the design of curves from scratch and the fitting of curves to predetermined constraints, such as positional constraints, tangency conditions, and specified curvatures, as well as others. In both cases the ultimate shape of the curve is often of paramount importance to the end user. Some shape aspects of curves are explicitly required; these include (local as well as global) convexity, monotonicity, the presence (or absence) of cusps, and monotone curvature. While it may not always be a simple matter to capture these shape characteristics in curves, which must simultaneously satisfy other design constraints, algorithms do exist for most cases. A key characteristic of these shape aspects, however, is that each has a precise mathematical formulation; there is no question whether the shape requirement has been satisfied for a given curve.

However, after all the design constraints have been met and the well-defined shape characteristics have been captured, a final judgment on the shape of the curve often remains. At this stage, the designer stops hearing precise requirements and begins hearing terms like "sweetness" and "aesthetics." The faired curve must not only satisfy all the precise shape requirements imposed on it, but it must also satisfy the subjective judgments of a wide range of critics.

This is the crux of the fairness problem. The notion of fairness is not simply subjective, ambiguous, and undefined—it is inherently undefinable. Yet this has not prevented researchers from attempting to formulate a mathematical definition for fairness. In order to make a real contribution to the designer who is required to create faired curves, debates over which is the best way to measure fairness should yield to the study of a host of fairness metrics and fairness algorithms. Researchers in CAGD should provide a wide selection of fairness metrics to the designers of curves—along with reliable descriptions of the ramifications of each of the metrics. This applies not just to fairness in the aesthetic sense but also to fairness as it relates to (imprecisely defined) functional requirements imposed on a design curve. The fairness metrics should be applicable locally as well as globally; they should be designed to give ultimate control over the shape of the curve to the designer himself—the final arbiter of the fairness of the designed curve.

5.1.1. Background. Shape control for curves in CAGD is a fertile research area. In this section we summarize some of the more prominent developments and techniques. However, we will deal mainly with topics connected with the subjective notion of fairness. There are several lines along which fairness research can be organized: the cubic spline and its descendants, variational approaches to fairness, subjective definitions of fairness, numerical approaches, and the definition of new fairness criteria.

The design of faired curves begins with the original spline curves, which were defined physically rather than mathematically. In this early scenario, the designer produced smooth curves through a series of points by placing metal weights at the designated points. The spline, a thin, elastic wooden beam, was then passed through the weights. The spline traced out a smooth, interpolating curve, which quite often was sufficiently pleasing in appearance. The resulting curve minimizes the strain energy of the spline, which is given by

$$\int (\kappa(s))^2 ds. \tag{5.1}$$

The term $\kappa(s)$ is the curvature as a function of arc length s of the curve under consideration.

The behavior induced by the minimization of the strain energy of the curve can be analyzed from (5.1). One tendency of this criterion is that the average curvature over the length of the curve tends to be minimized; however, spikes in the value of the curvature are penalized by the squaring of the curvature. In effect, the criterion tends to level the curvature function while keeping it, on the average, as low as possible.

The first mathematically formulated faired curve to gain widespread acceptance was the cubic interpolatory spline. The cubic spline emulated some of the desirable behavior of the original spline curve. The cubic spline is a piecewise polynomial which interpolates a series of data points with a single cubic segment between each pair of points. Moreover, the cubic spline maintains C^2 continuity at each joint. In addition to all these properties, the cubic spline also possesses the serendipitous property that among all C^2 functions which interpolate the data points at the proper parameter values and satisfy the given end conditions, the cubic spline $\mathbf{r}$ minimizes the value of

$$\int (\mathbf{r}''(t))^2 dt. \tag{5.2}$$

For a parametric curve **r**, the quantity (5.2) is a vector of the values that are obtained for each coordinate function of the curve. Thus, for a curve in R^3, (5.2) becomes

$$\int (\mathbf{r}''(t))^2 dt = \left(\int (x''(t))^2 dt, \int (y''(t))^2 dt, \int (z''(t))^2 dt \right).$$

Since this is a vector quantity, we minimize a norm of this vector—such as the Euclidean norm. This property is called the "minimum property" of the cubic spline, and it accounts for the characterization of the cubic spline as a faired curve. However, it should be noted that this property is, in a sense, spurious. The "minimum property" produces a faired curve only insofar as it simulates the minimum strain energy criterion, and this holds exactly only for curves that are parametrized by arc length. The "minimum property" is a simplification of the minimum strain energy property, which in real applications yields results that can be quite different from the results that would have obtained from the minimum strain energy criterion.

A significant amount of research has been devoted to improvements to the cubic spline. One shortcoming of the cubic spline is that it occasionally introduces inflection points that are not warranted by the data being interpolated. Oddly enough, this drawback is not a result of the discrepancy between the "minimum property" and the minimum strain energy criterion. Rather, it is a result of the fact that minimizing either of the quantities (5.1) or (5.2) fails to penalize inflections—changes in the sign of the signed curvature. The first attempt to deal with this problem was the spline in tension introduced by Schweikert in [38] and derived by an alternate approach by Cline in [6]. Schweikert characterizes a faired curve as a piecewise C^2 curve with at most one inflection point in each segment and with any straight line segments spanning entire segments. The spline in tension can be characterized as representing an elastic beam (as in the cubic spline)—but one that also responds to tension induced by pulling its ends. The degree of pulling at the ends is represented by the tension parameter α. The spline in tension exhibits the same smoothness conditions as the cubic spline but the fairness requirement is that the quantity $f'' - \alpha f'$ vary linearly in each segment. When α is zero, the spline in tension reduces to the cubic spline, and as α approaches infinity, the spline in tension approaches a piecewise linear interpolant of the data points.

The second major improvement to the cubic spline was the ν-spline, as described by Nielson in [26]. One problem with the original spline in tension was that its final representation entailed hyperbolic functions. Nielson developed an interpolating curve that preserved much of the functionality of the spline in tension yet had a piecewise polynomial representation that made it more suitable for CAGD. The added flexibility of the ν-spline was achieved by changing the smoothness constraints of the interpolant. In the parametric case the ν-spline provides an interpolant with G^2 continuity at the joins (as compared to the C^2 continuity of the previous splines). Note that G^2 continuity is the common term for second-order geometric continuity. This means that the curve is C^2 in the arc length parametrization but may not be C^2 in the current parametrization. Moreover, the ν-spline introduces a series of tension parameters, one for each segment. When all the tension parameters are zero, the ν-spline reduces to the cubic spline, and as all the tension parameters approach infinity, the ν-spline approaches a piecewise linear interpolant to the data points. Consequently, the ν-spline enforces a new fairness criterion

$$\int (\mathbf{r}''(t))^2 dt + \sum \nu_i [\mathbf{r}'(u_i)]^2 = \text{“as small as possible,”} \tag{5.3}$$

where the u_i are the parameter values of the joins and the ν_i are tension parameters.

The spline in tension approach provides an effective way to produce an interpolant to data points efficiently according to the fairness criteria described above. If there is a drawback to the approach, it is that the shape manipulation of the curve generally consists of the systematic flattening of the curve or of the offending segments. There may be some fairness applications in which flattening is not an acceptable option.

A second area of fairness research centers around the use of variational methods for the design of faired curves. One prominent development in this area is the nonlinear spline of Mehlum in [21]. Mehlum characterizes a nonlinear spline as a piecewise analytical function, with continuity constraints on its first few derivatives at the joins. The nonlinear spline interpolates to given data points, and in each segment presents some adjustable parameters that are available for shape control. Mehlum begins with a familiar fairness criterion, which he restates as

$$\int_0^L (\kappa(s))^2 ds = \text{“as small as possible,”} \tag{5.4}$$

where L is the total arc length of the curve and κ is a suitable measure of curvature. Mehlum imposes the constraints that (5.4) be minimized and that the resulting curve be parametrized by arc length. He uses the calculus of variations to derive formulations for the intrinsic equations for $\kappa(s)$ and $\tau(s)$ of the faired curve. In the process the value L becomes a user-provided quantity that can be regarded as a kind of tension parameter. Mehlum's method avoids the approximation in the fairness criterion that is implicit in the cubic spline approach. Because the results of this approach rely on the arc length parametrization, the difficult implementation problem for the nonlinear spline is the computation (or approximation) of the position vectors of the new curve from the intrinsic equations.

All the previous fairness criteria have captured the subjectivity of the phenomenon by virtue of the averaging implicit in the integration of the shape measures. The paper [8] by Farin and Sapidis endorses a subjective definition of fairness that avoids integration but concentrates instead on the curvature plot of the design curve: "*A curve is fair if its curvature plot consists of relatively few monotone pieces.*"

According to this criterion, one manifestation of a fairness violation is a slope discontinuity in the curvature plot of the design curve. This condition is especially apparent at the joins of spline curves which are not K^1 continuous. Note that K^1 is the notation for a parametric curve whose curvature plot is C^1 as a function of arc length. Farin and Sapidis describe a method for the fairing of B-spline curves (and surfaces) by which such discontinuities are discovered and remedied by a process of knot removal and new knot insertion. After this process the curvature plot typically consists of fewer monotone pieces. The Farin–Sapidis process is basically a local process, although the global fairness of the curve can in some sense be measured by a summation of the individual (absolute) slope discontinuities.

Another approach is to use numerical methods for the design of faired curves. Within this approach there are various avenues available. Nowacki, Liu, and Lu in [27] describe a method for the creation of faired, planar, parametric Bézier curves. Nowacki uses a variational approach to impose a fairness criterion and a set of constraints on the curve to be faired. The constraints include interpolation constraints, end condition constraints, and even constraints on the area under the curve. The most significant constraint, however, is that the resulting curve must be in Bézier form. In order to accomplish all this with variational methods, it is necessary to simplify the fairness criterion. Consequently, Nowacki uses the second derivative criterion from the "minimum property" of the cubic spline. The variational problem then becomes the optimization of a functional, using free parameters (beyond those needed for the explicit constraints) to minimize the fairness measure. The solution to the variation problem, thus posed, corresponds to the solution of a nonlinear system of equations—for which Nowacki relies upon traditional numerical methods.

A second numerical approach can be characterized as more of a brute force approach. Lott and Pullin in [19] describe such an approach (in the context of fairing B-spline surfaces). In this approach a curve is described by a combination of parameters, some constrained and some free, for shape adjustment. Automatic fairing of the curve is achieved by applying nonlinear optimization methods to the free parameters, using a pre-established fairness criterion as the objective function. In this approach, particular attention must be paid to implementation techniques needed to expedite the process. For example, [19] recommends the mathematical simplification of the fairness criterion—even to the point of sacrificing the parametric invariance of the criterion. In addition, [19] explores techniques to minimize the number of free parameters, efficient quadrature techniques, and suitable nonlinear optimization methods. Another significant issue is whether to use constrained or unconstrained optimization methods. Although constrained optimization is generally much slower, some problems require that the faired curve be constrained to remain close to the original design curve.

Finally, one last approach (see Moreton and Séquin [22], [23], Rando and Roulier [29], [30], and Rando [32]) is to derive and analyze new fairness criteria for curves (and surfaces). In this chapter we plan to expand upon this approach. We intend to establish guidelines and techniques for the derivation of suitable fairness criteria. The aim will be to derive a collection of new fairness criteria and to analyze the implications of the criteria both theoretically and empirically. Moreover, we plan to explore the issues that arise in the process of implementing these new fairness criteria using nonlinear optimization techniques. In fact, we have found that some implementation strategies have set the stage for the discovery of additional new fairness criteria.

5.2. Fairness Metrics for Curves

The construction of a system for the design of fair curves requires a method of assessing whether or not a curve is fair. Even with such a method, it may occur that many curves satisfy the criteria for fairness, and yet to the designer it may be obvious that some of these curves are "fairer" than others. For example, a designer may want to design a planar curve passing through given points and having specified unit tangent vectors at each of the points. The fairness criteria may be that the curve avoid extraneous inflection points. Many curves may satisfy these criteria, but some may exhibit undesirable properties such as regions with sharp bends and regions with nearly flat spots while others may not have these undesirable properties. In Fig. 5.1 the two curves pass through two given points with given unit tangent vectors. The curve on the top has undesirable properties as described above, while the lower curve has no areas like this. It might be judged then that the lower curve is "fairer" than the upper curve even though both of the curves are fair.

It is desirable for this reason that a method for discriminating levels of fairness be developed. That is, a numerical measure of the fairness of a

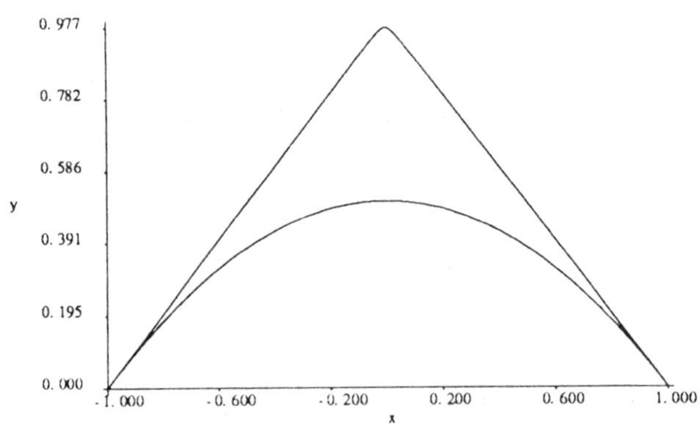

FIG. 5.1. *Lower curve "fairer" than upper curve.*

curve is needed so that algorithms to determine the level of fairness can be developed. Such a measure will be called a *fairness metric*. Strictly speaking, such measures of fairness will not be metrics in the mathematical sense of the word. We use the term loosely as a measure.

The development of such fairness metrics must take into account the fact that fairness criteria are very subjective. The perception of fairness will vary from application to application and from designer to designer. No one measure of fairness will suffice for every situation. Yet there are several widely accepted properties which are considered indicators of fairness. We will present several fairness metrics based on these properties as tools from which the designer may choose. In addition, we will present algorithms which use these metrics to construct the fairest curve satisfying a given fairness criterion in addition to other given constraints. Such constructions can be used to encourage various fairness tendencies in different situations and even in different regions of a composite curve.

It is desirable that any such metric should not depend on the parametrization used for a parametric curve. Rather, it should depend only on the geometric invariants of the curve. Thus, two different parametrizations of the same curve will have the same measure of fairness, and the measure of fairness for any parametric curve may be computed using the most convenient parametrization—usually the one inherent in the type of curve being used. In this situation, a reparametrization solely for the purpose of computing the fairness metric can be avoided.

The geometric invariants for curves which we will use are the concepts of curvature, radius of curvature, torsion, unit normal vector, unit tangent vector, unit binormal vector, and arc length. These concepts are independent of any allowable parametrization up to a sign. The sign will have no effect on the measure of fairness.

The Frenet–Serret formulas (see [9]) for parametric curves are useful in the analysis of the fairness metrics. Let ρ and τ be the radius of curvature and torsion, respectively, let κ be the curvature, and let $\mathbf{t}$, $\mathbf{n}$, and $\mathbf{b}$ be the unit tangent, normal, and binormal vectors, respectively, for $\mathbf{r}$. The Frenet–Serret formulas using arc length parametrization are

$$\mathbf{t} = \mathbf{r}', \tag{5.5}$$

$$\mathbf{t}' = \kappa\mathbf{n}, \tag{5.6}$$

$$\mathbf{n}' = -\kappa\mathbf{t} + \tau\mathbf{b}, \tag{5.7}$$

$$\mathbf{b}' = -\tau\mathbf{n}. \tag{5.8}$$

The arc length of a parametric curve $\mathbf{r}$ with parameter $t \in [a, b]$ is

$$S_r = \int_a^b |\mathbf{r}'(t)|dt. \tag{5.9}$$

We first derive a specific fairness metric to demonstrate the main approach we will take. Let $\mathbf{r}$ be a parametric curve. Let $\mathbf{c}$ be the parametric curve consisting of the centers of curvature of $\mathbf{r}$. Then $\mathbf{r}$ is called the *design curve*, and $\mathbf{c}$ is called the *derived curve*. Let S_c be the arc length of the curve $\mathbf{c}$. Note that S_c is independent of the parametrization of $\mathbf{r}$. We define the fairness measure of $\mathbf{r}$ to be

$$\mu = S_c. \tag{5.10}$$

Typically, $\mathbf{r}$ will depend on a finite number of parameters such as control points. Some of these parameters may be determined by the particular problem, such as end conditions involving interpolation, tangency, curvature, etc. The remaining free parameters may be chosen to minimize μ while ensuring that $\mathbf{r}$ satisfies the constraints inherent in the problem.

The effect of minimizing μ in this case is to make the arc of the centers of curvature as short as possible. Heuristically, this should "pull" the centers of curvature to a small region thus making $\mathbf{r}$ as near to being a circle as possible. The idea here is that a circle is widely considered to be a fair curve since it is devoid of any changes in curvature which could produce sharp bends and nearly flat regions. Let τ and ρ be the torsion and radius of curvature of $\mathbf{r}$, respectively, and let $\mathbf{n}$ be the unit normal vector. The formula for $\mathbf{c}$ is

$$\mathbf{c}(t) = \mathbf{r}(t) + \rho(t)\mathbf{n}(t). \tag{5.11}$$

We will show that this metric yields

$$\mu = \int_0^{S_r} [\rho^2\tau^2 + (\rho')^2]^{1/2} ds, \tag{5.12}$$

where S_r is the arc length of $\mathbf{r}$. (See also Rando and Roulier [29] and Roulier, Rando, and Piper [36].)

The derivation of (5.12) from (5.10) is presented in [29] and [36]. We reproduce it here for completeness. Note that from (5.11) we get

$$\mathbf{c}' = \mathbf{r}' + \rho'\mathbf{n} + \rho\mathbf{n}'.$$

Observe also that $\rho\kappa = 1$ if neither ρ nor κ is zero. A substitution using this observation, (5.5), and (5.7) gives

$$\mathbf{c}' = \rho'\mathbf{n} + \rho\tau\mathbf{b}.$$

It now follows from the orthogonality of unit vectors $\mathbf{n}$ and $\mathbf{b}$ that

$$|\mathbf{c}'|^2 = \mathbf{c}' \cdot \mathbf{c}' = (\rho')^2 + \rho^2\tau^2.$$

This last equation gives (5.12).

The integral in (5.12) is with respect to arc length s of $\mathbf{r}$, but μ is independent of the parametrization. We may evaluate μ as the arc length of the curve of centers of curvature $\mathbf{c}$ in whatever parametrization is used with $\mathbf{r}$. The equation (5.12) expresses analytically the effect of minimizing μ. That is, large values of ρ, τ, and ρ' are minimized. Thus the tendency is to bring the curve as much as possible into a plane by making τ small. The presence of the ρ^2 term enforces the tendency to bring the curve close to a circular arc rather than a section of a straight line, while the presence of the $(\rho')^2$ term tends to eliminate sharp bends and flat spots. If the original curve $\mathbf{r}$ is a planar curve, then $\tau = 0$ and (5.12) reduces to

$$\mu = \int_0^{S_r} |\rho'| ds, \tag{5.13}$$

the *total variation* of ρ (Rudin [37]). (See the next section for the definition.) In this case, the curve $\mathbf{c}$ is the *evolute* of $\mathbf{r}$.

In any event, minimizing μ here tends to minimize the magnitude of the rate of change of the radius of curvature ρ of $\mathbf{r}$. Obviously, one must avoid inflection points in the curves $\mathbf{r}$ to be considered since the arc length of the arc of the centers of curvature will be infinite for such a curve. To observe this, note that $\rho = 1/\kappa$ and that $\kappa = 0$ at an inflection point. If there is no curve from the given family which satisfies the given constraints without an inflection point, then this metric cannot be used.

We may use this approach to generate additional fairness metrics. The general approach then is to produce *derived curve* $\mathbf{c}$ depending only on the geometric invariants of the *design curve* $\mathbf{r}$ so that minimizing the arc length of $\mathbf{c}$ has the desired fairing effect on $\mathbf{r}$. The arc length of such curves $\mathbf{c}$ will therefore be independent of the parametrization of $\mathbf{r}$. The relationship of $\mathbf{c}$ to the fairness metrics which follow will not be as intuitive as in the previous case, but an analytical examination of the metric will reveal its properties. We will list a few such curves $\mathbf{c}$ in Table 5.1 with their metrics.

The derivation of each of these is accomplished in a similar manner as above for (5.12) using the Frenet–Serret formulas and the fact that $\mathbf{t}$, $\mathbf{n}$, and $\mathbf{b}$ are mutually orthogonal unit vectors.

The first, (5.14), is just the metric presented earlier. Note that $\mathbf{c} = \rho\mathbf{b}$ in (5.18) gives the same metric. The second (5.15) is perhaps the most useful of the metrics. It tends to minimize the rate of change of curvature and the magnitude of the torsion. Note that we will refer to the tendency to minimize the rate of change of the curvature or the rate of change of the radius of curvature as *smoothing*. Unlike the first metric, this metric is applicable even when $\mathbf{r}$ has an inflection point. In the planar case it reduces to the total variation of the curvature. In the general case, the tendency will be to "pull" the curve into a plane due to the τ^2 term, while the presence of the $(\kappa')^2$ term avoids sharp bends by minimizing the overall rate of change of curvature. The presence of the κ^2 term encourages a tendency toward a straight line segment rather than

TABLE 5.1

Metric	Derived Curve $\mathbf{c}$	Fairness metric $\mu = S_c$	R^2	R^3
(5.14)	$\mathbf{r} + \rho\mathbf{n}$	$\int_0^{S_r} [\rho^2\tau^2 + (\rho')^2]^{1/2} ds$	Smooths	Smooths Rounds
(5.15)	$\kappa\mathbf{b}$	$\int_0^{S_r} [\kappa^2\tau^2 + (\kappa')^2]^{1/2} ds$	Smooths	Smooths Flattens
(5.16)	$\kappa\mathbf{n}$	$\int_0^{S_r} [\kappa^2\tau^2 + (\kappa')^2 + \kappa^4]^{1/2} ds$	Smooths Straightens	Smooths Flattens
(5.17)	$\kappa\mathbf{t}$	$\int_0^{S_r} [\kappa^4 + (\kappa')^2]^{1/2} ds$	Smooths Straightens	Smooths Flattens
(5.18)	$\rho\mathbf{b}$	$\int_0^{S_r} [\rho^2\tau^2 + (\rho')^2]^{1/2} ds$	Smooths	Smooths Rounds
(5.19)	$\mathbf{n}$	$\int_0^{S_r} [\tau^2 + \kappa^2]^{1/2} ds$	Straightens	Flattens
(5.20)	$\rho^2\mathbf{n}$	$\int_0^{S_r} [4\rho^2(\rho')^2 + \rho^4\tau^2 + \rho^2]^{1/2} ds$	Smooths Rounds	Smooths Rounds
(5.21)	$\rho^2\mathbf{t}$	$\int_0^{S_r} [4\rho^2(\rho')^2 + \rho^2]^{1/2} ds$	Smooths Rounds	Smooths Rounds

a circular arc as in the first metric. In the planar case both of the metrics tend to smooth the design curve when minimized, while in the nonplanar case the first tends to round the curve and the second tends to flatten the curve in addition to smoothing it. For these reasons, we may refer to the first metric as a *rounding metric* and to the second metric as a *flattening metric* in the nonplanar case. In the planar case both are *smoothing metrics.* The third metric (5.16) and fourth metric (5.17) enhance the tendency toward a straight line in the planar case due to the presence of the κ^4 term. The fourth eliminates the tendency to force the curve into the osculating plane because the τ^2 term is not present. Nevertheless, both of these can be considered to be flattening or straightening, and smoothing in both the planar and nonplanar cases. The sixth metric (5.19) tends to force the curve into a plane and to a straight line segment without enforcing the avoidance of sharp changes in the curvature. The integrand in (5.19) is called the *total curvature* in Mehlum [21]. Finally, the last two metrics given by (5.20) and (5.21) encourage smoothing and rounding in both the planar and nonplanar cases. In addition, (5.20) tends to pull the design curve into a plane due to the presence of the τ^2 term.

We now discuss another metric inspired by the notion of *spherical curvature.* The *centers of spherical curvature* of a parametric curve $\mathbf{r}(s)$ parametrized by arc length is given in [18] as

$$\mathbf{c}(s) = \mathbf{r} + \rho\mathbf{n} + \rho'\sigma\mathbf{b}, \tag{5.22}$$

where $\rho = 1/\kappa$ is the radius of curvature and $\sigma = 1/\tau$ is called the *radius of torsion.*

The *radius of spherical curvature* is given by

(5.23) $$\zeta = |\rho\mathbf{n} + \rho'\sigma\mathbf{b}| = [\rho^2 + (\rho'\sigma)^2]^{1/2}.$$

It is well known that *a parametric curve* $\mathbf{r}$ *lies on a sphere if either* $\kappa' \neq 0$ *and* $\zeta =$ constant, *or* $\kappa' = \tau = 0$ $(\kappa \neq 0)$ (cf. [12] pp. 160–161). In the latter case, the curve is planar and the metrics (5.14) and (5.15) given above apply. In the first case, if the total variation of the radius of spherical curvature is zero, then the curve lies on a sphere. Thus, the next metric is the total variation of ζ,

(5.24) $$\mu = \int_0^{S_r} |\zeta'(s)| ds,$$

where from (5.23) we see that

(5.25) $$\zeta' = \frac{\rho'\sigma((\rho/\sigma) + \rho''\sigma + \rho'\sigma')}{(\rho^2 + (\rho'\sigma)^2)^{1/2}}.$$

A simpler metric related to this is based on the fact that $\zeta' = 0$ if the term in the right side of the numerator above is zero. Thus, if the minimum of the following metric is zero, then so is the previous one. In fact, if $\kappa' \neq 0$ and all of the terms are defined, then the minimum of the previous metric is zero if and only if the minimum of the next is zero.

(5.26) $$\mu = \int_0^{S_r} |\nu(s)| ds,$$

where

(5.27) $$\nu = (\rho/\sigma) + \rho''\sigma + \rho'\sigma'.$$

An examination of the centers of spherical curvature (5.22) reveals that

$$\mathbf{c}' = \mathbf{t} + \rho'\mathbf{n} + \rho\mathbf{n}' + (\rho''\sigma + \rho'\sigma')\mathbf{b} + \rho'\sigma\mathbf{b}'.$$

We may now apply the Frenet–Serret formulas (5.5)–(5.8) and the facts that $\rho\kappa = 1$ and $\tau\sigma = 1$ to this to obtain

$$\mathbf{c}' = ((\rho/\sigma) + \rho''\sigma + \rho'\sigma')\mathbf{b}.$$

This last observation together with the fact that $\mathbf{b}$ is a unit vector gives $\mathbf{c}' \cdot \mathbf{c}' = [(\rho/\sigma) + \rho''\sigma + \rho'\sigma']^2 = \nu^2$. Thus, (5.26) also represents the arc length of the centers of spherical curvature. This means that if $\kappa' \neq 0$ and all of the terms are defined, *the total variation of the radius of spherical curvature is zero if and only if the arc length of the centers of spherical curvature is zero.* Therefore, *minimizing the metric* (5.26) *tends to "pull" the curve to the surface of a sphere.*

We also note that combinations of the metrics given here may be used to create new metrics which take advantage of the beneficial properties of each. For example, if we want to enhance the tendency to flatten and straighten which is implied in metric (5.15) while retaining the tendency to smooth, we can combine it with metric (5.19). This can be accomplished by minimizing the quantity

$$\mu = \int_0^{S_r} [\kappa^2\tau^2 + (\kappa')^2]^{1/2} ds + \int_0^{S_r} [\tau^2 + \kappa^2]^{1/2} ds.$$

This is equivalent to minimizing the sum of the arc lengths of the derived curves $\mathbf{c} = \kappa\mathbf{b}$ and $\mathbf{c} = \mathbf{n}$ for these two metrics. This concept will be discussed further in the next section.

5.3. Implementation Issues

As noted in the previous section, the definition of fairness metrics for the design of parametric curves is a necessary first step toward facilities for the automatic generation of faired curves; however, difficult implementation issues remain to be addressed. The proposed metrics are designed to be amenable with the most general parametric curves. Nevertheless, implementation improvements are possible for many popular CAGD curve types. In this section we will describe both the generic algorithms as well as some specialized algorithms that are suitable for specific types of curves. Moreover, we examine the ramifications of applying these algorithms not just to single segment parametric curves but to spline curves as well.

Although the most salient implementation issues center around the algorithms that are chosen for any particular curve type, the software engineering aspects of the automatic generation of faired curves also entail some interesting problems. Our approach to the design of faired curves relies on the use of standard unconstrained optimization. Our implementation strategy has concentrated on an object-oriented programming approach to computer-aided geometric design.

The automatic fairness problem is as follows: a design curve $\mathbf{r}$ is created, subject to certain design constraints. The design curve is a parametric curve that is defined by a finite set of values, some of which are fixed by the design requirements, some of which are free to act as shape adjustment variables. The automatic generation of a faired curve in this scenario consists of the choice of a set of values for the free variables such that the resulting curve minimizes one of the fairness metrics described previously.

The algorithm follows.

Algorithm 5.1.

1. Choose a fairness metric μ to use for fairing.
2. Determine the free variables for the problem at hand, and express μ as a function of these free variables.

3. Choose a method for the evaluation of the function μ determined above.
4. Choose a minimization algorithm and suitable starting values for the algorithm to minimize μ, and perform the minimization using the evaluation method chosen in the previous step.

Note that step 4 can be terminated at any time that the user is satisfied with the fairness of the curve in question since the process results in progressively fairer curves. This can be done by setting loose stopping criteria or by allowing a user interrupt in the implementation.

In the basic case, the design curve $\mathbf{r}$ is a simple, bounded regular curve. The design curve thus represents a mapping $\mathbf{r}$ from a closed interval $[a, b]$ in R to R^2 or to R^3. The design curve is assumed to be analytic—this is not an unreasonable imposition because in the basic case, the curve is assumed to be a single segment. The design curve is described by the arbitrary parameter t over the closed interval $[a, b]$. The function

$$s = s(t) = \int_a^t |\mathbf{r}'(u)| du, \quad a \leq t \leq b$$

describes the arc length of the design curve and provides an allowable change of parameters on the interval $[a, b]$. S_r represents the total arc length of the design curve over the interval $[a, b]$ as in (5.9). Specific fairness metrics may impose further restrictions on the allowable shape characteristics of the design curve.

In a typical problem the design curve $\mathbf{r}$ is a cubic Bézier curve segment that is constrained to start at P_0 and to end at P_1. The tangent directions at the ends are also fixed by the design requirements. The only free variables available for shape adjustment, then, are the magnitudes d_1 and d_2 of the end tangent vectors. The fairness metric is a functional defined on the mapping $\mathbf{r}$, that takes the form

$$\mu = \int_0^{S_r} m(s) ds$$

and since the design curve $\mathbf{r}$ is a function of the free shape variables, $\mathbf{v} = (d_1, d_2)$, the fairness metric becomes a function of these variables. The faired curve in this problem is the curve for which $\mu(\mathbf{v})$ attains a minimum. The function $m(s)$ is a real valued function, whose domain corresponds to the arc length of the design curve.

As discussed in the previous section, the fairness of the design curve should be related exclusively to the geometry of the curve. It should not be affected by changes to nongeometric aspects of the design curve. Specifically, since curves in CAGD are often parametric curves, the fairness measure for a curve must remain constant irrespective of changes to the parametrization to the curve. Consequently, the fairness description function, $m(s)$, will consist exclusively of the geometric invariants κ, ρ, and τ of the design curve and their derivatives. As seen earlier, these values describe the shape aspects of the curve and provide an insight into the shape influences of each of the fairness metrics.

When the fairness metric is equivalent to the arc length of a derived curve **c** as described in the previous section, this relationship can be exploited for the efficient evaluation of these fairness metrics. If the derived curve is evaluated at multiple, evenly spaced parametric intervals, the resulting total chord length offers a rough approximation of the arc length of the derived curve and, consequently, of the fairness metric. This technique provides a useful starting technique for automatic generation of the faired curve because of its speed of execution and because of the simplicity of the programming—all the computations can be carried out in the arbitrary parametrization of the design curve. Because the technique is an uncontrolled approximation, however, it is generally not suitable for exacting fairness problems. Our experience has indicated that the use of the chord length evaluation technique results in curves with curvature plots that for the most part maintain a steady, low curvature value but which exhibit spikes in the curvature values between the evaluation points. These spikes represent significant, local deviations of the derived curve from its chord segment approximation in occasional segments of the derived curve. The resultant design curves are nearly linear for most of their length—with intermittent, brief regions of very high curvature.

An alternate approach to the evaluation of fairness metrics relies on adaptive quadrature. (One example is Romberg integration as implemented in [28].) This approach can be used to avoid the problems attendant with the chord length method. Adaptive quadrature can be programmed to attain more accurate approximations to the fairness metric. In some cases, for example, when the fairness metric is utilized as an averaging of the shape quantities across the curve, a very high degree of accuracy is not strictly required. However, when the fairness metric is used to reproduce exact representations of known shape features (for example, circular arcs, curves on spheres, conic sections), a high degree of accuracy in the evaluation of the fairness metric is a necessity. The problem with the adaptive quadrature approach is that it is significantly slower than the chord length approach; and this deficiency is exacerbated in the nonlinear optimization code, which depends on many function evaluations. For fairness metrics that are associated with derived curves, adaptive quadrature can, of course, be applied to the computation of the arc length of the derived curve; in addition, it is possible to employ an adaptive quadrature technique directly for an approximation of the fairness function μ.

The selection of a nonlinear optimization code is itself a difficult implementation issue. For many problems it is necessary to impose constraints on the faired design curve. One common constraint used (cf. Ferguson [10]) is that the faired curve remains "close" to the original curve being faired. This constraint is especially important for fairness metrics that minimize curvature and torsion; if the geometric constraints imposed by the fixed parameters in the problem definition permit new curves that are of unlimited arc length, an unconstrained optimization code may fail to find a reasonable solution. Because the curvature and torsion of the new curve of unbounded arc length may approach zero, the fairness metric of such curves may be misleadingly small. Nevertheless, it is often possible to define the basic geometric constraints of the problem so that such cases are prohibited. In this case it is possible to choose an unconstrained optimization code. The advantage of such a choice is that the unconstrained optimization code is an order of magnitude faster than the constrained optimization code. We have tried a number of unconstrained optimization algorithms in our implementation. The pattern search technique of Hooke and Jeeves (cf. [14]) is easy to code and does not require the computation of gradients; the same advantages apply to the simplex method of Nelder and Meade in [25]. For the most part, however, we have used the Fletcher-Reeves method of conjugate gradient descent as described in [28].

The approaches described above have the advantage of generality; they can be employed for any design curve that can be described parametrically with the first few derivatives available. However, our experience with these algorithms has demonstrated the need for faster and/or more accurate techniques for the evaluation of the fairness metrics. We have had success here by exploiting characteristics of some fairness metrics in conjunction with the characteristics of some design curve representation forms. These improvements are made, of course, at the expense of the generality of the algorithms.

A new fairness evaluation technique arises when the fairness metric can be expressed as the total variation of some invariant shape function of the design curve. The total variation of a function f defined on an interval $[a, b]$ follows from the establishment of a partition δ of $[a, b]$ where $\delta = (x_0, x_1, \ldots, x_n)$, that is, $a = x_0 < x_1 < \cdots < x_n = b$.

DEFINITION 5.3.1. *Let*

$$T(\delta; f) = \sum_{i=1}^{n} |f(x_i) - f(x_{i-1})|. \tag{5.28}$$

The total variation *of f on $[a, b]$ is given by*

$$V_a^b(f) = \sup_{\delta} T(\delta; f). \tag{5.29}$$

Moreover, if f is of bounded variation on $[a, b]$ and f is absolutely continuous, then

$$V_a^b(f) = \int_a^b |f'(x)| dx. \tag{5.30}$$

The crucial property of the total variation, in the present context, is that f is monotone on $[a, b]$ if and only if

$$V_a^b(f) = |f(b) - f(a)|. \tag{5.31}$$

The notions of total variation and derived fairness curves come together in the study of planar evolutes (see, for example, [39]). As noted above, the derived curve (5.11) (also (5.14)) for a planar design curve is the planar evolute of the design curve. The planar evolute is a regular curve if for the design curve $\rho' \neq 0$, that is, the radius of curvature is a monotone function. Moreover, since $\kappa = 1/\rho \neq 0$, the curvature of the design curve is also a monotone function. The element of arc length of the evolute is $|\rho'|ds$, where s is the arc length parameter of the design curve. Consequently, for each segment of the design curve $[c, d]$ with monotone curvature, the arc length of the evolute is given by

$$L_c^d = |\rho(d) - \rho(c)|. \tag{5.32}$$

By a similar argument, it can be shown that for planar design curves with monotone curvature, the arc length of the derived curve $\mathbf{c} = \kappa\mathbf{b}$ is given by

$$L_c^d = |\kappa(d) - \kappa(c)|. \tag{5.33}$$

The points on the curve at which $\kappa' = 0$ are called the *vertices* of the curve. For the evaluation of these metrics, then, the problem becomes one of locating the vertices of the curve—between which each segment has monotone curvature. Roulier, Rando, and Piper in [36] show that for any Bézier curve the problem of finding the vertices of the curve can be reduced to the problem of finding the roots of a nonparametric Bézier curve on a closed interval. In fact, it is not difficult to show that the same holds for rational Bézier curves. For a robust root-finding algorithm, it is possible to exploit the properties of non-parametric Bézier curves, such as end interpolation, the variation diminishing property, the convex hull property, and to use the deCasteljau subdivision algorithm, in order to bracket regions containing single zero points. With this information, a numerical method such as Brent's method (cf. [28]) can be used to find the root quickly.

Finally, the total variation evaluation technique suggests a justification for the study of other fairness metrics. There are fairness metrics that recommend themselves first because they lend themselves to evaluation by the total variation approach. Their utility in a CAGD environment may be established if examination of the metrics reveals shape behaviors that prove useful to the designers of faired curves. The idea is to begin with a function of the shape invariants of the design curve such as

$$h = \kappa^2\tau^2. \tag{5.34}$$

The total variation of (5.34) is

$$V_0^{S_r}(h) = 2\int_0^{S_r} |\kappa'\kappa\tau^2 + \tau\tau'\kappa^2|ds. \tag{5.35}$$

This quantity contains "averaged" information about the shape of the design curve; it is geometrically invariant and can be computed by finding the zeros of the derivative of (5.34). Examination of the metric reveals that in the plane, the metric vanishes; consequently, it is applicable only to twisted curves. By a similar process, the shape function

$$h = \kappa^2 + \tau^2 \tag{5.36}$$

yields the following fairness metric:

$$V_0^{S_r}(h) = 2\int_0^{S_r} |\kappa\kappa' + \tau\tau'|ds. \tag{5.37}$$

This metric is applicable even when the torsion of the design curve vanishes.

All the preceding metrics and algorithms have been described in terms of single segment, parametric curves; however, these approaches can be easily extended to spline curves. Some additional considerations come into play. One concern is the behavior of the fairness metric at the knots. For the fairness metrics themselves, if the design curve is not of sufficient smoothness, the fairness metric function will contain discontinuities. In general, this will cause problems in the evaluation of the fairness metric. For fairness metrics associated with derived curves, it is possible to expand the definition of the derived curve to include the linear segments that join the discontinuous segments of the derived curve. Minimization of this function tends to diminish the discontinuities in the derived curve while it minimizes the overall arc length of the derived curve segments.

We have also found that it is possible to extend the vertices method of evaluation for fairness metrics to B-spline curves. For B-splines, the function that characterizes the roots of the applicable shape function can be stated as a nonparametric B-spline curve. This is possible because the curves in each knot segment are well defined and the continuity requirements at each knot are also known. Consequently, it is possible to transform from the piecewise polynomial representation of the function to the B-spline representation. The resulting nonparametric B-spline is equivalent to the piecewise function; however, it can be represented with fewer control ordinates.

5.3.1. Software engineering considerations. One final implementation issue involves the software engineering of a system for the automatic fairing of curves. We have chosen to implement our algorithms using object-oriented design and object-oriented programming in the C++ language. As a result of this decision, we have developed a library of curve (and surface) classes with a myriad of computational methods, which is potentially of value to CAGD researchers (among others). A primary benefit of the object-oriented approach is that a problem domain (such as the classes of curve representations) can be implemented in various levels of abstraction. At any level in the inheritance hierarchy, curve classes can be reused by others, who will be relieved of the burden of reproducing and testing the fundamental operations associated with that class of curves. Moreover, the basic curve functionality can be extended by future users of libraries so that specialized classes can be produced for ad hoc purposes without disturbing the basic, generic functionality.

In attempting to use an object-oriented approach in a CAGD domain, we have encountered several interesting implementation issues. The object-oriented approach begins with the definition of a hierarchy of object classes for the problem domain. In the realm of CAGD, the notion of a class of curves being a generalization of another class is often discussed; however, there is no consensus on the nature of such an inheritance hierarchy (cf. [1]). Moreover, specialization of curve classes is usually achieved by the restriction of certain classes. For example, a Bézier curve is a rational Bézier curve with equal weights. Object-oriented languages such as C++ do not easily support this type of specialization. Another issue surrounds the need for efficiency in the computational methods. The object-oriented approach stresses high level programming constructs before speed of execution. However, we have found that it is possible to exploit the hybrid nature of the C++ programming language to implement classes that exploit the computational efficiency of procedural programming languages.

On the other hand, the object-oriented approach offers some structural affinities to the CAGD domain. We found that it was straightforward to implement our curve classes underneath a parametric evaluator. In this architecture, many of the operations that can be applied to any parametric curve are collected in a high-level parametric curve class. The operations include such things as the computation of curvature and torsion and the signatures for the evaluation and differentiation of the parametric curve. In fact, nearly all the computation of fairness quantities has been centralized in this class. As a result, a new curve representation can be added to the library in such a way that it automatically inherits all the machinery for the generation of faired curves.

Finally, we designed our CAGD curve classes with an eye to the standardized sharing of geometric data. The incipient STEP (cf. [16]) standard for the digital exchange of product data has specified an object-oriented representation of geometric data—including curves, surfaces, and solids. When this standard is in place, it will be possible to share geometric data between applications not only by means of text files (similar to an IGES [17] data exchange) but also by means of a standard interface to a database representation of the geometry. Our curve classes have been designed to be compatible with the STEP geometry representations.

In some areas of software engineering (such as the development of graphical user interfaces), object-oriented programming has already demonstrated its efficacy. Object-oriented programming has not yet been widely accepted for CAGD applications. This is unfortunate because the classes of objects in CAGD are relatively stable. Consequently, object-oriented programming offers the promise of providing high-level programming constructs with many benefits. A properly constructed object-oriented architecture could provide the CAGD developer with reusable and extensible code; it could provide the CAGD researcher with a programming environment in which the low-level, machine-dependent details of CAGD algorithms could be avoided; it could provide structures through which CAGD data could be shared among diverse users.

5.3.2. Monotone curvature. As was indicated earlier, a function f of bounded variation on closed interval $[a, b]$ is monotone if and only if the total variation of f satisfies

$$V_a^b(f) = |f(b) - f(a)|. \tag{5.38}$$

The right side of (5.38) represents the absolute minimum attainable for the total variation of f on $[a, b]$. Therefore, this minimum can be attained if and only if f is monotone on $[a, b]$, and no lower value of $V_a^b(f)$ can ever be attained.

With these comments in mind, it is obvious from (5.14) and (5.15) that for planar curves, the fairness metrics are the total variations of the radius of curvature and curvature, respectively. If a parametric curve $\mathbf{r}$ satisfying the constraints imposed (including fixed values $\mathbf{r}(a)$ and $\mathbf{r}(b)$ at the endpoints a and b) exists that has monotone radius of curvature (or curvature), then minimizing (5.14) (or (5.15)) will produce a curve with monotone radius of curvature (or curvature). The main difference between the two metrics is that (5.14) produces monotone radius of curvature which requires that the curvature be bounded away from zero, while (5.15) allows an inflection point if the signed curvature is used in the planar case.

These ideas have been discussed extensively in [36]. In addition, a theorem stating exactly when a parametric curve $\mathbf{r}$ satisfying endpoint constraints on position, tangency direction, and curvature value is a convex curve with monotone curvature has been stated there as well. A theorem was also stated regarding Bézier curves. That is, given end conditions as above, which satisfy the necessary and sufficient conditions for a convex curve with monotone curvature for n sufficiently large, there is a Bézier curve of degree n satisfying the constraints and which is also a convex curve with monotone curvature. Additional results on monotone curvature can be found in the paper by Frey and Field [11].

The point to remember then, is that if a curve from the family under consideration exists with monotone curvature, then minimizing these metrics will find one. Moreover, if the constraints involve specifying the curvature at the endpoints, then one can determine immediately at the end of the minimization process if the resulting curve has monotone curvature. That is the minimum value of the fairness metric must equal the absolute value of the difference of the radius of curvature (or curvature) at the endpoints.

The facts discussed here also suggest an alternate approach. That is, instead of setting up a minimization problem, set up a root-finding problem. For example, if a curve with monotone curvature is sought with curvatures κ_1 and κ_2 specified at the endpoints, solve the equation

$$\mu = |\kappa_2 - \kappa_1| \tag{5.39}$$

as a function of the free parameters. If such a curve exists then this will find it, otherwise, the process will not converge to a solution.

Finally, for convex curves with monotone curvature and given end tangent directions and radii of curvature ρ_1 and ρ_2, a third approach is suggested by Roulier in [34]. That is, first produce a curve $\mathbf{c}$ which is convex, passes through the centers of curvature, and whose arc length is $|\rho_2 - \rho_1|$, then produce curve $\mathbf{r}$ as the planar involute of $\mathbf{c}$ which passes through the given endpoints. It is known that $\mathbf{r}$ will have monotone curvature and be convex. The difference between this and the previous approaches is that the curve $\mathbf{c}$ will be the first curve constructed—for example, a Bézier curve, and the curve $\mathbf{r}$ will be an involute of $\mathbf{c}$—that is not a Bézier curve.

Theory and algorithms are presented in [34] which produce a Bézier curve satisfying such end constraints and having a prescribed arc length. These algorithms are fast and require finding the unique root of a function of a single variable. The starting values for the iterative solution are chosen automatically, and convergence is guaranteed. Thus the curve $\mathbf{c}$, and therefore the curve $\mathbf{r}$, are found quickly without the guesswork involved in selecting starting values for multivariable problems. The negative aspect to this is that the design curve $\mathbf{r}$ produced is not a Bézier curve or other standard construct. The reader is referred to [34] and [35] for further discussion.

The three approaches to monotone curvature discussed in this section each have advantages and disadvantages as described above. The first approach which involves minimizing the fairness metrics (5.14) or (5.15) has the advantage over the other two approaches in that even if the family of curves has no member with monotone curvature, the minimization process will produce a curve with curvature as close as possible to being monotone in the sense that the total variation of the curvature is minimized. The second approach will fail to converge in such a case since it will be attempting to solve a nonlinear system of equations for which there is no solution. The third approach will produce a curve with monotone curvature if one exists at all, but if no such curve exists this approach will fail. If the user has verified that a curve with monotone curvature exists for the given constraints, then the third approach will produce the fastest solution since it reduces the problem to solving a single nonlinear equation of a single variable and automatically chooses appropriate starting values. In summary then, the first and third approaches appear to be the most useful. We will present some examples of these in the next section.

Since the first approach is just Algorithm 5.1 presented above using one of the two specific metrics (5.14) or (5.15), we will not list it here. We simply remind the user that the constraints for such a problem usually involve specifying curvatures at the endpoints in addition to interpolation and tangency conditions. We list the steps of the third approach below in Algorithm 5.2. Recall that this approach uses the algorithm for generating a Bézier curve of a specified arc length and constructing an involute of this curve.

The second algorithm follows.

Algorithm 5.2.

1. Set up the conditions constraining the problem. That is, specify the endpoints P_1 and P_2, the unit tangent vectors $\mathbf{u}_1$ and $\mathbf{u}_2$, and the curvatures $0 < \kappa_1 < \kappa_2$ of the design curve $\mathbf{r}$ to be constructed.
2. Construct the normal vectors $\mathbf{n}_1$ and $\mathbf{n}_2$ perpendicular to the unit vectors $\mathbf{u}_1$ and $\mathbf{u}_2$, respectively, and the corresponding centers of curvature C_1 and C_2. (C_1 is the point on the directed line through P_1 in the direction $\mathbf{n}_1$ distance $\rho_1 = 1/\kappa_1$ from P_1, and C_2 is similarly defined.)
3. Find a convex Bézier curve B_n passing through C_1 and C_2 tangent to $\mathbf{n}_1$ and $\mathbf{n}_2$, respectively, with arc length equal to $\rho_1 - \rho_2 = 1/\kappa_1 - 1/\kappa_2$.
4. Construct the design curve $\mathbf{r}$ as an involute of B_n. That is, start at C_2 and for each point $B_n(t)$ (which is parametrized so that $B_n(0) = C_2$) calculate $\mathbf{r}(t) = B_n(t) - (\rho_2 + S(t))\mathbf{u}(t)$ where

$$\mathbf{u}(t) = \frac{B_n'(t)}{|B_n'(t)|} = \mathbf{n}_2$$

and $S(t)$ is the arc length of the segment of B_n from $B_n(0) = C_2$ to $B_n(t)$.

5.3.3. Examples. Although it is possible to predict the behavior of many of the new fairness metrics based solely on an analysis of the shape quantities involved, it is also informative to see how the fairness criteria work on actual cases. We have selected a small group of examples which illustrate some of the salient tendencies of the various metrics. It is important to recognize that while some pairs of fairness criteria result in curves that are visually quite distinct, other pairs of fairness criteria result in curves that are visually indistinguishable. In these cases we rely on shape interrogation tools—such as plots of curvature, torsion, or total curvature—to identify the subtle shape differences in the faired curves.

For the first set of examples we used a two-segment, cubic NURB (which has a uniform knot vector (0.0,1.0,2.0)) with the control points and weights given in Table 5.2:

TABLE 5.2

Index i	Point P_i	Weight w_i
0	(0.0,0.0,0.0)	1.0
1	(1.0,0.0,0.0)	free (positive)
2	free	free (positive)
3	(2.0,0.5,0.5)	free (positive)
4	(1.0,1.0,1.0)	1.0

In effect, the endpoints and the end tangent directions are fixed, the degree of the curves is fixed. However, there is considerable flexibility for shape alteration provided by a free control point and three free weights. In the examples that follow, the starting values for the free weights was 1.0, and the starting value for the free control point was (2.0, 0.25, 0.333).

Within the geometric constraints of the problem, it is possible that the faired curve turn out to be planar—however, it is not mandatory. Under these circumstances, an effective implementation strategy would be to further constrain the geometry of the problem so that the result will of necessity lie in a plane. The simplified problem will consist of fewer free parameters and may be in a form that is easier to solve. The planar faired result can then be calculated and evaluated for suitability. If necessary, the original problem can then be solved for a possible alternative (nonplanar) result.

The examples that follow implement only the primitive fairness metrics described in the previous section; however, from these results it is possible to envision the effects of building composite metrics. Specifically, the results of the logical ANDs and logical ORs of shape quantities are quite apparent. Figures 5.2 and 5.3 depict the faired curves that are produced by applying metrics (5.14) and (5.15), respectively. Even a visual examination of the curves reveals that metric (5.14) yields a faired curve that is not quite as smooth—

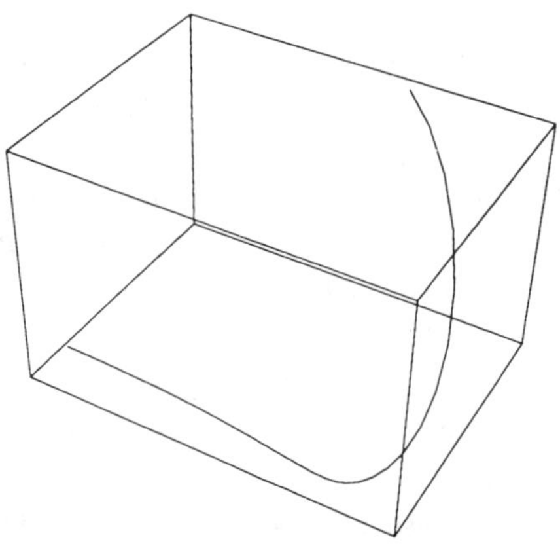

FIG. 5.2. NURB *faired by* (5.14).

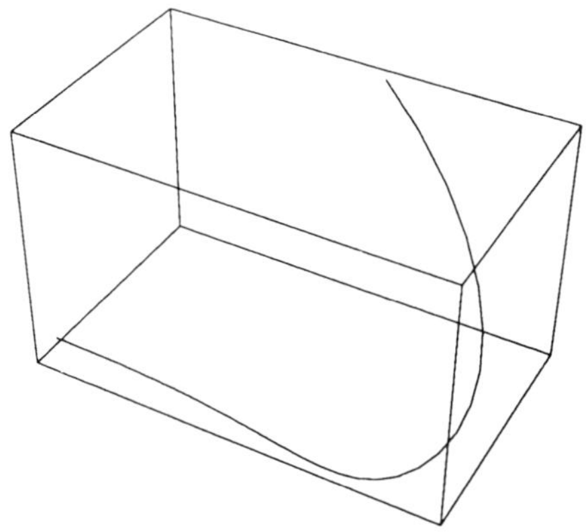

FIG. 5.3. NURB *faired by* (5.15).

that has sharper turns. Metric (5.15) produces a curve with a smaller curvature overall.

Figures 5.4–5.7 are the curvature and torsion plots of the two faired curves. These plots indicate that the curvature of the faired curve of metric (5.14) never gets very close to zero; this would, of course, result in a large value for ρ. Accordingly, the torsion for this curve oscillates but exhibits no extreme spikes. In contrast, the curvature for the faired curve of metric (5.15) does nearly vanish at one point, and in the neighborhood of this point, there is a surge in the torsion of the curve. The multiplication of curvature and torsion in the fairness metric suggests a logical OR of the two quantities. In regions of small curvature, large torsion is tolerated, and in regions of small torsion, large curvature is tolerated.

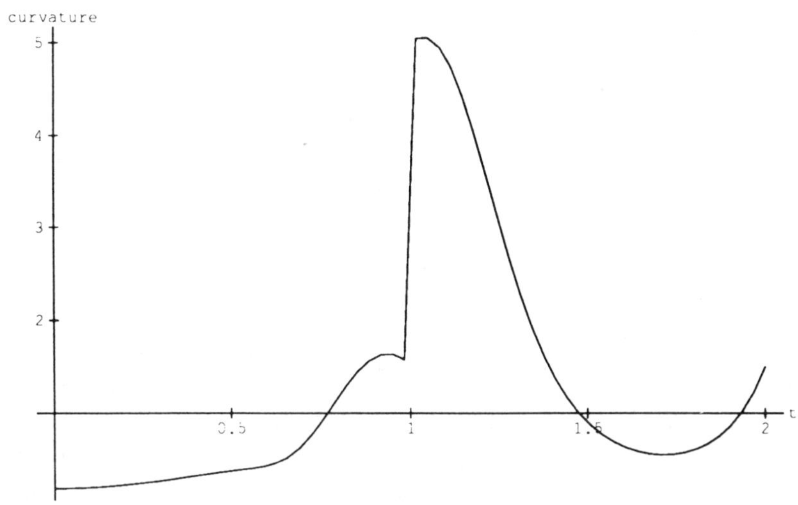

FIG. 5.4. *Curvature plot for Fig.* 5.2.

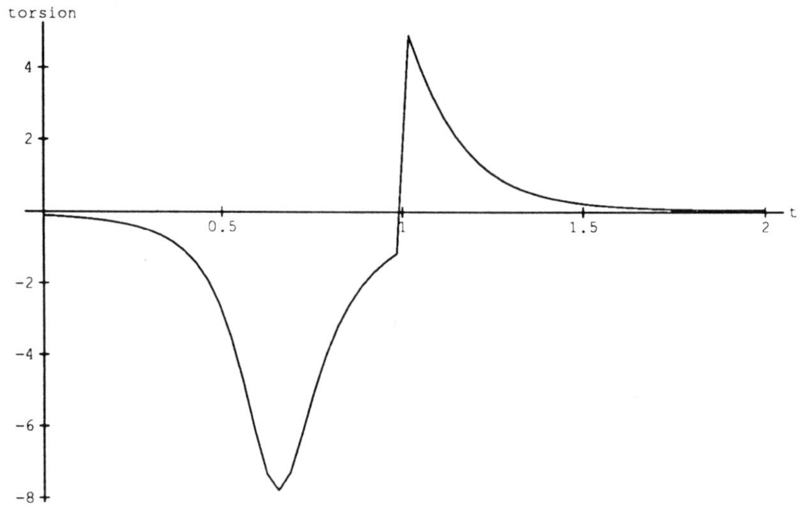

FIG. 5.5. *Torsion plot for Fig. 5.2.*

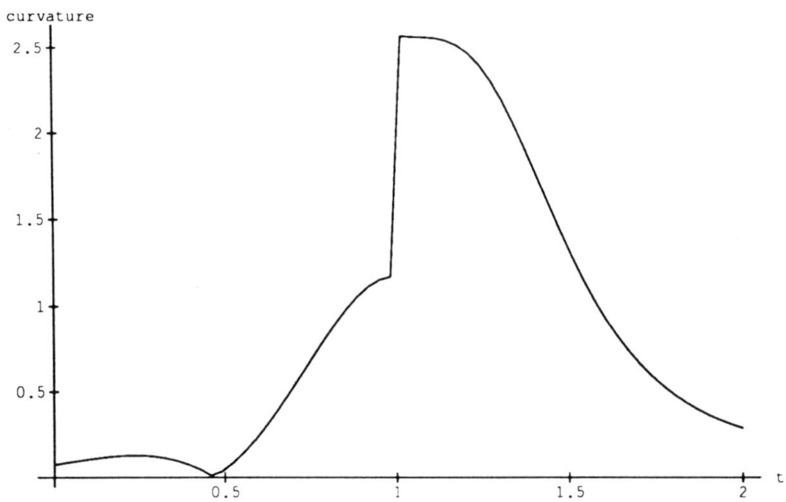

FIG. 5.6. *Curvature plot for Fig. 5.3.*

Figures 5.8 and 5.9 depict the faired curve and its curvature plot for metric (5.16). The analysis in the previous section suggests that this faired curve should be similar to the faired curve of metric (5.15). In this example, the plots of the two faired curve are visually similar. In fact, the curvature plots of the two curves have analogous profiles. Both plots nearly vanish near the parameter value 0.5; both have concomitant surges in the torsion in that region. Moreover, the torsion for both curves nearly vanishes in the parameter range (1.0, 2.0), as the curve is nearly flattened into a plane. However, the faired curve of metric (5.16) consistently exhibits a lower curvature—the effect of the κ^4 term in metric (5.16).

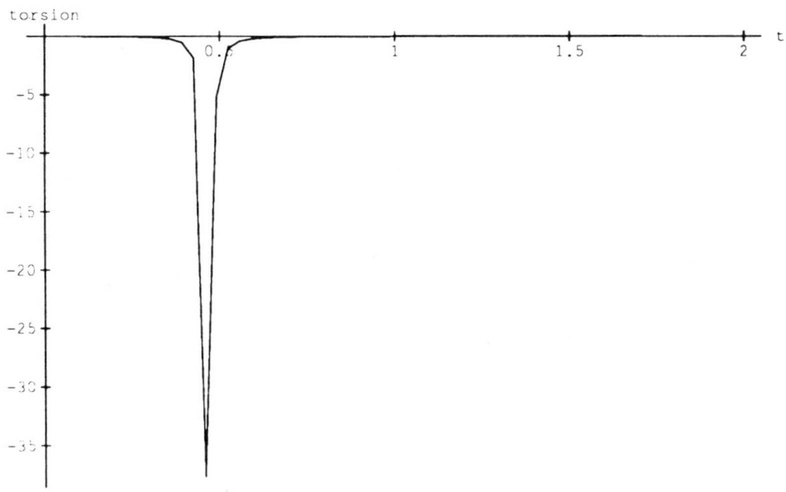

FIG. 5.7. *Torsion plot for Fig.* 5.3.

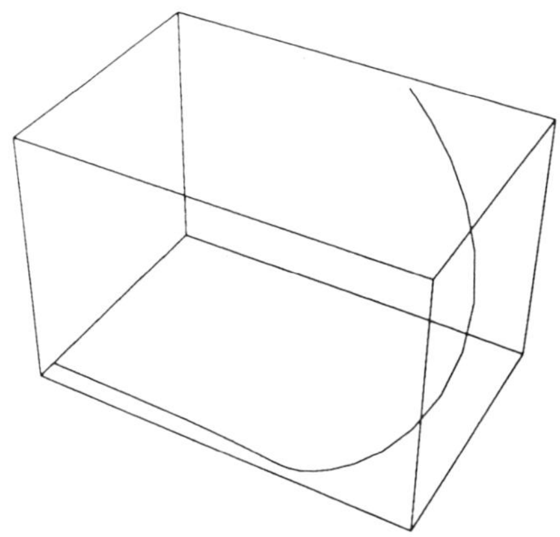

FIG. 5.8. NURB *faired by* (5.16).

Figures 5.10 and 5.11 depict the faired curve from metric (5.19), and the plot of its total curvature, respectively. From the total curvature plot it is apparent that the curvature for the faired curve remains consistently small. The torsion virtually vanishes for the length of the curve—indicating that the faired curve is planar. Consequently, the total curvature essentially corresponds to the curvature and is small for the most of the length of the curve. This curve does not exhibit the surges in total curvature that appear in the previous metrics in the regions in which low curvature compensates for high torsion. This metric illustrates the effect of a logical AND of the values of curvature and torsion: both values are held low for the length of the faired curve.

Figures 5.12–5.14 depict the faired curve, the curvature plot, and the torsion plot for metric (5.20). This curve is visually distinct from the previous curves. The curvature is consistently low for the faired curve; however, it never

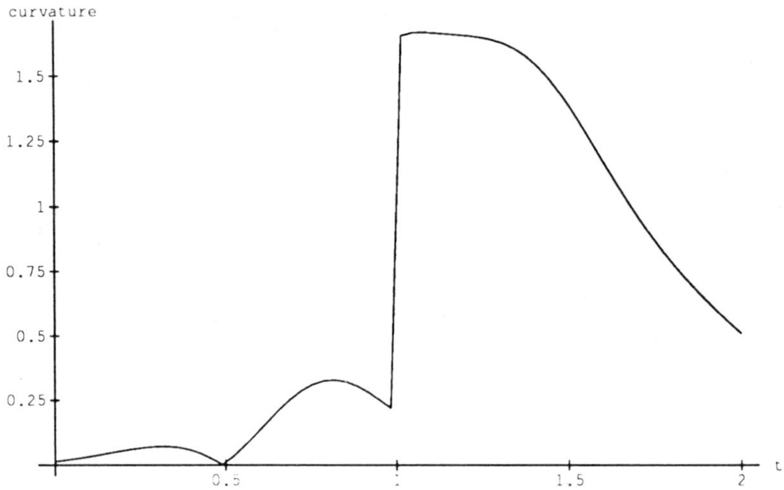

FIG. 5.9. *Curvature plot for Fig.* 5.8.

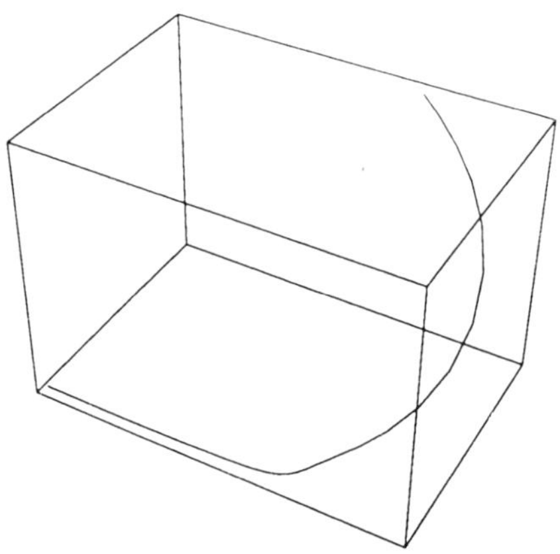

FIG. 5.10. NURB *faired by* (5.19).

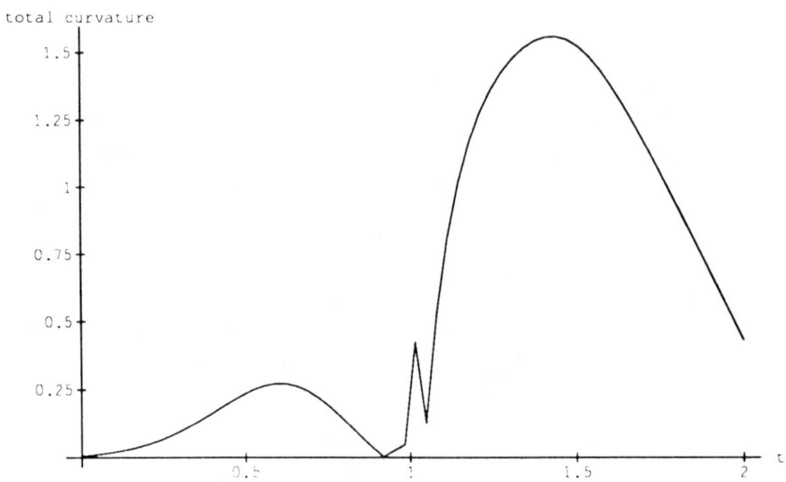

FIG. 5.11. *Total curvature plot for Fig.* 5.10.

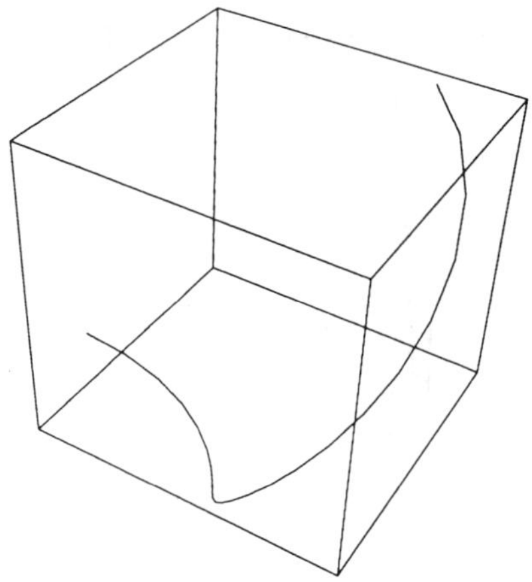

FIG. 5.12. NURB *faired by* (5.20).

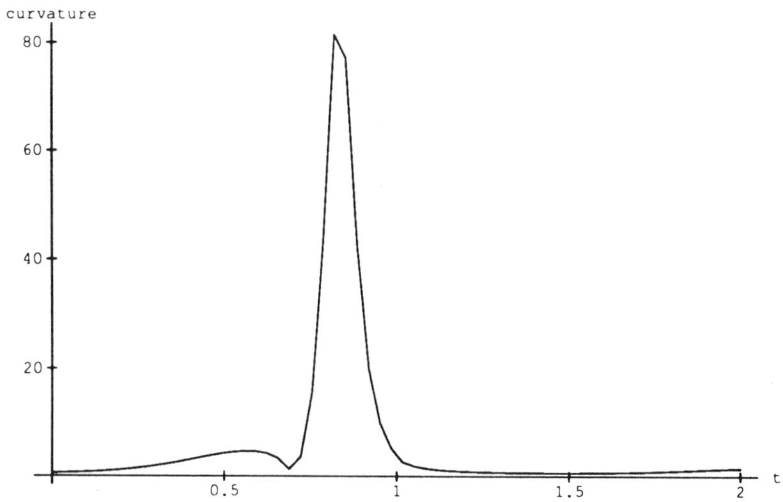

FIG. 5.13. *Curvature plot for Fig.* 5.12.

approaches very close to zero (since this would result in large values for the radius of curvature). In contrast to the preceding metrics, regions of high curvature (i.e., low radius of curvature) coincide with regions of high torsion. This phenomenon manifests itself in the plot of the faired curve as a quick turn near the middle of the curve.

Figure 5.15 (based on new geometric constraints) depicts a curve faired according to the spherical curvature metric (5.26). The constraints for this problem are in Table 5.3.

The starting values for the minimization process are (0.5, 1.0, 0.75) for the free point, and 1.0 for each free weight. The fairness value for the faired curve is 2.1 so the curve does not lie completely on a sphere; however, the plot indicates that this is the tendency of this faired curve.

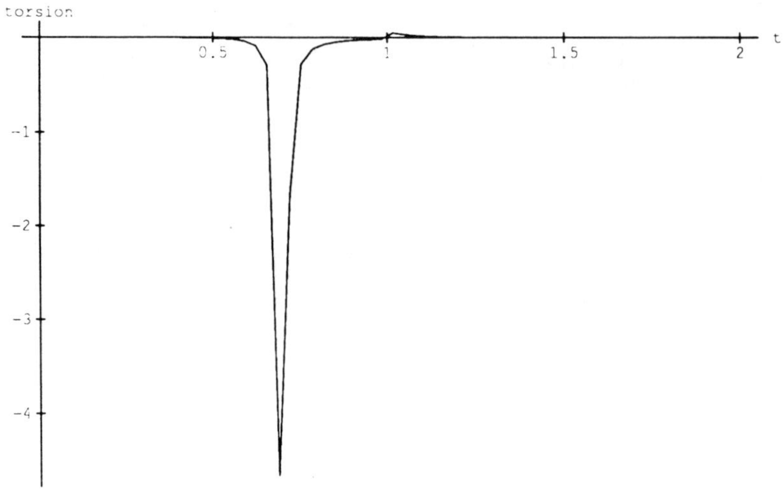

FIG. 5.14. *Torsion plot for Fig.* 5.12.

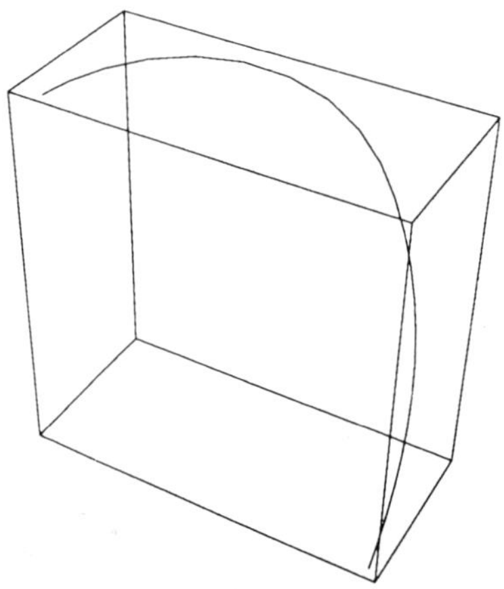

FIG. 5.15. NURB *with different constraints faired by* (5.26).

The last two examples for curves illustrate the use of the algorithms for monotone curvature. The end conditions for interpolation tangency and curvature are given as follows.

$$P_1 = (0,0), \quad P_2 = (5,0),$$
$$\mathbf{u}_1 = (0.92106, -.389420), \quad \mathbf{u}_2 = (0.070737, 0.99749),$$
$$\kappa_1 = 0.1, \quad \kappa_2 = 3.0.$$

Figure 5.16 shows a seventh-degree Bézier curve (the lower curve) together with its planar evolute (the upper curve). The Bézier curve has monotone curvature, and the planar evolute has arc length $\rho_1 - \rho_2 = 9.666\ldots$ which is the minimum attainable. The actual metric used in the minimization process was

TABLE 5.3

Index i	Point P_i	Weight w_i
0	(1.0,0.0,0.0)	1.0
1	(1.0,0.25,0.25)	free (positive)
2	free	free (positive)
3	(0.0,1.0,1.0)	free (positive)
4	(0.0,0.0,1.0)	1.0

(5.15). Figure 5.17 shows the curvature plot of the Bézier curve which clearly has monotone curvature. The eight parameters are two free parameters at each endpoint and the coordinates of the two middle control points C_3 and C_4. One of the two free parameters at each endpoint is the length of the end segment of the control polygon, and the other is a distance along a line (determined by the curvature and the first parameter) parallel to the end segment. Specifically, for an nth degree Bézier curve, the end parameters are α_i and β_i for $i = 1, 2$. The control points C_0, C_1, C_2, C_{n-2}, C_{n-1}, C_n are determined by

$$C_0 = P_1, \quad C_n = P_2,$$
$$C_1 = P_1 + \alpha_1 \mathbf{u}_1, \quad C_{n-1} = P_2 - \alpha_2 \mathbf{u}_2,$$
$$C_2 = C_1 + \beta_1 \mathbf{u}_1 + d_1 \mathbf{n}_1, \quad C_{n-2} = C_{n-1} - \beta_2 \mathbf{u}_2 + d_2 \mathbf{n}_2,$$

where

$$d_1 = \frac{n}{n-1} |\Delta C_0|^2 \kappa_1 \quad \text{and} \quad d_2 = \frac{n}{n-1} |\Delta C_{n-1}|^2 \kappa_2.$$

A more detailed discussion of these parameters is presented in [36]. It should be made clear that we could have found a fifth-degree Bézier curve with monotone curvature for these conditions using the four free end parameters alone, but we decided to use a seventh-degree Bézier curve to illustrate more clearly the types of parameters which can be used.

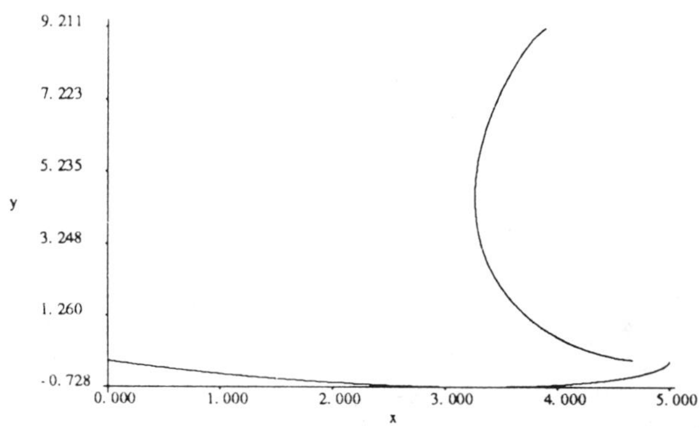

FIG. 5.16. *Bézier curve with monotone curvature and its evolute produced by metric* (5.15).

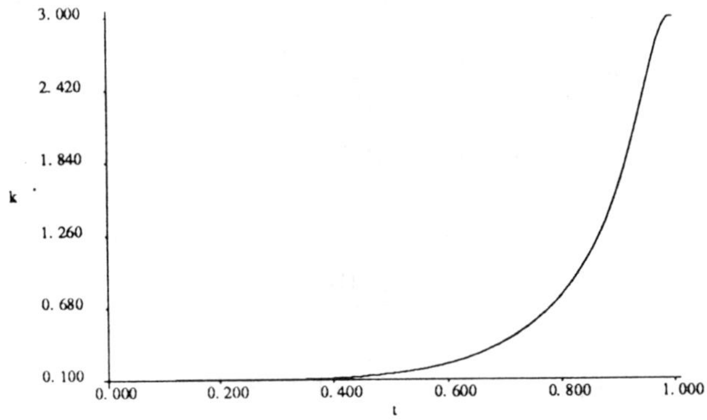

FIG. 5.17. *Curvature plot for Fig.* 5.16.

Figure 5.18 shows a convex Bézier cubic curve (the upper curve) connecting the two centers of curvature and tangent to the normal vectors as described above. It has arc length $\rho_1 - \rho_2 = 9.666\ldots$. The lower curve is its involute as described above. It passes through the given endpoints, is tangent to the given unit vectors, has the given curvatures, and has monotone curvature. Figure 5.19 shows the curvature plot of the involute curve, which clearly has monotone curvature. The plot shows the curvature decreasing from 3.0 to 0.1 because of the parametrization of the evolute which passes from C_2 to C_1 as the parameter t goes from 0 to 1. The actual evaluation of the arc length was accomplished using composite 15 point Gaussian quadrature with four subintervals. This resulted in rapid but accurate evaluation.

5.4. Fairness Metrics for Surfaces

Measures of fairness for surfaces may be designed in a manner similar to those for curves. As in the case of curves, it is desirable that any such measure depends only on the geometric invariants of the design surface so that a

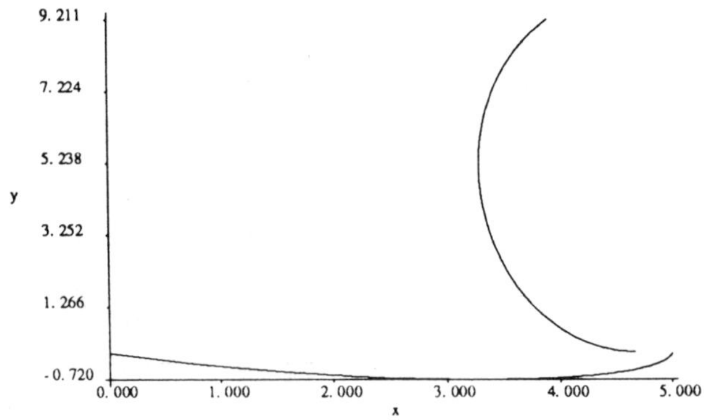

FIG. 5.18. *Curve with monotone curvature as involute of Bézier curve of specified arc length.*

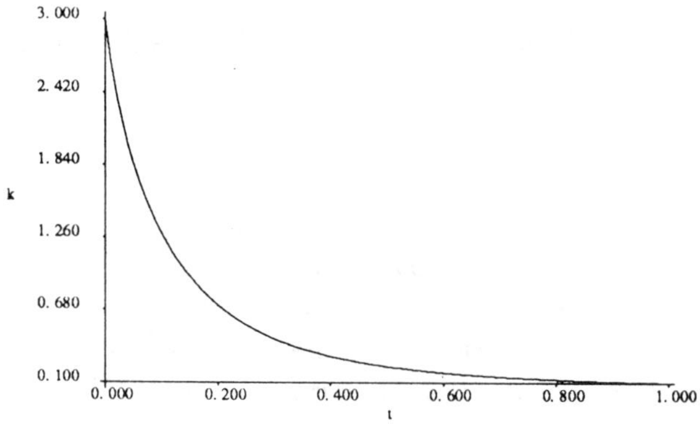

FIG. 5.19. *Curvature plot for Fig.* 5.18.

reparametrization will not produce a different value of the measure. As one would expect, the analysis required to study and develop such fairness measures is considerably more difficult than it was for curves.

The technique of using derived curves which depend only on the geometric invariants of the design curve has been applied to surfaces by Rando and Roulier in [30]. Let $\mathbf{r}(u, v)$ be a parametric *design surface* with $a \leq u \leq b$ and $c \leq v \leq d$. We assume that $\mathbf{r}$ is analytic. This is generally not a problem, since most of the surfaces used in CAGD are locally analytic. Therefore, any algorithm based on the measures of fairness developed here may be applied locally—patch by patch. In a manner similar to that for curves we may construct a *derived surface* $\mathbf{c}(u, v)$ which depends only on the geometric invariants of $\mathbf{r}$. The surface area Γ of the derived surface $\mathbf{c}$ is the fairness metric for $\mathbf{r}$. An analysis similar to that for curves must be performed to determine the effect of minimizing such a metric.

The geometric invariants for $\mathbf{r}$ are the *Gaussian curvature* K and the *mean curvature* H. These may be expressed in terms of the *principal curvatures* k_1 and k_2:

$$K = k_1 k_2, \tag{5.40}$$

$$H = (k_1 + k_2)/2. \tag{5.41}$$

These terms may also be expressed in terms of the *first and second fundamental forms* of $\mathbf{r}$. The reader is referred to any standard text in differential geometry, such as [5], for further details.

The desirable properties of a fairness metric for surfaces are similar to those for curves. Such a metric should provide the surface designer with a set of tools which will allow predictable modifications to the design surface locally or globally. Generally, the designer should be able to round, flatten, or roll the surface and minimize the magnitude of changes in the appropriate geometric invariants by minimizing the fairness metric as a function of the free parameters of the design surface.

It is observed in [30] that a reparametrization $\mathbf{r}(s,t)$ of the surface $\mathbf{r}$ exists so that at each point

$$\left|\frac{\partial \mathbf{r}}{\partial s} \times \frac{\partial \mathbf{r}}{\partial t}\right| = 1. \tag{5.42}$$

This is the equivalent of the reparametrization of a curve by arc length, and is used to analyze the effect of minimizing the surface area Γ of the derived surface $\mathbf{c}$. The actual evaluation of Γ may be accomplished by minimizing the surface area of $\mathbf{c}$ in the original parametrization of $\mathbf{r}$ or any convenient parametrization.

The following three derived surfaces are presented in [30]:

The *flattening metric* is the surface area of the derived surface

$$\mathbf{c}(u,v) = K(u,v)\mathbf{n}(u,v). \tag{5.43}$$

The *rounding metric* is the surface area of the derived surface

$$\mathbf{c}(u,v) = \mathbf{r}(u,v) + [H(u,v)/K(u,v)]\mathbf{n}(u,v). \tag{5.44}$$

The *rolling metric* is the surface area of the derived surface

$$\mathbf{c}(u,v) = [K(u,v) + H^2(u,v)]\mathbf{n}(u,v). \tag{5.45}$$

Here $\mathbf{n}(u,v)$ is the unit normal vector to the surface at $\mathbf{r}(u,v)$.

It is given by

$$\mathbf{n}(u,v) = \frac{(\partial \mathbf{r}(u,v)/\partial u) \times (\partial \mathbf{r}(u,v)/\partial v)}{|(\partial \mathbf{r}(u,v)/\partial u) \times (\partial \mathbf{r}(u,v)/\partial v)|}. \tag{5.46}$$

If $\mathbf{r}$ is reparametrized as indicated above, then it is obvious that

$$\mathbf{n} = \frac{\partial \mathbf{r}}{\partial s} \times \frac{\partial \mathbf{r}}{\partial t}. \tag{5.47}$$

Let G represent the parameter domain of $\mathbf{r}(u,v)$, and let G^* be the parameter domain of $\mathbf{r}(s,t)$ reparametrized as above. The surface area Γ of $\mathbf{c}$ is given by

$$\Gamma = \int\int_G \left|\frac{\partial \mathbf{c}}{\partial u} \times \frac{\partial \mathbf{c}}{\partial v}\right| dudv = \int\int_{G^*} \left|\frac{\partial \mathbf{c}}{\partial s} \times \frac{\partial \mathbf{c}}{\partial t}\right| dsdt. \tag{5.48}$$

Since

$$\left|\frac{\partial \mathbf{r}}{\partial s} \times \frac{\partial \mathbf{r}}{\partial t}\right| = 1$$

by (5.42), it is obvious that the area of the parameter domain G^* is equal to the surface area of the design surface $\mathbf{r}$.

The analysis of the effect of minimizing the surface area of a given derived surface depends on the representation of Γ as given in (5.48) in terms of the geometric invariants given above. Note that in addition to (5.42), the reparametrization of $\mathbf{r}$ may be done in such a way that the parameter curves are lines of curvature and so that the arc lengths of the two families of curvature lines may serve as invariant parameters. See [5] and [12]. The analysis is carried out on the reparametrized form of the problem since in addition to (5.42) we also have the nice properties that

$$\frac{\partial \mathbf{r}}{\partial s}, \quad \frac{\partial \mathbf{r}}{\partial t} \quad \text{and} \quad \frac{\partial \mathbf{r}}{\partial s} \times \frac{\partial \mathbf{r}}{\partial t}$$

are mutually orthogonal unit vectors. In addition, the formulas of Rodrigues are valid for this parametrization:

$$\begin{aligned} \frac{\partial \mathbf{n}}{\partial s} &= -k_1 \frac{\partial \mathbf{r}}{\partial s}, \\ \frac{\partial \mathbf{n}}{\partial t} &= -k_2 \frac{\partial \mathbf{r}}{\partial t}. \end{aligned} \tag{5.49}$$

For the flattening metric (5.43), we have $\mathbf{c} = K\mathbf{n}$. An application of (5.49) to the two partial derivatives of $\mathbf{c}$ gives

$$\frac{\partial \mathbf{c}}{\partial s} = \frac{\partial K}{\partial s}\mathbf{n} - Kk_1 \frac{\partial \mathbf{r}}{\partial s}$$

and

$$\frac{\partial \mathbf{c}}{\partial t} = \frac{\partial K}{\partial t}\mathbf{n} - Kk_2 \frac{\partial \mathbf{r}}{\partial t}.$$

Thus, the normal vector to $\mathbf{c}$ is

$$\frac{\partial \mathbf{c}}{\partial s} \times \frac{\partial \mathbf{c}}{\partial t} = -Kk_1 \frac{\partial K}{\partial t}\left[\frac{\partial \mathbf{r}}{\partial s} \times \mathbf{n}\right] - Kk_2 \frac{\partial K}{\partial s}\left[\mathbf{n} \times \frac{\partial \mathbf{r}}{\partial t}\right] + k_1 k_2 K^2 \frac{\partial \mathbf{r}}{\partial s} \times \frac{\partial \mathbf{r}}{\partial t}.$$

The fact that the vectors in the right-hand side of the above equation are mutually orthogonal unit vectors and (5.40) gives the integrand of the fairness metric (5.48) for derived curve $\mathbf{c}$ as given in (5.43):

$$\left|\frac{\partial \mathbf{c}}{\partial s} \times \frac{\partial \mathbf{c}}{\partial t}\right| = |K|\left[\left(k_1 \frac{\partial K}{\partial t}\right)^2 + \left(k_2 \frac{\partial K}{\partial s}\right)^2 + K^4\right]^{1/2}. \tag{5.50}$$

An examination of (5.50) reveals that minimizing the surface area of $\mathbf{c}$ given by (5.43) tends to encourage minimizing the magnitude of the Gaussian curvature K and extreme changes in the Gaussian curvature along the lines of curvature. Thus, the tendency is to *flatten* the design surface $\mathbf{r}$.

A similar analysis of the other two derived surfaces (5.44) and (5.45) yields expressions which reveal their properties as well. The rounding metric

$\mathbf{c}(u,v) = \mathbf{r}(u,v) + [H(u,v)/K(u,v)]\mathbf{n}(u,v)$ given by (5.44) gives:

$$\left|\frac{\partial \mathbf{c}}{\partial s} \times \frac{\partial \mathbf{c}}{\partial t}\right| = \Big[(1-Qk_1)^2(1-Qk_2)^2 + (1-Qk_1)^2\left(\frac{\partial Q}{\partial t}\right)^2 + (1-Qk_2)^2\left(\frac{\partial Q}{\partial s}\right)^2\Big]^{1/2}, \tag{5.51}$$

where $Q = H(s,t)/K(s,t)$.

The rolling metric $\mathbf{c}(u,v) = [K(u,v) + H^2(u,v)]\mathbf{n}(u,v)$ given by (5.45) gives

$$\left|\frac{\partial \mathbf{c}}{\partial s} \times \frac{\partial \mathbf{c}}{\partial t}\right| = |W|\left[\left(k_1\frac{\partial W}{\partial t}\right)^2 + \left(k_2\frac{\partial W}{\partial s}\right)^2 + W^2K^2\right]^{1/2}, \tag{5.52}$$

where $W = K + H^2$.

An analysis of (5.51) reveals that minimizing the metric generated by the derived surface given by (5.44) tends to minimize the magnitudes of $1-Qk_1$ and $1-Qk_2$. This minimum is zero when $k_1 = k_2$. Thus, a tendency of minimizing this metric is to pull the surface to a sphere. A similar analysis of (5.52) reveals that minimizing it tends to make the surface more cylindrical or conical. For further details on these see [30].

A new metric which also is a rolling metric is generated by the derived surface given by

$$\mathbf{c}(u,v) = H(u,v)\mathbf{n}(u,v). \tag{5.53}$$

A similar analysis of this derived surface gives the integrand

$$\left|\frac{\partial \mathbf{c}}{\partial s} \times \frac{\partial \mathbf{c}}{\partial t}\right| = |H|\left[\left(k_1\frac{\partial H}{\partial t}\right)^2 + \left(k_2\frac{\partial H}{\partial s}\right)^2 + H^2K^2\right]^{1/2}. \tag{5.54}$$

Minimizing this metric tends to force $H = 0$, or $K = 0$ and (for example) $k_1 \neq 0$ and

$$\frac{\partial H}{\partial t} = 0.$$

Thus, the tendency is to force the design surface either to a planar, cylindrical, or possibly a conical shape.

We introduce another new metric with derived surface

$$\mathbf{c}(u,v) = \frac{1}{2(2H^2(u,v) - K(u,v))}\mathbf{n}(u,v). \tag{5.55}$$

A similar analysis of this derived surface gives the integrand

$$\left|\frac{\partial \mathbf{c}}{\partial s} \times \frac{\partial \mathbf{c}}{\partial t}\right| = |R|\left[\left(k_1\frac{\partial R}{\partial t}\right)^2 + \left(k_2\frac{\partial R}{\partial s}\right)^2 + R^2K^2\right]^{1/2}, \tag{5.56}$$

where

$$R(s,t) = \frac{1}{2(2H^2(s,t) - K(s,t))}.$$

Note that it is easy to see that

$$R = \frac{1}{k_1^2 + k_2^2}.$$

Also, observe that

$$\frac{\partial R}{\partial s} = -2\frac{k_1\frac{\partial k_1}{\partial s} + k_2\frac{\partial k_2}{\partial s}}{(k_1^2 + k_2^2)^2}$$

and

$$\frac{\partial R}{\partial t} = -2\frac{k_1\frac{\partial k_1}{\partial t} + k_2\frac{\partial k_2}{\partial t}}{(k_1^2 + k_2^2)^2}.$$

Minimizing this metric tends to force small values of R and small values of $k_2\frac{\partial R}{\partial s}$ and $k_1\frac{\partial R}{\partial t}$ while not allowing both $k_1 = 0$ and $k_2 = 0$. The tendency of minimizing this metric is to roll the design surface if the constraints and the initial conditions (such as k_1 = constant) warrant, but it will not flatten the surface. For example, if $k_1 = 0$ everywhere, then the metric minimizes to zero if $\frac{\partial k_2}{\partial s} = 0$ everywhere, that is, if k_2 is constant as a function of s for fixed t. This can be observed from the equations above. If the initial conditions and constraints do not allow one of the principal curvatures to be a constant, then minimizing this metric will tend to make a slight bulge or rounding but not a spherical shape. Unlike the rounding metric given by the derived surface (5.44) which is undefined if either $k_1 = 0$ or $k_2 = 0$, this metric is defined as long as the principal curvatures are not both zero to start with. In practice, this metric is applicable in most surface fairing situations as long as the starting conditions for the minimization are not too severe. Some of the examples to follow will illustrate these comments.

Examples of the application of the first three metrics to surfaces appear in [30]. The examples which we will present here will concentrate mainly on the flattening metric given by derived surface (5.43) and the two new metrics given by derived surfaces (5.53) and (5.55). The general algorithm for surface fairing is the same as Algorithm 5.1 for curves.

Algorithm 5.3.

1. Choose a fairness metric Γ and its associated derived surface to use for fairing.
2. Determine the free variables for the problem at hand, and express Γ as a function of these free variables.
3. Choose a method for the evaluation of the function Γ determined above. One relatively easy way to estimate the value of Γ is to calculate an array of points on the derived surface $\mathbf{c}(u,v)$ in the original parametrization and estimate the surface area from these.

4. Choose a minimization algorithm and suitable starting values for the algorithm to minimize Γ, and perform the minimization using the evaluation method chosen in the previous step.

5.4.1. Examples. The examples included here have some common aspects. All of the examples involve fairing a single Bézier patch with certain constraints. In all cases, the fairness metric Γ is estimated by calculating an 11×11 point grid using the original parametrization on the derived surface $\mathbf{c}(u_i, v_j)$, where $u_i = i/10$ and $v_j = j/10$ for $i = 0, 1, \ldots, 10$ and $j = 0, 1, \ldots, 10$. The fairness metric is then approximated as the sum of the areas of the triangles with vertices $P_{i,j}$, $P_{i+1,j}$, $P_{i+1,j+1}$ and with vertices $P_{i,j}$, $P_{i,j+1}$, $P_{i+1,j+1}$, respectively, for $i = 0, 1, \ldots, 9$ and $j = 0, 1, \ldots, 9$. This is admittedly somewhat crude, but it has the advantage of being relatively easy to implement since it requires no calculations or special functions beyond those required to evaluate $\mathbf{c}(u, v)$ and no reparametrizations. As we will see, the results are quite satisfactory for the examples presented below.

The first group of examples apply some of the metrics to the generation of a bicubic Bézier patch for grid control points of the form $P_{i,j} = (x_i, y_j, z_{i,j})$ where $x_i = i/3$ and $y_j = j/3$ for $i = 0, \ldots, 3$ and $j = 0, \ldots, 3$. The eight end control points $P_{i,j}$ for $j = 0$ and $j = 3$ and $i = 0, \ldots, 3$ are fixed, and the x and y coordinates of the remaining eight control points are also fixed. The free variables then will be the eight z coordinates $z_{i,j}$ for $j = 1$ and $j = 2$ and $i = 0, \ldots, 3$. This will allow us to easily observe the results which follow from the fairing with the individual metrics.

The first figure is an initial bicubic Bézier patch. The graph of the patch is shown in Fig. 5.20, and the graphs of the Gaussian curvature and mean curvature are shown in Figs. 5.21 and 5.22, respectively. Figure 5.23 shows the graph of this surface faired with the flattening metric induced by the derived surface (5.43). Figures 5.24 and 5.25 show the Gaussian and mean curvatures, respectively, for the faired surface. The flattening effect is evident when the faired surface and the original are compared. An examination of the Gaussian

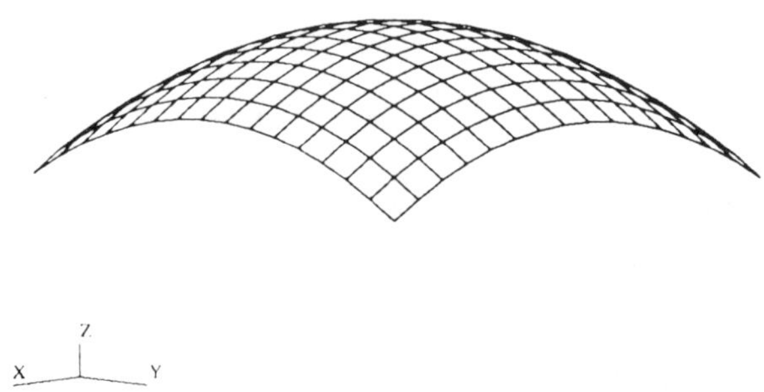

FIG. 5.20. *Unfaired bicubic Bézier patch.*

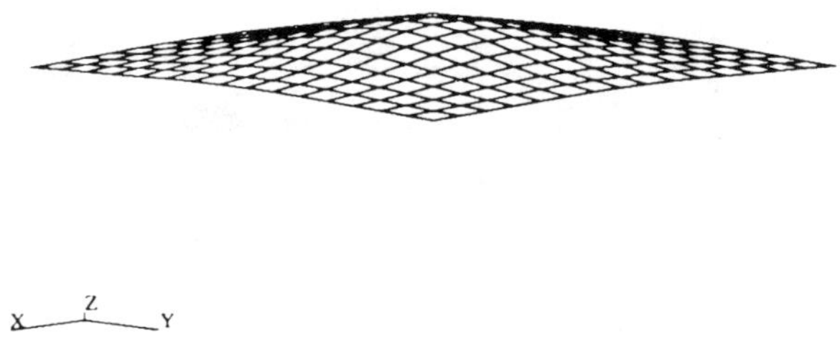

FIG. 5.21. *Gaussian curvature for Fig.* 5.20.

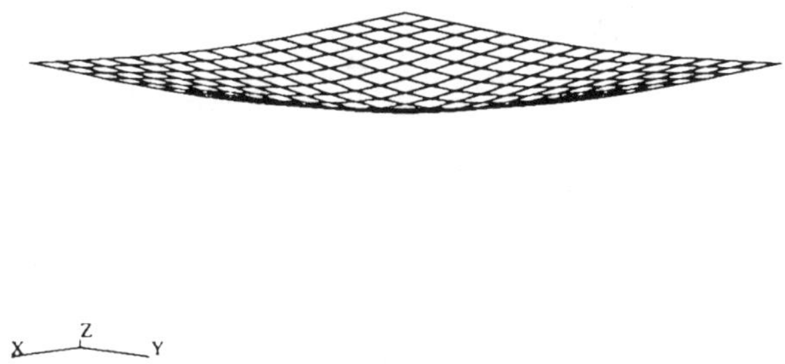

FIG. 5.22. *Mean curvature for Fig.* 5.20.

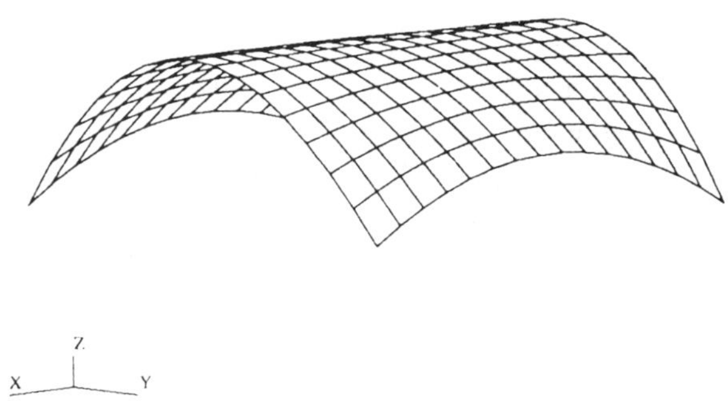

FIG. 5.23. *Faired with flattening metric* (5.43).

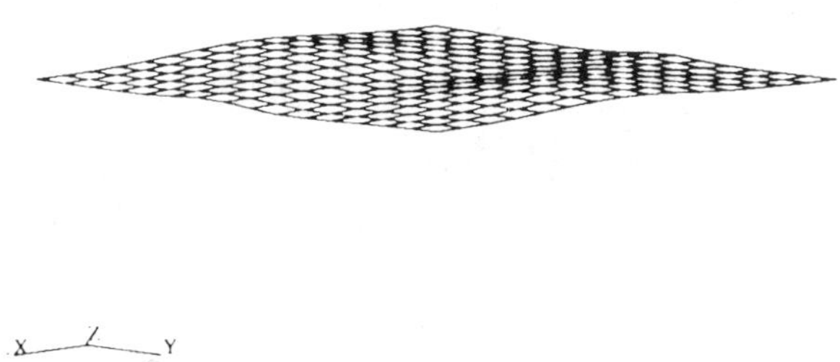

FIG. 5.24. *Gaussian curvature for Fig.* 5.23.

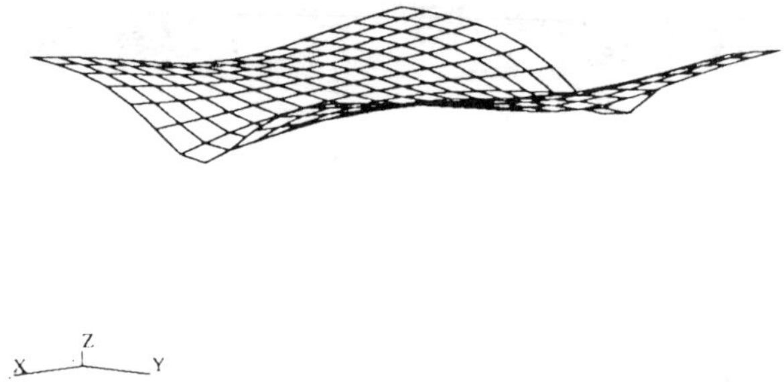

FIG. 5.25. *Mean curvature for Fig.* 5.23.

curvature plots shows that the faired surface has a flatter Gaussian curvature closer to zero overall.

The next group of figures shows the result of fairing the same initial surface as above with the new rolling metric induced by the derived surface given by (5.53). Figure 5.26 shows the faired patch, and Figs. 5.27 and 5.28 show the Gaussian and mean curvatures, respectively, for this patch. The rolling effect of fairing with this metric is clearly evident from these figures. The Gaussian curvature is zero everywhere, and the mean curvature is constant along lines parallel to the y axis. Thus, one of the principal curvatures is zero everywhere, and the other is constant along lines parallel to the y axis.

We next applied the new "bulging" metric induced by the derived surface given by (5.55) to the original patch. Figure 5.29 shows the faired patch, and Figs. 5.30 and 5.31 show the Gaussian and mean curvatures, respectively. The "bulging" effect is evident near the top of Fig. 5.29 as a slightly sharper bend. The Gaussian and mean curvature plots clearly show a higher magnitude near the center than those in Figs. 5.21 and 5.22 for the original patch.

To show the rolling ability of the metric induced by (5.55), we modified the original patch so that the two edges parallel to the y axis are straight and parallel. The plot of this patch is shown in Fig. 5.32, and its Gaussian and mean curvatures are shown in Figs. 5.33 and 5.34, respectively. The faired patch is shown in Fig. 5.35, and its Gaussian and mean curvatures are shown in Figs. 5.36 and 5.37, respectively. The rolling effect is clearly evident here. The

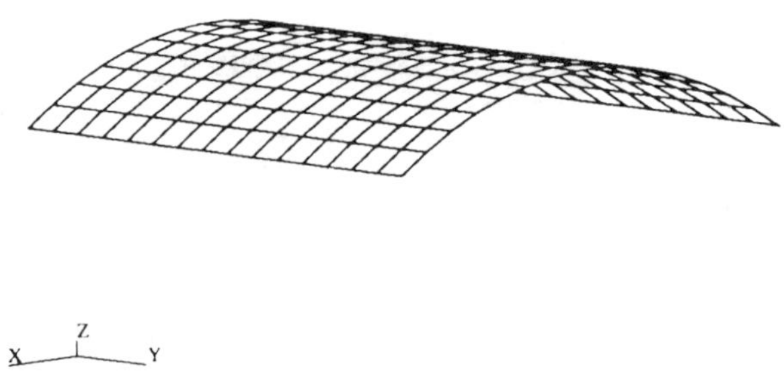

FIG. 5.26. *Faired with new rolling metric* (5.53).

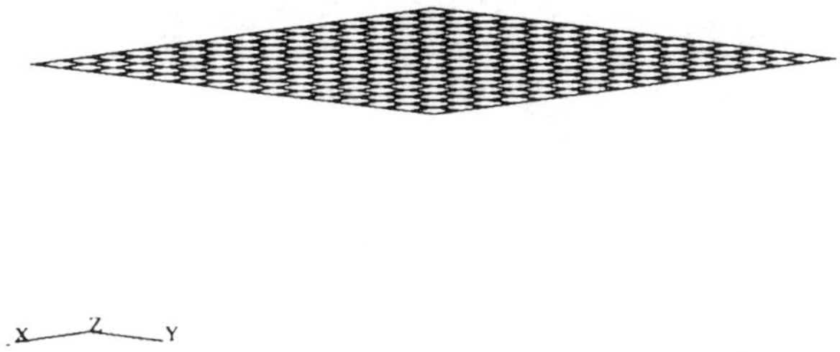

FIG. 5.27. *Gaussian curvature for Fig.* 5.26.

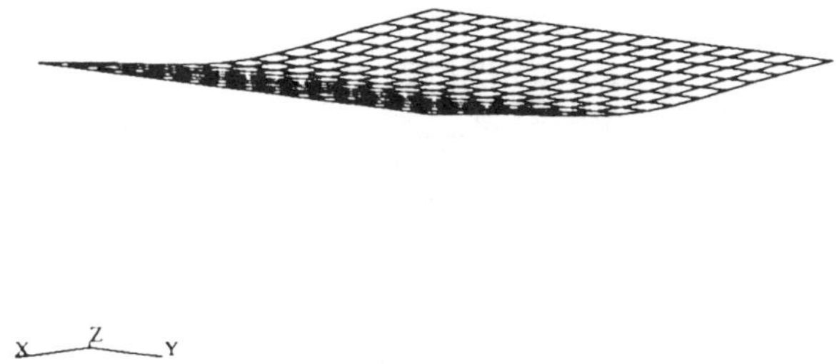

FIG. 5.28. *Mean curvature for Fig.* 5.26.

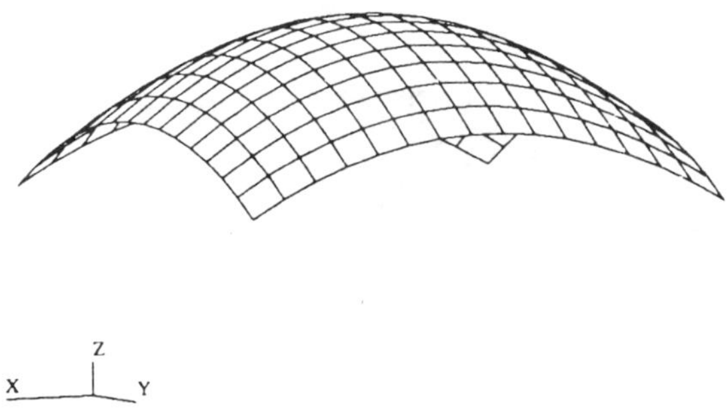

FIG. 5.29. *Faired with bulging metric* (5.55).

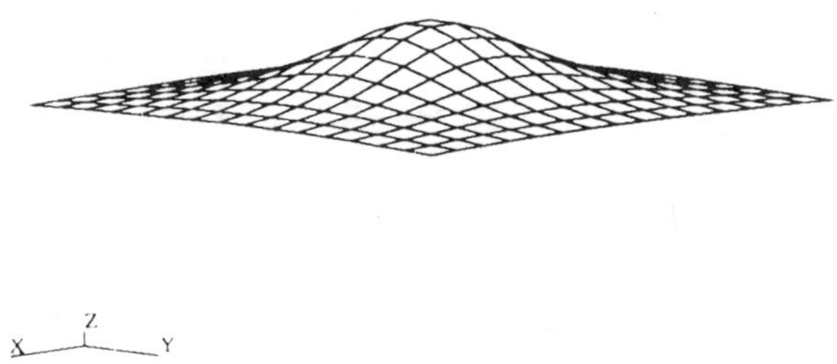

FIG. 5.30. *Gaussian curvature for Fig.* 5.29.

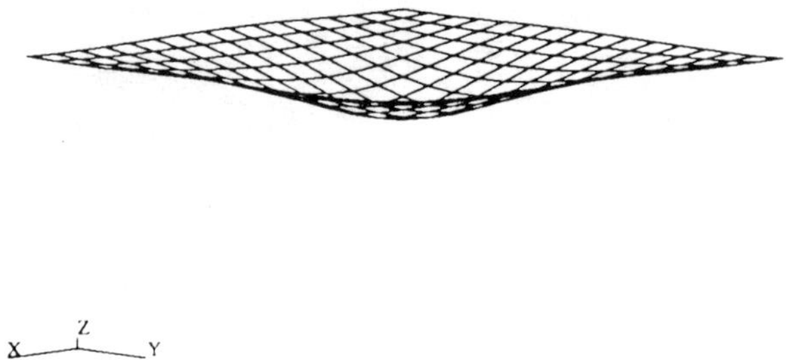

FIG. 5.31. *Mean curvature for Fig. 5.29.*

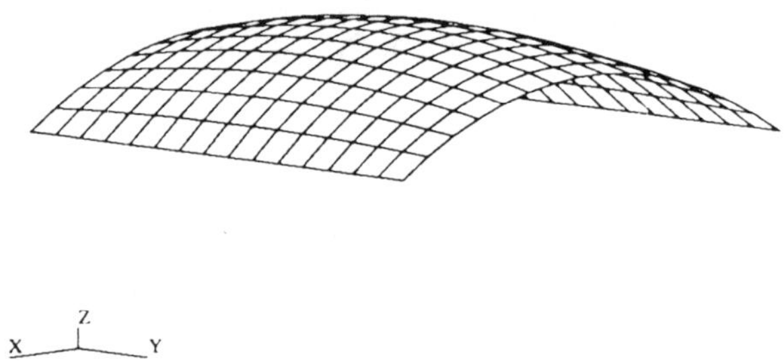

FIG. 5.32. *Modified bicubic patch.*

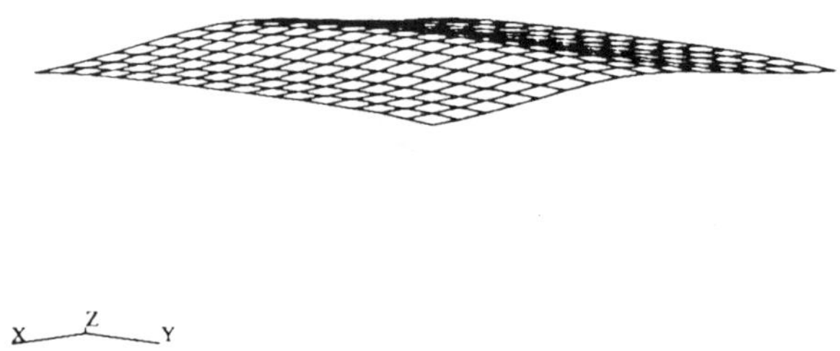

FIG. 5.33. *Gaussian curvature for Fig. 5.32.*

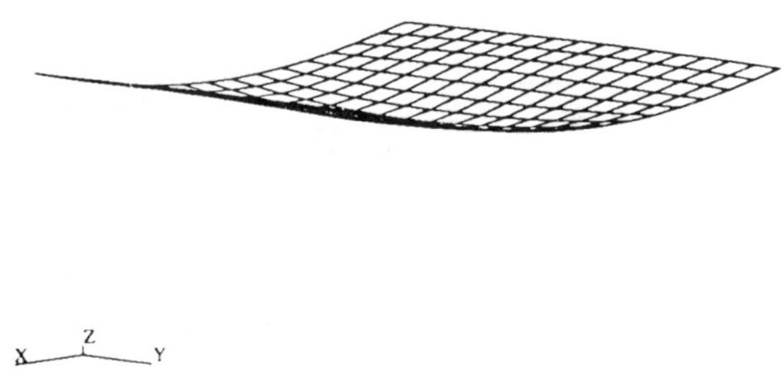

FIG. 5.34. *Mean curvature for Fig. 5.32.*

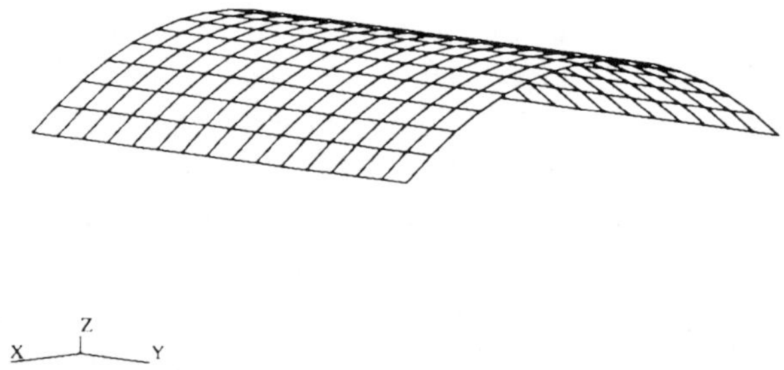

FIG. 5.35. *Faired patch from Fig. 5.32 using bulging metric* (5.55)—*demonstrates rolling capability of this metric.*

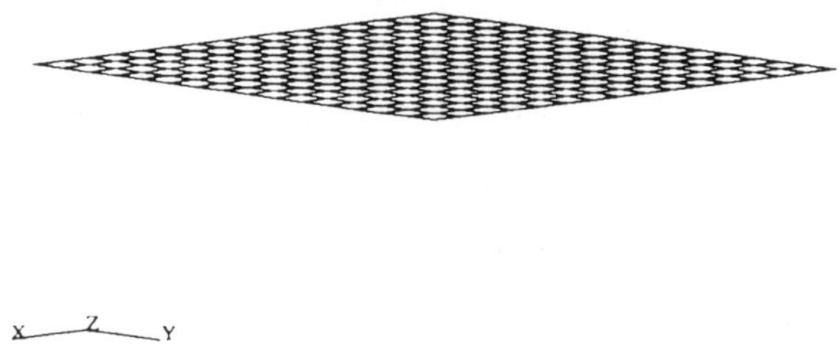

FIG. 5.36. *Gaussian curvature for Fig.* 5.35.

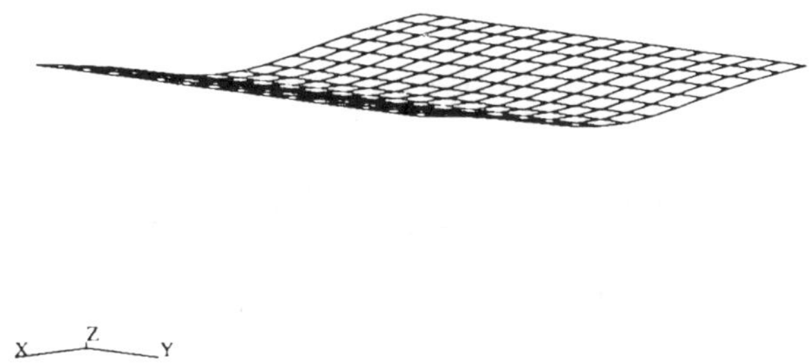

FIG. 5.37. *Mean curvature for Fig.* 5.35.

reason why this figure faired to a rolled surface is that the two edges parallel to the y axis were forced to remain straight. This encouraged one of the principal curvatures to tend to zero. As discussed earlier, this forces the other principal curvature to be constant in one direction.

The next group of figures involves a biquintic patch. The control points are on a grid as above, but all of their coordinates are fixed except for the z coordinates of the four middle control points $P_{2,2}$, $P_{2,3}$, $P_{3,2}$, and $P_{3,3}$. The patch then has all boundary curves and first partial derivatives on the boundary determined. The four parameters for the minimization process are the z coordinates of the four control points indicated above. The initial configuration forces a small "dimple" in the center of the patch. This is shown in Fig. 5.38. The Gaussian and mean curvatures are shown in Figs. 5.39 and 5.40, respectively. The next three figures, Figures 5.41–5.43, show the patch, Gaussian curvature, and mean curvature, respectively, after fairing with the flattening metric (5.43). The "dimple" is removed, and the magnitudes and ripples of both the Gaussian and mean curvatures are sharply reduced.

Figure 5.44 shows the patch which results from applying the "bulge" metric induced by (5.55) to the patch of Fig. 5.41 produced by the flattening metric. Figures 5.45 and 5.46 show the Gaussian and mean curvatures, respectively, for this patch. The result is a slightly rounder look on the top left of the figure. If this metric had been applied to the original patch, the "dimple" would have been enhanced rather than removed. The figures for this are not included here.

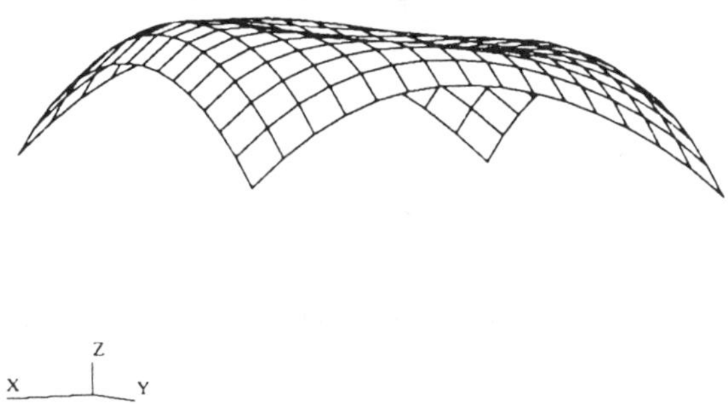

FIG. 5.38. *Biquintic Bézier patch.*

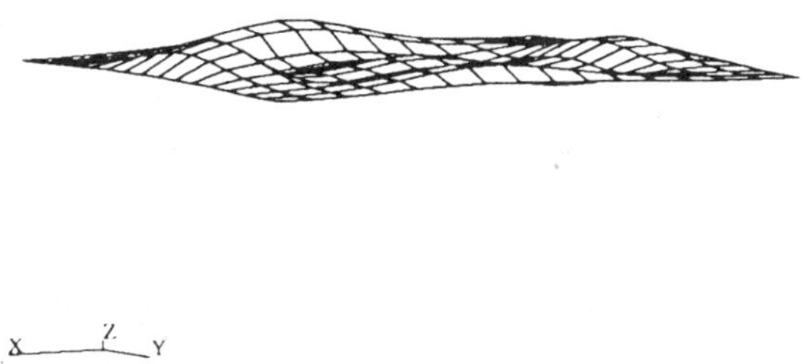

FIG. 5.39. *Gaussian curvature for Fig.* 5.38.

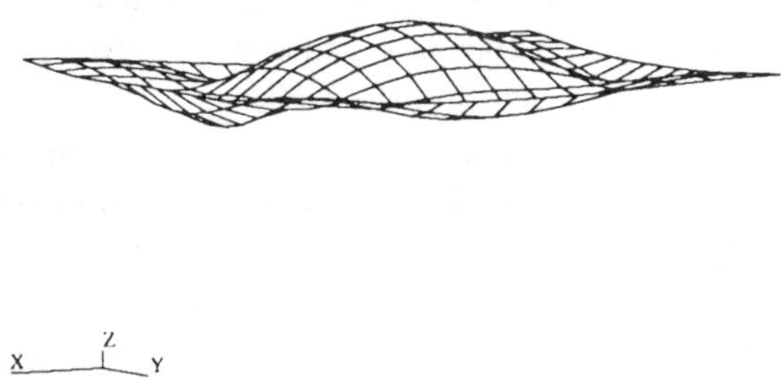

FIG. 5.40. *Mean curvature for Fig.* 5.38.

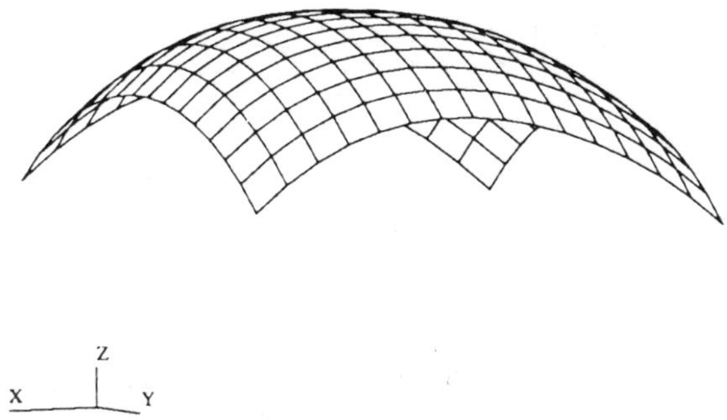

FIG. 5.41. *Faired patch of Fig. 5.38 with flattening metric* (5.43).

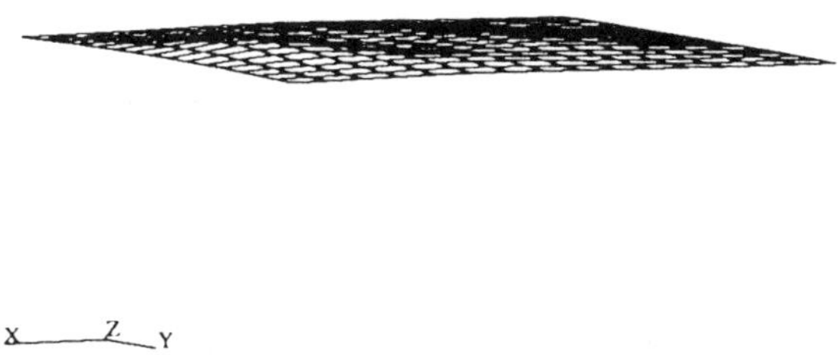

FIG. 5.42. *Gaussian curvature for Fig.* 5.41.

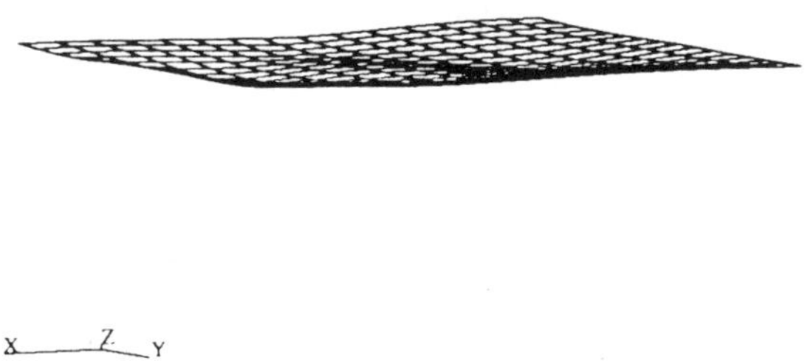

FIG. 5.43. *Mean curvature for Fig.* 5.41.

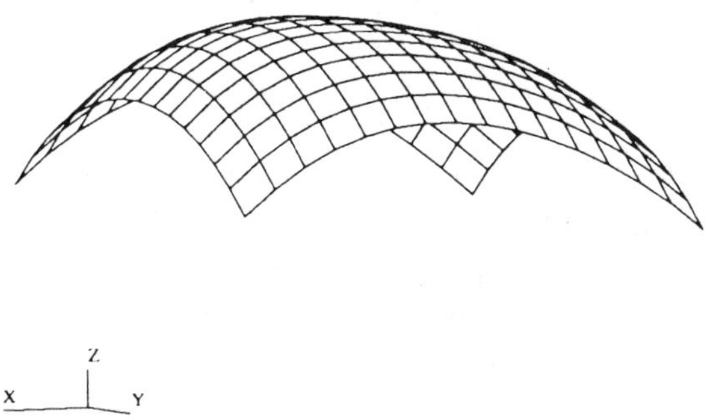

FIG. 5.44. *Faired patch of Fig. 5.41 with bulging metric* (5.55).

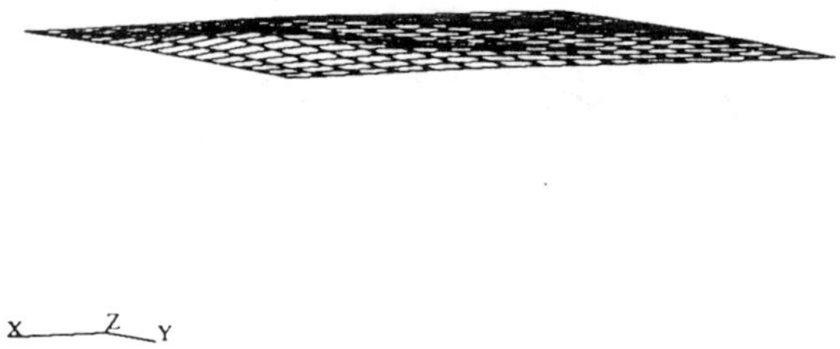

FIG. 5.45. *Gaussian curvature for Fig.* 5.44.

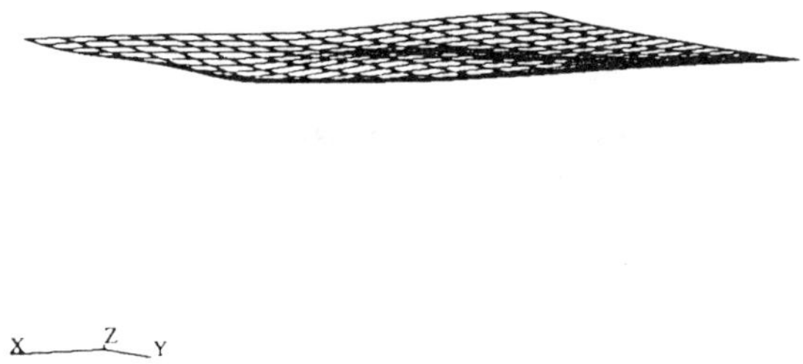

FIG. 5.46. *Mean curvature for Fig.* 5.44.

The final group of figures illustrates the use of these metrics in cases involving nongrid control points and with parameters other than the z coordinates of some of the control points. The initial data here are the control points of a Bézier bicubic. The first and last rows of control points are the same, and the resulting patch resembles a "wing" with a bulge on the front edge. The initial patch is shown in Fig. 5.47. The first example fairs the patch with the new rolling metric induced by (5.53). The four free parameters are the lengths of the segments $P_{0,1}P_{1,1}$, $P_{0,2}P_{1,2}$, $P_{2,1}P_{3,1}$, and $P_{2,2}P_{3,2}$. All control points are fixed except $P_{1,1}$, $P_{1,2}$, $P_{2,1}$, $P_{2,2}$, which are constrained to lie on the initial four line segments. The partial derivatives in the u direction all have common direction at the back side top and at the back side bottom. The result of the fairing process is shown in Fig. 5.48. The rolling effect is obvious. The "bulge" in the patch is removed and the lines along the "wing" are all straight.

If the patch is faired using the new "bulge" metric (5.55) and the same parameters, the "bulge" collapses to the back of the "wing." The reason for the problem is that both principal curvatures are close to zero at points on the surface away from the leading edge of the "wing." The parameter constraints will exacerbate this problem if the parameters are increased. Therefore, the only way to minimize is to shrink the parameters. If we change the parameters appropriately, a better patch results. For this, the six parameters are now the x and z coordinates of $P_{1,1}$ and $P_{1,2}$, and the x coordinates of $P_{2,1}$ and $P_{2,2}$.

FIG. 5.47. *Bicubic Bézier "wing."*

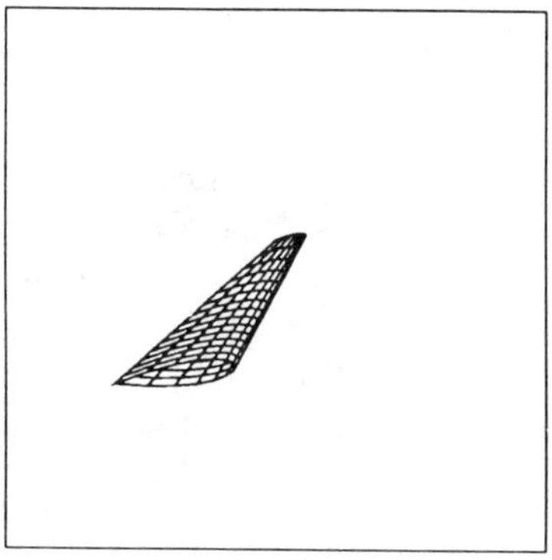

FIG. 5.48. *Bicubic Bézier "wing" of Fig. 5.47 faired with new rolling metric using end segment lengths and fixed tangent directions at back of "wing."*

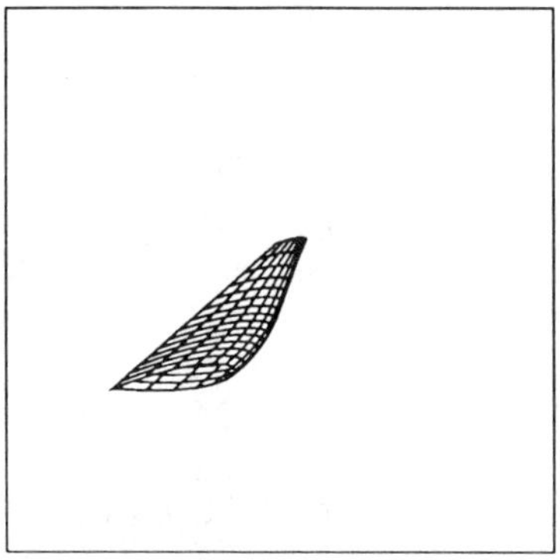

FIG. 5.49. *Bicubic Bézier "wing" of Fig. 5.47 faired with bulging metric using x and z coordinates of certain control points with relaxed tangent directions at back of "wing."*

The resulting patch is shown in Fig. 5.49. Note that the "bulge" is smoothed out and less severe than in Fig. 5.47.

5.5. Conclusion

A wide variety of fairness metrics for parametric curves and surfaces have been presented and discussed. A successful application of these metrics to practical problems requires an understanding of the effect of minimizing the metric under the given constraints as well as an understanding of the minimization algorithm used as well as the technique for evaluating the metric. The results can be quite good when the metrics are used properly. The derived curve and derived surface approach is particularly useful in that it allows the use of the original parametrization for the evaluation of the metric as well as an easy way to evaluate it.

References

[1] P. J. Barry and R. Goldman, *What is the natural generalization of a Bézier curve?* in Mathematical Methods in Computer Aided Geometric Design, T. Lyche and L. Schumaker, eds., Academic Press, New York, 1989, pp. 71–86.

[2] W. Boehm, G. Farin, and J. Kahmann, *A survey of curve and surface methods in CAGD*, Comput. Aided Geom. Design, 1 (1984), pp. 1–60.

[3] C. de Boor, *A Practical Guide to Splines*, Springer-Verlag, New York, 1978.

[4] H. G. Burchard, J. A. Ayers, W. H. Frey, and N. S. Sapidis, *Approximation with aesthetic constraints*, Research Report GMR-7814, General Motors Research Lab, Warren, MI, 1992.

[5] M. P. do Carmo, *Differential Geometry of Curves and Surfaces*, Prentice-Hall, Englewood Cliffs, NJ, 1976.

[6] A. K. Cline, *Scalar- and planar-valued curve fitting using splines under tension*, Comm. ACM, 17 (1974), pp. 219–220.

[7] G. Farin, *Curves and Surfaces for Computer Aided Geometric Design*, Academic Press, New York, 1988.

[8] G. Farin and N. Sapidis, *Curvature and the fairness of curves and surfaces*, IEEE Comput. Graph. Appl., 3 (1989), pp. 52–57.

[9] I. D. Faux and M. J. Pratt, *Computational Geometry for Design and Manufacture*, Ellis Horwood, Chichester, England, 1979.

[10] D. Ferguson, P. Frank, and A. Jones, *Surface slope control using constrained optimization on the B-spline representation*, Comput. Aided Geom. Design, 5 (1988), pp. 87–103.

[11] W. H. Frey and D. A. Field, *Designing Bézier conic segments with monotone curvature*, Tech. Rept. GMR-7485, General Motors Research Laboratories, Warren, MI, 1991.

[12] H. W. Guggenheimer, *Differential Geometry*, Dover, New York, 1977.

[13] D. Hill, E. Passow, and L. Raymon, *Approximation with interpolatory constraints*, Illinois J. Math., 20 (1976), pp. 65–71.

[14] R. Hooke and T. A. Jeeves, *Direct search solution of numerical and statistical problems*, J. Assoc. Comput. Mach., 8 (1961), pp. 212–229.

[15] J. Hoschek, *Detecting regions with undesirable curvature*, Comput. Aided Geom. Design, 1 (1984), pp. 183–192.

[16] ISO/DIS 1 10303-42, Industrial Automation Systems and Integration—Product Data Representation and Exchange—Part 42: Integrated Resourses, Geometric and Topological Representation, February 1993, available from International Standards Organization (ISO).

[17] Initial Graphics Exchange Specification (IGES), Version 4.0, NBSIR 88-3813, P. R. Kennicott, ed., U.S. Dept. of Commerce, NBS, Gaithersburg, MD, 1989.

[18] E. Kreyszig, *Differential Geometry*, Dover, New York, 1959.

[19] N. J. Lott and D. Pullin, *Method for fairing B-spline surfaces*, Comput. Aided Design, 20 (1988), pp. 597–604.

[20] D. McAllister and J. A. Roulier, *An algorithm for computing a shape preserving osculatory quadratic spline*, ACM Trans. Math. Software, 7 (1981), pp. 331–347.

[21] E. Mehlum, *Nonlinear splines*, in Computer Aided Geometric Design, R. E. Barnhill and R. F. Riesenfeld, eds., Academic Press, New York, 1974, pp. 173–208.

[22] H. Moreton and C. Séquin, *Minimum variation curves and surfaces for computer-aided geometric design*, in Designing Fair Curves and Surfaces, N. S. Sapidis, ed., Society for Industrial and Applied Mathematics, Philadelphia, 1994, pp. 123–159.

[23] ——, *Functional optimization for fair surface design*, Comput. Graph., 26 (1992), pp. 167–176.

[24] F. Munchmeyer, *Mathematical ship lines and surfaces*, Marine Technology, 19 (1982), pp. 219–227.

[25] J. A. Nelder and R. Mead, *Simplex method for function minimization*, Computer J., 7 (1965), pp. 308–313.

[26] G. Nielson, *Some piecewise polynomial alternatives to splines under tension*, in Computer Aided Geometric Design, R. E. Barnhill and R. F. Riesenfeld, eds., Academic Press, New York, 1974, pp. 209–235.

[27] H. Nowacki, D. Liu, and X. Lu, *Fairing Bézier curves with constraints*, Comput. Aided Geom. Design, 7 (1990), pp. 43–55.

[28] W. Press, *Numerical Recipes in C*, Cambridge University Press, Cambridge, UK, 1988.

[29] T. Rando and J. A. Roulier, *Fair curves and surfaces*, in Approximation Theory VI: Volume 2, C. K. Chui, L. L. Schumaker, and J. D. Ward, eds., Academic Press, New York, 1989, pp. 553–556.

[30] ——, *Designing faired parametric surfaces*, Comput. Aided Design, 23 (1991), pp. 492–497.

[31] ——, *Algorithms for local convexity of Bézier curves and surfaces* in Curves and Surfaces, P. J. Laurent, A. LeMéhauté, and L. L. Schumaker, eds., Academic Press, New York, 1991, pp. 403–406.

[32] T. Rando, *Automatic Fairness in Computer Aided Geometric Design*, Ph.D. dissertation, University of Connecticut, Storrs, CT, 1990.

[33] J. Roulier, *Bézier curves of positive curvature*, Comput. Aided Geom. Design, 5 (1988), pp. 59–70.

[34] ——, *Specifying the arc length of Bézier curves*, Comput. Aided Geom. Design, 10 (1993), pp. 25–56.

[35] ——, *Curves with monotone curvature from Bézier curves of specified arc length*,

in Geometric Modeling for Product Realization, P. Wilson, M. Wozny, and M. Pratt, eds., IFIP Transactions B-8, North-Holland, Amsterdam, 1993.

[36] J. A. Roulier, T. Rando, and B. Piper, *Fairness and monotone curvature*, in Approximation Theory and Functional Analysis, C. K. Chui, ed., Academic Press, Boston, 1990, pp. 177–199.

[37] W. Rudin, *Principles of Mathematical Analysis*, McGraw-Hill, New York, 1974.

[38] D. G. Schweikert, *An interpolation curve using a spline in tension*, J. Math. Phys., 45 (1966), pp. 312–317.

[39] J. Stoker, *Differential Geometry*, Wiley, New York, 1969.

CHAPTER

6

Minimum Variation Curves and Surfaces for Computer-Aided Geometric Design

Henry P. Moreton and Carlo H. Séquin

6.1. Introduction

In this paper we introduce curvature variation as a fairness metric, the Minimum Variation Curve (MVC), the Minimum Variation Network (MVN), and the Minimum Variation Surface (MVS). These curves and surfaces are computed to satisfy a set of interpolation conditions while minimizing a fairness functional that measures the variation of curvature.

A traditional measure of the fairness of curves is the arc length integral of the squared magnitude of curvature,

$$\int_0^l \|\vec{\kappa}(s)\|^2 \, ds. \tag{6.1}$$

This integral is proportional to the strain energy of an idealized thin beam forced into the shape of the curve. A curve minimizing functional (6.1) is called a Minimal Energy Curve (MEC). In this work we present a new fairness metric based on the derivative or variation of curvature. Here we minimize the arc length integral of the squared magnitude of the derivative of curvature with respect to arc length,

$$\int_0^l \left\| \frac{d\vec{\kappa}(s)}{ds} \right\|^2 ds. \tag{6.2}$$

The curve minimizing this functional is a Minimum Variation Curve (MVC). There are several advantages of an MVC over an MEC. First, the MVC exhibits higher-order continuity, where the MEC is G^3 and the MVC is G^4 [19]. Second, the MVC naturally forms circular arcs. Third, the MVC allows for the specification of tangent and/or curvature at all points, where the MEC only provides for the specification of tangent direction. Fourth, the MVC is naturally *convexity preserving*: it is guaranteed not to have any extraneous points of inflection [29].

Analogous to curves, a common measure of surface fairness is the strain energy of a thin plate; this is proportional to the area integral of the sum of the principal curvatures squared,

$$\int \kappa_1^2 + \kappa_2^2 \, dA.$$

Again we have developed a fairness metric that minimizes the variation of curvature. In this case we minimize the variation of the normal curvature in the principal directions,[1]

$$\int \left(\frac{d\kappa_1}{d\hat{e}_1}\right)^2 + \left(\frac{d\kappa_2}{d\hat{e}_2}\right)^2 dA. \tag{6.3}$$

The *normal curvature* at a point on a surface in a direction specified by a surface tangent vector is determined from the intersection curve of the surface with the plane spanned by the surface normal and the given tangent vector. The principal directions, $\hat{e}_1$ and $\hat{e}_2$, and the principal curvatures, κ_1 and κ_2, at a point on a surface are the directions and magnitudes of the minimum and maximum of all possible normal curvatures at that point [5] (Fig. 6.1).

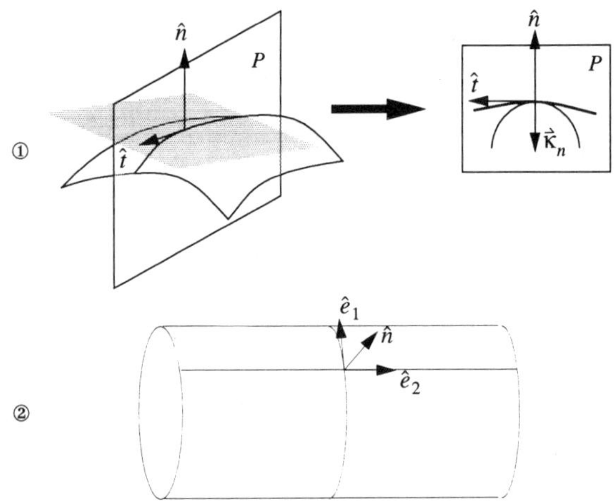

FIG. 6.1. (1) *Normal curvature in direction t: The curvature of the curve formed by the intersection of the surface and a plane P containing the normal and tangent.* (2) *Principal directions and principal curvatures: The directions and magnitudes of the maximum and minimum normal curvature.*

Similar to the MVC functional, the MVS functional evaluates to zero for cyclides, e.g., spheres, cylinders, cones, and tori. The MVS also exhibits superior fairness and higher-order continuity.

Minimum variation curves are specified by an ordered set of interpolation conditions, each consisting of position and, optionally, tangent and curvature (Fig. 6.2). Discontinuities can also be introduced where necessary; in Fig. 6.2(4)

[1] In this paper we use the convention that $\hat{x}$ indicates that x is a unit vector.

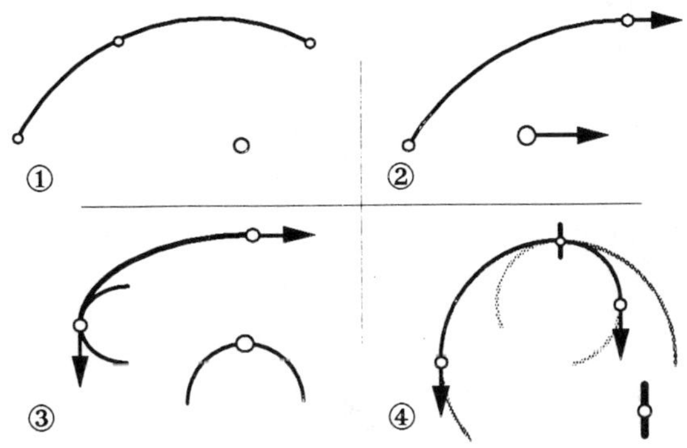

FIG. 6.2. *Curve specification through interpolation conditions and corresponding graphical symbols.* (1) *Three points.* (2) *A point and a point tangent.* (3) *A point tangent and a point curvature.* (4) *A curvature discontinuity.*

a curvature discontinuity is introduced to allow circular arcs to be formed. In general, the minimum variation curve resulting from a given set of interpolation conditions has no closed form representation. The MVC discussed here is approximated using piecewise parametric quintic polynomial curves. This choice permits simple communication with existing modeling systems whose curves are also based on a polynomial representation, e.g., Bézier and B-spline. Once the interpolation conditions are specified, an initial curve approximation satisfying the interpolation conditions is computed using heuristics that are designed to produce an initial shape close to the final MVC. Numerical techniques are then used to iteratively minimize the curvature variation functional while maintaining compliance with the given set of interpolation conditions and G^2 continuity. Note that G^2 continuity must be maintained because of the order of the terms in the functional (6.2). An intuitive explanation for requiring G^2 continuity is that if G^2 continuity were not enforced, piecewise circular curves would be formed.

Surface modeling and calculation is carried out as a three-step process (Fig. 6.3). First, the interpolation conditions of the surface are specified. Surface interpolation conditions consist of positions, and optionally, tangents (surface normals), and principal curvatures and directions. These specification tuples are interconnected, forming a graph of geometric specifications. Second, a G^2 continuous network of MVC, the minimum variation network (MVN), is computed matching the given surface geometries. Finally, quadrilateral or triangular quintic surface patches are placed in the openings of the network and used to approximate the MVS. The patches are initialized to interpolate the edges of the network, to meet with G^2 continuity at the vertices, and to satisfy the specified geometric interpolation conditions. Similar to the computation of the MVC, the MVS is computed iteratively using numerical techniques to simultaneously minimize the fairness functional plus a penalty function that

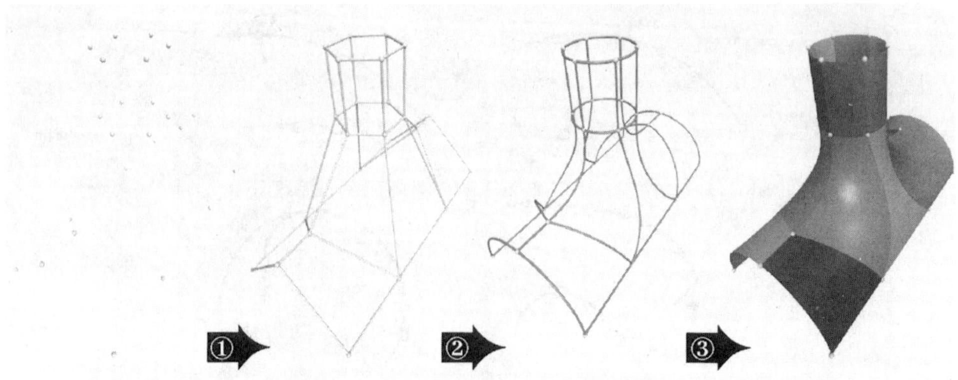

FIG. 6.3. MVS *computation as a three-step process.* (1) *Formation of graph of specified interpolation conditions.* (2) *Computation of* MVC *network.* (3) *Calculation of interpolating* MVS *patches.*

increases with cross-boundary tangent and/or curvature discontinuities. The resulting surface spans the edges of the network forming a G^1 or G^2 continuous surface with G^2 continuity at the vertices of its constituent patches. Even without a penalty for G^2 discontinuity, G^2 continuity tends to propagate along patch boundaries because of the nature of the surface functional, thus creating an approximately G^2 surface. We observe results superior to those produced by methods that focus strongly on patch–patch continuity but neglect the distribution of curvature across the interior of the patches [14].

Section 6.2 reviews previous related work, discussing approaches, advantages, and shortcomings. Sections 6.3, 6.4, and 6.5 describe, respectively, the computation of minimum variation curves, the construction of networks of the MVC, and the calculation of the MVS from geometric specifications. Section 6.6 presents a comparison of the use of minimum variation with other methods for curve and surface modeling. Finally, §6.7 discusses the factors effecting the efficiency of computing MVS.

6.2. Previous Work

The work described in this paper is naturally broken into curve, network, and surface computation. Correspondingly, we first discuss minimum variation curves in comparison to the closely related minimum energy curves. In §6.2.2 we review work on computing networks of curves for the purpose of creating a surface interpolating scattered or irregularly spaced, ordered/connected data. We then discuss the surface computation phase in §6.2.3, reviewing results in the computation of collections of patches meeting with G^1 continuity. In section 6.2.4 we reference work on functional minimizations, constrained optimization, and finite element analysis—all applied to surface design.

6.2.1. Minimum variation curves—nonlinear splines. The computation of an MEC and an MVC both require the solution of nonlinear equations, thus they are called *nonlinear* splines. Most algorithms for the computation of an MEC calculate discrete points on the curve, as in [13], for example. In contrast, Mehlum's algorithm [15] computes a G^1 approximation composed of circular arcs.

Independent of our work, Roulier, Rando, and Piper [29] outline an algorithm for the computation of individual Bézier segments that minimize a fairness metric. They mention equation (6.2) as one such metric. Their algorithm uses a simple "direct search method" to find the Bézier segment that minimizes the fairness metric. They also present a theorem stating the natural convexity of MVCs:

> If the constraints involve given unit tangent vectors and curvature values at the first and last points, and if the family contains at least one curve with monotone curvature satisfying these conditions, then any curve from the family which minimizes S_c (functional (2)) will have monotone curvature.

In other independent work Ohlin [19], [18] discusses the computation of the MEC and the MVC using a $\theta(s)$ representation. In this scheme, θ is the angle formed with the x axis, and s is the arc length parameter. Once a solution is computed in $\theta(s)$ form, it is approximated by a conventional, parametric quintic representation.

Our work computes the MVC that satisfies a set of interpolation conditions. We find a differentiable curve that minimizes (6.2) while interpolating specified positions, tangent directions, and curvatures. The curve is computed and approximated with a piecewise polynomial representation.

6.2.2. Minimum norm networks. A minimum norm network (MNN) is a network of curves meeting with specified continuity that minimize an energy norm. Such networks are typically used as a stepping stone or building block in the construction of a scattered data interpolant. Nielson [16] introduced the MNN using linear energy terms to produce a C^1 network and a resulting G^1 surface. Nielson and Franke [17] compare several techniques that operate on triangulated points. Pottmann [24] presents a generalization of the MNN to produce a C^2 surface.

In our surface modeling scheme, the network is an MNN where the energy norm is the curvature variation functional (6.2). The MVN is G^2 continuous and satisfies the given set of interpolation conditions. This network is either used as a fixed framework, into which MVS patches are fit, or it is used to initialize the surface for a more global minimization during which the network is allowed to flex and deform.

6.2.3. G^1 continuous patch assemblies. Peters [22] provides a good classification and review of G^1 interpolation techniques. All of the methods discussed are constructive, using heuristics to set those degrees of freedom that are neither fixed by continuity constraints nor set as side effects of the construction method. These methods rely on the computation of a network of curves that interpolate the data, subject to varying continuity and connectivity constraints. Peters has done a great deal of work on the construction of geometrically continuous surfaces. His most recent work outlines a method for creating "C^k" surfaces. Most relevant to this discussion, Peters [21] has shown that a curve network maintaining G^2 continuity is necessary and sufficient for the construction of a G^1 surface. This result assumes that a single polynomial patch is placed in each network opening, and that the opening boundaries are fixed. In addition, there are no restrictions on the order of (i.e., the number of edges joining) the network nodes. Our work combines the construction of a G^1 continuous surface with the setting of the unconstrained degrees of freedom to form a fair G^1 continuous surface. No explicit G^1 or G^2 construction is used; rather, a suitable penalty function is incorporated into the objective function.

6.2.4. Optimization, minimization, and finite element analysis. In [36] a system using finite difference methods for the computation of smooth surfaces is described. The system minimizes the total energy of a fictitious elastic plate. In [25] Pramila describes techniques for ship hull design that employ finite element analysis to minimize a quadratic functional approximating strain energy. Celniker and Gossard [4] present a free-form design system that uses finite element analysis to simulate physical models. Interactive deformation is carried out by simulating forces applied to the subject model. Surfaces are represented by triangular patches meeting with C^1 continuity. Linearized approximations are used to model physical deformations.

Rando and Roulier [28] propose several specialized geometrically based fairness functionals. These functionals are referred to as "flattening," "rounding," and "rolling." They apply these functionals to Bézier patches. Some of the Bézier control points are fixed in order to guarantee continuity, while others are varied to minimize the functionals.

Hagen and Schulze [9] use the calculus of variations to fit generalized Coons patches to three-dimensional data. The resulting patches minimize a strain energy fairness criterion. The analysis uses simplifying approximations to limit the complexity of calculations.

Most recently, Kallay and Ravani [12] discuss a method for determining "optimal" twist vectors of the surface formed by a rectangular mesh of cubic curves. In their work, twist vectors are computed minimizing a quadratic energy term.

Our work uses higher-order patches and the full nonlinear expression of functional (6.3) to achieve the highest possible surface quality with the fewest underlying patches.

6.3. Curve Computation

We cast the problem of computing the MVC as a nonlinear optimization/finite element problem. The curve is broken into a series of quintic polynomial elements satisfying the given geometric and continuity constraints. The curve integral minimized using a gradient descent optimization procedure. A heuristically chosen starting curve greatly accelerates convergence toward minimum variation.

6.3.1. Curve representation through Hermite elements. There are several possible representations for a quintic polynomial element (e.g., B-spline and Bézier). We chose the Hermite form because of the ease with which the geometric specifications can be mapped to the defining parameters of the Hermite segments. Also, a simple change of basis converts this form to other polynomial representations that are typically used in geometric modeling systems. Quintic Hermite curves are specified by the position of the endpoints and by the first two parametric derivatives at these locations. In vector notation this can be expressed as

$$\vec{C}(t) = \begin{bmatrix} \vec{C}(0) \\ \vec{C}'(0) \\ \vec{C}''(0) \\ \vec{C}''(1) \\ \vec{C}'(1) \\ \vec{C}(1) \end{bmatrix}^T \cdot \begin{bmatrix} H_0(t) \\ H_1(t) \\ H_2(t) \\ H_3(t) \\ H_4(t) \\ H_5(t) \end{bmatrix},$$

where $H_j(t)$ are quintic blending functions. The computation of $H_j(t)$ is described in [6]. The mapping from the geometric to the parametric description of the curve is carried out using the following equations for element i:

$$\vec{C}_i(0) = \vec{p}_i,$$
$$\vec{C}_i(1) = \vec{p}_{i+1},$$
$$\vec{C}'_i(0) = m_{(i,0)}\hat{t}_i, \tag{6.4}$$
$$\vec{C}'_i(1) = m_{(i,1)}\hat{t}_{i+1},$$
$$\vec{C}''_i(0) = m^2_{(i,0)}\vec{\kappa}_i + \alpha_{(i,0)}m_{(i,0)}\hat{t}_i, \tag{6.5}$$
$$\vec{C}''_i(1) = m^2_{(i,1)}\vec{\kappa}_{i+1} + \alpha_{(i,1)}m_{(i,1)}\hat{t}_{i+1},$$

where $\vec{p}_i$, $\hat{t}_i$, $\vec{\kappa}_i$ are the position, tangent direction, and curvature vector at one end of the curve. The tangent direction is computed to be

$$\hat{t}_i = \vec{C}'_i / \|\vec{C}'_i\|,$$

and the curvature is

$$\vec{\kappa}_i = \frac{(\vec{C}'_i \times \vec{C}''_i) \times \vec{C}'_i}{\|\vec{C}'_i\|^4}.$$

Note that in the context of the MVC functional (6.2) curvature is defined as

$$\vec{\kappa}_i = \frac{\vec{C}'_i \times \vec{C}''_i}{\|\vec{C}'_i\|^3}.$$

During the minimization, the scalar $m_{(i,0)}$ and $m_{(i,1)}$ in equations (6.4) and (6.5) must be constrained to be positive; this is because if either m_i became negative then the corresponding $\vec{C}'_i$ would reverse direction. We impose this constraint by using m_i^2 rather than m_i:

$$\begin{aligned}
\vec{C}'_i(0) &= m^2_{(i,0)}\hat{t}_i, \\
\vec{C}'_i(1) &= m^2_{(i,1)}\hat{t}_{i+1}, \\
\vec{C}''_i(0) &= m^4_{(i,0)}\vec{\kappa}_i + \alpha_{(i,0)} m^2_{(i,0)}\hat{t}_i, \\
\vec{C}''_i(1) &= m^4_{(i,1)}\vec{\kappa}_{i+1} + \alpha_{(i,1)} m^2_{(i,1)}\hat{t}_{i+1}.
\end{aligned}$$

A curve is made up of a sequence of vertices connected by quintic elements. G^2 curves are pieced together from these elements by sharing geometric specifications at the common points. Data structures associated with the vertices hold the point, tangent, and curvature information, while data structures associated with the connecting elements hold the parameters $\alpha_{(i,0)}$, $\alpha_{(i,1)}$, and $m_{(i,0)}, m_{(i,1)}$ (Fig. 6.4). Each element is defined by three data structures; two structures are associated with the vertices at the element's endpoints and one structure is associated with the element itself. By distributing the curve/element specification in this way, adjacent elements share vertex structures and are guaranteed to meet with curvature continuity. Note that discontinuities can be introduced by giving adjacent elements independent geometric specifications (Fig. 6.4).

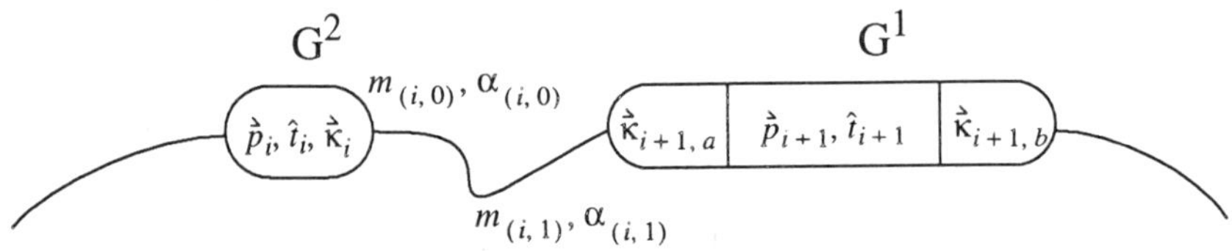

FIG. 6.4. *Schematic view of curve representation. On the left is a G^2 joint where incident curves share the full geometric specification. On the right is a G^1 joint where incident curves have independent curvatures.*

6.3.2. Parametric functionals. The MVC curvature functional (6.2) is defined in terms of an integral of a function over arc length. To evaluate this functional and its gradient in the context of the parametric piecewise polynomial curves described in §6.3.1, the functional must be converted to a compatible form and evaluated in a piecewise fashion. The value of the functional for the curve as a whole is computed as the sum of the values of the functional for each element. In the first conversion step the arc-length-based definition

$$\int_0^l \left(\frac{d\vec{\kappa}}{ds}\right)^2 ds$$

is changed to an integral of a function of the curve $\vec{C}(t)$ parameterized by t of the form

$$\int_{\alpha}^{\beta} f(\vec{C}(t))\, dt.$$

The bounds, α, β, of the integral are set to 0 and 1, since the Hermite representation is parameterized with t varying from 0 to 1. The differential with respect to s is converted to a differential in t. Since

$$\frac{dt}{ds} = \frac{1}{\|\vec{C}'(t)\|} \qquad \text{then } ds = \|\vec{C}'(t)\|\, dt,$$

where

$$\|\vec{C}'(t)\| = (\vec{C}'(t) \cdot \vec{C}'(t))^{1/2}.$$

Finally, the derivative of curvature with respect to arc length $d\vec{\kappa}/ds$ transforms to $(d\vec{\kappa}/dt)(dt/ds)$. These two steps yield

$$\int_0^l \frac{d\vec{\kappa}}{ds}^2 ds = \int_0^l \left(\frac{d\kappa}{dt}\frac{dt}{ds}\right)^2 \|\vec{C}'(t)\|\, dt, \qquad \int_0^l \frac{d\vec{\kappa}}{ds}^2 ds = \int_0^1 \frac{d\vec{\kappa}}{dt}^2 \frac{1}{\|\vec{C}'(t)\|}\, dt.$$

Finally, we find the expression for $d\vec{\kappa}/dt$ in term of the parametric derivatives of the curve, $\vec{C}(t)$. The expression for curvature is

$$\text{(6.6)} \qquad \vec{\kappa}(t) = \frac{\vec{C}'(t) \times \vec{C}''(t)}{\|\vec{C}'(t)\|^3} \equiv \frac{u}{v}.$$

Taking the derivative of (6.6) with respect to t yields

$$\frac{d\kappa}{dt} = \frac{\nu du - u d\nu}{\nu^2},$$

where

$$\begin{aligned} u &= \vec{C}'(t) \times \vec{C}''(t), \\ du &= \vec{C}'(t) \times \vec{C}'''(t), \\ \nu &= \|\vec{C}'(t)\|^3, \\ d\nu &= 3\|\vec{C}'(t)\|^2(\vec{C}'(t) \cdot \vec{C}''(t)). \end{aligned}$$

6.3.3. Minimization. We use conjugate gradient descent to iteratively minimize the fairness functional [27]. The curve is represented as a point in $\mathcal{R}^n$ corresponding to its n degrees of freedom. There is one degree of freedom for each variable or coefficient not fixed by geometric or continuity constraints. Each iteration step consists of computing the gradient with respect to all the degrees of freedom and then computing a conjugate direction using the Polak–Ribiere method [27]. Finally, a line minimization is performed to find the minimum of the functional in the conjugate direction. Iteration continues until the gradient has very small magnitude indicating a (local) minimum. This optimization algorithm requires the evaluation of the functional and its partial derivatives with respect to all the available degrees of freedom. Because it has no closed form solution, the evaluation of the functional must be performed using numerical techniques. This is accomplished using 20 point Gauss–Legendre quadrature [27]. Similarly, the partial derivatives are in integral form and are computed using the same quadrature rules.

6.3.4. Initialization. The descent scheme described in the previous section starts with an initial curve and iteratively refines that curve until the optimal curve is achieved. In this section we discuss the problem of finding a suitable initial curve, i.e., one that lies in the same "valley" of the solution space in which the minimum lies. The optimization procedure requires that initial values be provided for all parameters not explicitly set. Several researchers have studied the problem of finding an interpolating curve for an ordered set of points, e.g., [32] and [33]. The formulas that we use for the initial guess are a combination of some of these approaches. The values of m_{i0} and m_{i1} of equations (6.4) and (6.5) and in Fig. 6.4 are related to the arc length of the curve and are initialized to the chord length $\|\vec{p}_i - \vec{p}_{i+1}\|$. The parameters α_{i0} and α_{i1} specify the nature of the parameterization at the endpoints; we set them to 0, causing the curve to be arc length parameterized at its endpoints; the first and second derivatives are orthogonal. The tangent directions are set to the weighted average of the incident chord directions. The chords are inversely weighted relative to their length (Fig. 6.5):

$$\grave{t}_i = \frac{\vec{p}_{i-1} - \vec{p}_i}{\|\vec{p}_{i-1} - \vec{p}_i\|^2} + \frac{\vec{p}_i - \vec{p}_{i+1}}{\|\vec{p}_i - \vec{p}_{i+1}\|^2}, \qquad \hat{t}_i = \frac{\grave{t}_i}{\|\grave{t}_i\|}.$$

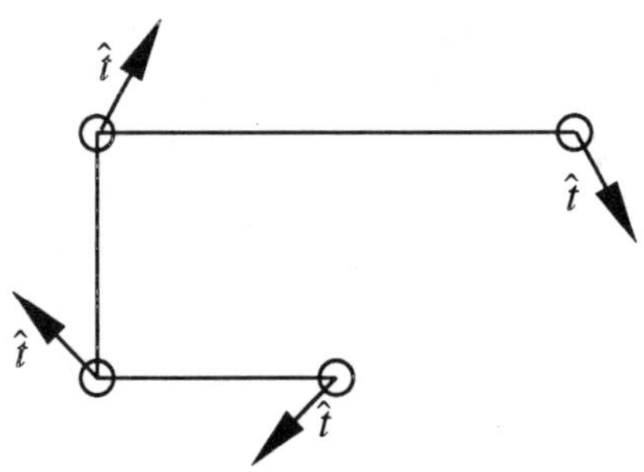

FIG. 6.5. *The tangent computed as the average of incident chords.*

The magnitude of the curvature vector $\vec{\kappa}_i$ at each interpolation is set equal to the curvature of the circle that interpolates the given point and its two neighbors (Fig. 6.6). The direction of curvature is perpendicular to the previously determined tangent direction and is in the plane of the three points defining the circle. Note that curvature vector is not necessarily in the direction

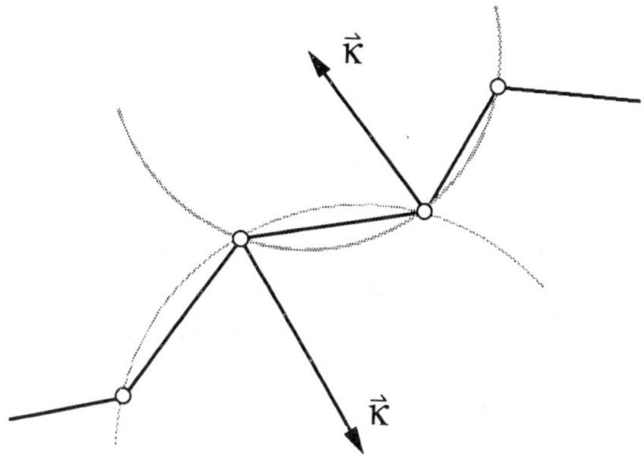

FIG. 6.6. *Curvature vectors initialized by fitting circles through adjacent points.*

of the radius of this circle. Infinite radii result from three colinear points and indicate zero curvature.

In the case where the subject curve is not closed and the elements at the ends of the curve have endpoints that do not have two neighbors, a construction is used that will lead to a symmetrical curve segment over this end element. The tangent direction is set to the tangent direction at the other endpoint mirrored about the perpendicular plane through the midpoint of the intervening chord (Fig. 6.7). The curvature vector has length equal to the magnitude of curvature of the single initialized neighbor and direction perpendicular to the tangent in the plane defined by the tangent and chord.

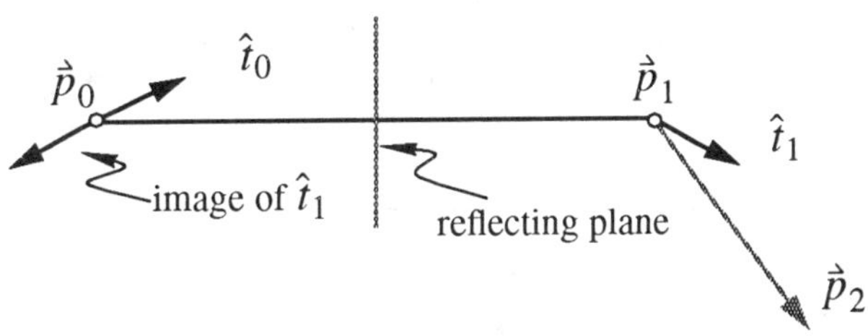

FIG. 6.7. *Construction of tangent direction at the open endpoint,* p_0*, of a curve.*

Our heuristic technique for selecting an optimization starting point has proven to be efficient and robust. Given a set of geometric constraints to interpolate, our approach to initialization places the starting point of the optimization near a desirable local minimum.

6.3.5. Existence, uniqueness, and sensitivity. Because the MEC has been much more widely studied (since the 17th century), we discuss results from these studies and relate them to the relatively new MVC. In [10] Jerome discusses the necessary and sufficient conditions for the existence of an MEC.

To the authors' knowledge there has been no study of the uniqueness of the MVC.

In [7] Fisher and Jerome discuss the stable and unstable equilibrium of an MEC. An MEC is said to be at stable equilibrium if it holds its shape in the absence of an arc length constraint. In most practical applications we seek the stable equilibrium solution that has the shortest length. This suggests the use of the continuum method, the approach of computing the MVC by initially placing the curve under tension (by penalizing arc length) and gradually relaxing the tension until stable equilibrium is achieved and no tension remains. Figure 6.8 illustrates two MVCs for the same set of geometric constraints. In addition to the positional constraints, the curvature at the central point is fixed. In comparing the two solutions, note that while the solution without the loop is shorter, it has a larger MVC functional value.

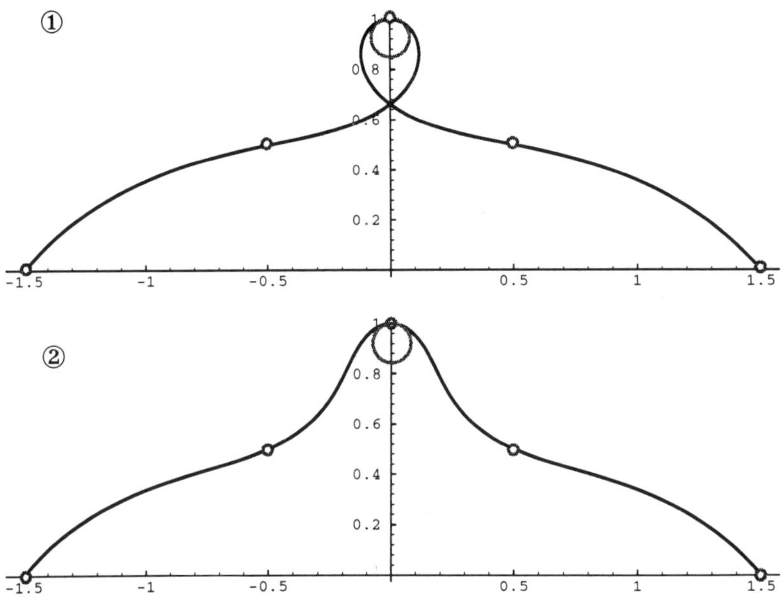

FIG. 6.8. *Multiple MVCs from one specification.* (1) *The MVC resulting from an initial curve containing a loop.* (2) *The MVC resulting from an initial curve without a loop. Note that while curve* (2) *is probably the desired curve, curve* (1) *has a lower MVC functional value. The curves are specified by five positional constraints (small circles) and a single curvature constraint (large circle) at the center of the curve. Note how the MVC distributes the high curvature at its midpoint.*

6.3.6. Multielement segments. Because of the limited descriptive power of polynomial elements, a single element can only approximate the ideal minimal variation curve. To improve the approximation, multiple elements can be inserted between vertices. In practice, a single element per constraint pair is normally sufficient. For many applications requiring high quality curves it is not so important that the minimal variation curve is accurately approximated, only that its desirable curvature properties are realized.

6.4. Minimum Variation Network Computation

Minimum variation networks (MVN) are used to initialize the boundaries of the patches from which the surface is composed. Since we can only use quadrilateral or triangular patches, we have to make sure that the network derived from the graph connecting the given constraint points meets these requirements; if necessary, openings with more than four sides can be subdivided. Many of the techniques used in computing the network are used in the computation of an MVS, where the initial shape of the curve network is allowed to change. In this section we outline the methods used to compute the G^2 curve network.

6.4.1. Network representation and continuity. The network of curves is defined via the parameters of a virtual second-order "fitting" surface at each vertex of the network and a description of how each curve segment emerges from within the fitting surfaces specified at its endpoints. Each fitting surface is defined by the vertex position $\vec{p}$, the conjugate directions $\hat{w}_1, \hat{w}_2$, and the normal curvatures in these directions, $\kappa_{w_1}, \kappa_{w_2}$. Conjugate directions are equivalent to principal directions in that coupled with the associated curvatures, they fully characterize the curvature of a surface at a point [5]. Conjugate directions are more amenable to optimization because they do not have to be constrained to mutual orthogonality. The network is represented by quintic Hermite curves. These Hermite curves are defined by the positions and first two parametric derivatives at their endpoints. Each curve in the network is defined by the position $\vec{p}$, tangent direction $\hat{t}$, and three scalar parameters, m, α, c, at each endpoint. The mapping from these values to the parameters defining the corresponding Hermite curve is

$$
\begin{aligned}
&\vec{P} = \vec{p}, \qquad \vec{P}' = m^2\hat{t}, \qquad \vec{P}'' = m^4\vec{\kappa} + \alpha m^2\hat{t}, \\
&\vec{\kappa} = \kappa_n\hat{n} + c\hat{b}, \qquad \hat{b} = \hat{n}\times\hat{t}, \\
&\kappa_n = \hat{t}\cdot\Big[K\Big]\cdot\hat{t}, \\
&\Big[K\Big] = \begin{bmatrix}\hat{w}_1\\ \hat{w}_2\\ \hat{n}\end{bmatrix}^{-1}\cdot\begin{bmatrix}\kappa_{w_1} & 0 & 0\\ 0 & \kappa_{w_2} & 0\\ 0 & 0 & 0\end{bmatrix}\cdot\left(\begin{bmatrix}\hat{w}_1\\ \hat{w}_2\\ \hat{n}\end{bmatrix}^{-1}\right)^T.
\end{aligned}
\tag{6.7}
$$

Note the the curvature of the curve is the sum of two orthogonal components; κ_n, the component in the normal direction, is a function of the fitting surface and the tangent direction of the curve at its endpoint; c, the component in the binormal direction, is independent of the fitting surface and represents the curvature of the curve "within the surface."

Using the same techniques employed in curve computation, the MVC functional is minimized using conjugate gradient descent to iteratively adjust those variables not fixed by constraints. At each iteration step, $\hat{w}_1, \hat{w}_2$ are renormalized, and $\hat{t}$ is projected onto the plane spanned by $\hat{w}_1, \hat{w}_2$ and also renormalized. It is this normalization step in combination with the construction outlined in equation (6.7) which guarantees that G^2 continuity is maintained and the specified interpolation conditions are satisfied.

6.4.2. Network initialization. The curve network must be initialized to some reasonable values before optimization may proceed. First, a vertex normal vector is initialized, then the tangent vectors of the incident curves are computed; next, the principal directions and curvatures are defined, and finally each curve's scalar coefficients are initialized.

The vertex normal is initialized as a weighted average of the incident face normals (weighted inversely proportional to the area of the incident face, i.e., the smaller the face the greater the weight [3]) (Fig. 6.9). The tangent vectors of curves incident to a vertex are set to the direction of the incident chords protected onto the plane defined by the vertex position $\vec{p}_i$ and normal $\hat{n}$ (Fig. 6.10).

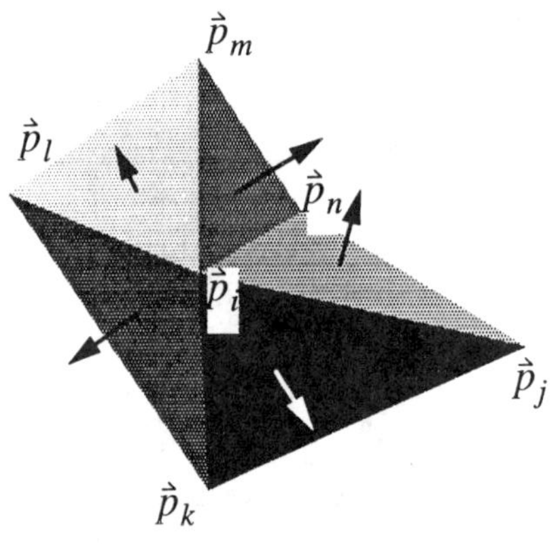

FIG. 6.9. *Vertex normal vector calculation. The normal is initialized to the average of the incident face normals inversely weighted by area.*

Once vertex normal vectors and incident tangent directions have been computed, the principal curvatures and principal directions at a vertex are calculated. Both Calladine [3] and Todd and McLeod [35] describe approaches for estimating the curvature of polyhedral surfaces. Calladine's method only estimates Gaussian curvature. Todd and McLeod require that a pairing be established among the points neighboring a point; this is a nontrivial task at vertices of odd order. At even order vertices, it remains problematic since the results vary greatly depending on the pairing chosen. Our approach uses a least squares fit of sample tangent directions and normal curvatures to compute the principal directions and curvatures. The initialization of these values is very important to the speed of convergence. First consider the situation shown in Fig. 6.10. A junction is shown with a number of incident edges. For each

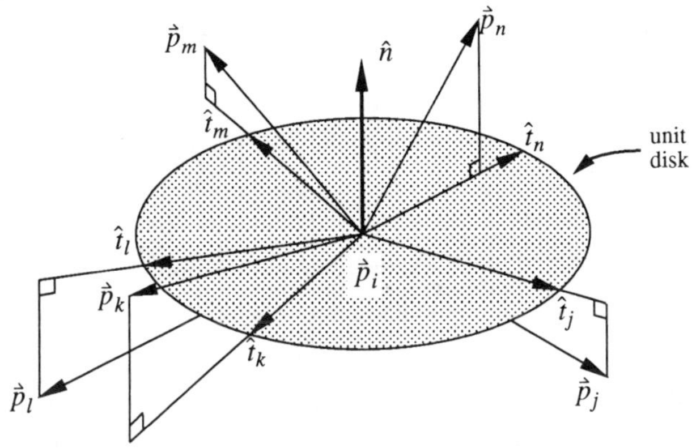

FIG. 6.10. *Tangent initialization. The projection of incident chords onto the plane defined by the normal.*

edge, we calculate the curvature implied by that edge emanating from the vertex. Starting with edge $\overline{\vec{p}_n\vec{p}_i}$ we reflect $\vec{p}_n$ through the normal and fit a circle through $\vec{p}'_n$, $\vec{p}_i$, and $\vec{p}_n$. The radius of the resulting circle is the desired radius of curvature in that normal plane (Fig. 6.11). Repeating this procedure for each of the incident edges provides a set of sample tangent directions and normal curvatures. This set is used to compute a least squares fit for the principal directions and principal curvatures of the fitting surface at the vertex.

We start with the expression for normal curvature expressed with respect to any convenient orthonormal basis in the plane defined by the normal,

$$\kappa_n = \hat{t} \cdot \left[K \right] \cdot \hat{t},$$

$$\left[K \right] = \begin{bmatrix} \hat{e}_{1,x} & \hat{e}_{1,y} \\ -\hat{e}_{1,y} & \hat{e}_{1,x} \end{bmatrix} \cdot \begin{bmatrix} \kappa_1 & 0 \\ 0 & \kappa_2 \end{bmatrix} \cdot \left(\begin{bmatrix} \hat{e}_{1,x} & \hat{e}_{1,y} \\ -\hat{e}_{1,y} & \hat{e}_{1,x} \end{bmatrix}^{-1} \right)^T,$$

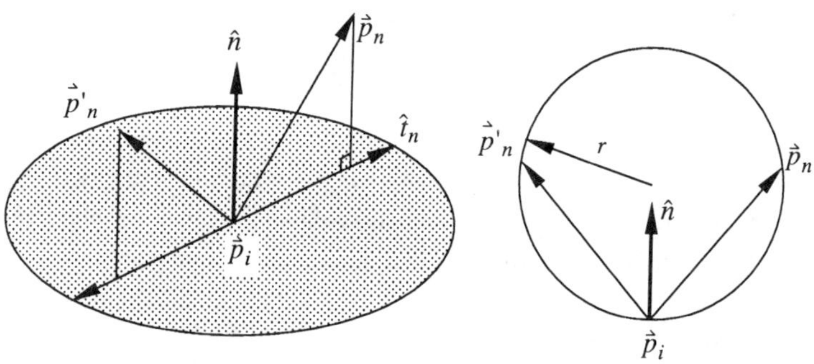

FIG. 6.11. *Calculating an approximate radius of curvature in the direction of t_n.*

and extract the tangent components, to produce an overdetermined set of linear equations:

$$\text{(6.8)} \qquad \begin{bmatrix} \hat{t}^2_{0,x} & \hat{t}_{0,x}\hat{t}_{0,y} & \hat{t}^2_{0,y} \\ \hat{t}^2_{1,x} & \hat{t}_{1,x}\hat{t}_{1,y} & \hat{t}^2_{1,y} \\ \cdot & \cdot & \cdot \\ \hat{t}^2_{m,x} & \hat{t}_{m,x}\hat{t}_{m,y} & \hat{t}^2_{m,y} \end{bmatrix} \begin{bmatrix} \hat{e}^2_{1,x}\kappa_{\hat{e}_1} + \hat{e}^2_{1,y}\kappa_{\hat{e}_2} \\ 2\hat{e}_{1,x}\hat{e}_{1,y}(\kappa_{\hat{e}_1} - \kappa_{\hat{e}_2}) \\ \hat{e}^2_{1,x}\kappa_{\hat{e}_2} + \hat{e}^2_{1,y}\kappa_{\hat{e}_1} \end{bmatrix} = \begin{bmatrix} \kappa_{n,0} \\ \kappa_{n,1} \\ \cdot \\ \kappa_{n,m} \end{bmatrix},$$

$$Ax = b.$$

The general formula for computing the least squares solution to this type of system is $A^T A\bar{x} = A^T b$, where $\bar{x}$ is the least squares solution for "x" in equation (6.8). Having solved for $\bar{x}$ we have three equations and four unknowns,

$$\begin{bmatrix} \hat{e}^2_{1,x}\kappa_1 + \hat{e}^2_{1,y}\kappa_2 \\ 2\hat{e}_{1,x}e_{1,y}(\kappa_1 - \kappa_2) \\ \hat{e}^2_{1,x}\kappa_2 + \hat{e}^2_{1,y}\kappa_1 \end{bmatrix} = \begin{bmatrix} \bar{x}_0 \\ \bar{x}_1 \\ \bar{x}_2 \end{bmatrix}.$$

Adding the fact that $\hat{e}^2_{1,x} + \hat{e}^2_{1,y} = 1$ allows us to solve for the principal directions and principal curvatures.

Finally the scalars associated with each curve are set as follows: m is set to chord length, and α, c are set to zero.

6.4.3. Optional network constraints. Since the quality of the network has a direct impact on the quality of the resulting surface, we present an optional heuristic constraint. A very successful method for improving the quality of a curve network is to force curve segments entering a vertex from "opposite" directions to form a G^2 continuous curve through the vertex [31]. During initialization, the joint tangent direction for the two opposite curve segments is initialized to the average of the individual tangents computed by chord projection. The curvature parameters then follow from the initial second-order fitting surface, as above. Full G^2 continuity is imposed by forcing the two curves to also share the binormal component, c, from (6.7). This technique can be applied automatically to vertices with an *even* number of segments joining. Alternatively, selected pairs of curve segments could be chosen manually, e.g., based on a priori knowledge that the sample points should lie on a continuous curve.

Figure 6.12 illustrates the two different procedures applied to 16 regularly spaced points on the surface of a torus. Figure 6.12(2) illustrates the changes to the network when G^2 continuity is imposed; it is obvious that this will lead to a much better approximation of a torus.

6.5. Surface Computation

As described in §1, the computation of an MVS satisfying a set of given constraints is broken into several steps. The focus of this selection is on the last

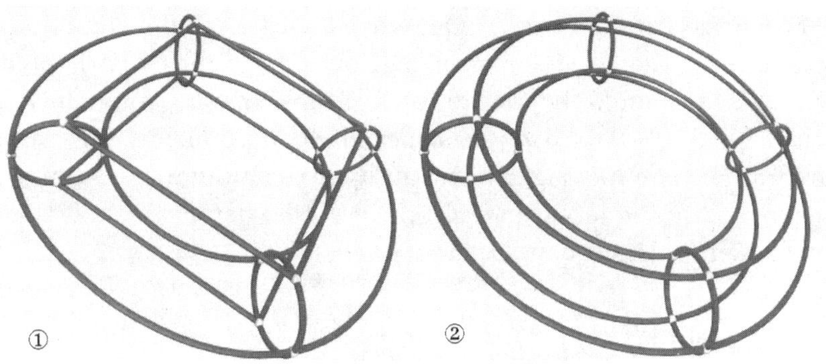

FIG. 6.12. *A network through* 16 *points on a torus* (1) *without continuous curves through vertices.* (2) *A network with the continuous curves.*

phase of the algorithm where surface patches are fit to an MVN. The curves may remain fixed or they may be used simply as a starting point for a more global optimization. The choice between a fixed or deformable network is up to the designer and does not affect the algorithms described here.

The MVS is approximated by a quilt of parametric polynomial patches that interpolate the curve network, satisfy the geometric constraints, and meet with approximate tangent continuity. The surface functional is then minimized by varying the remaining surface parameters that are not fixed by geometric constraints.

6.5.1. Bézier patches. The curves of the network are represented by quintic Hermite polynomial segments; one segment replaces each edge of the network of constraints. Consequently, the patches making up the interpolatory surface are (bi-)quintic patches. Peters [21] has demonstrated that quintics are sufficient to achieve tangent continuity for all triangular–quadrilateral patch–patch combinations. One patch is used for each opening in the network. Though we have found single patches to have sufficient descriptive power, it is simple to subdivide network patches, creating multiple patches per opening. The use of multiple patches improves the approximation of the theoretical MVS surface, which in general has no closed form representation. Note that while Peters' construction requires the curve network being interpolated to have G^2 continuity, the interpolatory surface resulting from his construction is only G^1 continuous across boundaries and at the vertices of the network. Our surfaces, additionally, may maintain either true G^2 continuity at the vertices of the network (see §6.5.6), or G^2 continuity across boundaries, resulting in superior overall curvature distribution.

Even though the boundary curves are in the Hermite form, we have chosen to use Bézier patches because of their superior numerical characteristics. Also Bézier patches are more amenable to rendering, and may be rendered directly by subroutines found in the graphics library of workstations such as the Silicon Graphics IRIS®.

6.5.2. Fairness functionals. Our choice of functional for minimization was prompted by the need for very high quality surfaces with predictable, intuitive behavior, and the desire to capture shapes commonly used in geometric modeling. This can be achieved by minimizing the *variation* of curvature, rather than its magnitude. Our approach minimizes the area integral of the sum of the squared magnitudes of the derivatives of the normal curvatures taken in the principal directions:

$$\int \left(\frac{d\kappa_n}{d\hat{e}_1}\right)^2 + \left(\frac{d\kappa_n}{d\hat{e}_2}\right)^2 dA. \tag{6.9}$$

Like the MVC functional, the MVS functional has associated shapes that are optimal in the sense that the functional evaluates to zero. In the case of the MVS functional, those shapes belong to a special family of curved surfaces called cyclides [2], [26] which includes spheres, cylinders, cones, and tori. They all have *lines of principal curvature* where the associated normal curvature remains constant. Lines of principal curvature are the paths of minimum and maximum normal curvature across a surface.

6.5.3. Parametric functionals. The fairness functional for surfaces (6.9) is defined in terms of an area integral. It is computed for each patch, and the value of the functional for the whole surface is the sum of the values for each patch. To evaluate the functional and its gradient in the context of the parametric polynomial patches described in §6.5.1, the functional must be converted to a compatible form. Here we outline the calculations necessary to evaluate the functional. The area-based definition is converted to integrals of functions of the independent parameters u and v in $\vec{S}(u, v)$. For quadrilateral patches, the bounds of the integrals are set to vary over the unit square, and the differential with respect to area is converted to differentials in u and v,

$$\int_0^1 \int_0^1 \left(\left(\frac{d\kappa_n}{d\hat{e}_1}\right)^2 + \left(\frac{d\kappa_n}{d\hat{e}_2}\right)^2 \right) \|S_u \times S_v\| \, du \, dv,$$

where

$$\|S_u \times S_v\| = \sqrt{EG - F^2}$$

and

$$\begin{aligned} E &= \vec{S}_u \cdot \vec{S}_u, \\ F &= \vec{S}_u \cdot \vec{S}_v, \\ G &= \vec{S}_v \cdot \vec{S}_v. \end{aligned} \tag{6.10}$$

The variables E, F, and G, are from the first fundamental form from differential geometry [5]. The principal curvatures κ_1 and κ_2 are the normal curvatures in the principal directions. Thus the problem of computing $d\kappa_n/d\hat{e}_1$ and $d\kappa_n/d\hat{e}_2$

becomes one of computing $d\kappa_1/d\hat{e}_1$ and $d\kappa_2/d\hat{e}_2$. First we find expressions for these in terms of derivatives taken in the parametric directions.

$$\frac{d\kappa_1}{d\hat{e}_1} = \frac{d\kappa_1}{d\hat{u}}(\hat{e}_1 \cdot \hat{S}_u) + \frac{d\kappa_1}{d\hat{v}}(\hat{e}_1 \cdot \hat{S}_v),$$

$$\frac{d\kappa_2}{d\hat{e}_2} = \frac{d\kappa_2}{d\hat{u}}(\hat{e}_2 \cdot \hat{S}_u) + \frac{d\kappa_2}{d\hat{v}}(\hat{e}_2 \cdot \hat{S}_v),$$

where

$$\hat{S}_u = S_u/\|S_u\|, \qquad \hat{S}_v = S_v/\|S_v\|.$$

Next we define the derivatives of κ_1, κ_2 taken in the *parametric directions* using simple parametric derivatives:

$$\frac{d\kappa_i}{d\hat{u}} = \frac{d\kappa_i}{du}\frac{1}{\|\vec{S}_u\|}, \qquad \frac{d\kappa_i}{d\hat{v}} = \frac{d\kappa_i}{dv}\frac{1}{\|\vec{S}_v\|}.$$

Finally, the parametric derivatives of κ_1 and κ_2 are computed from an expression derived from the fact that the principal directions are the eigenvalues of the curvature tensor. The expression for the curvature tensor is

$$\begin{bmatrix} a_{11} & a_{21} \\ a_{12} & a_{22} \end{bmatrix},$$

where

$$a_{11} = \frac{fF - eG}{EG - F^2}, \qquad a_{21} = \frac{eF - fE}{EG - F^2},$$

$$a_{12} = \frac{gF - fG}{EF - F^2}, \qquad a_{22} = \frac{fF - gE}{EG - F^2},$$

$$e = \hat{n} \cdot \vec{S}_{uu},$$

$$f = \hat{n} \cdot \vec{S}_{uv}. \tag{6.11}$$

$$g = \hat{n} \cdot \vec{S}_{vv},$$

E, F, and G are defined as in equation (6.10) and e, f, and g are the terms of the second fundamental form from differential geometry [5].

Since κ_1, κ_2 and $\hat{e}_1$, $\hat{e}_2$ are the eigenvalues and eigenvectors of the curvature tensor, we get the following expressions:

$$\kappa_i = \frac{a_{11} + a_{22} \pm \sqrt{a_{11}^2 + 4a_{12}a_{21} - 2a_{11}a_{22} + a_{22}^2}}{2},$$

$$\vec{e}_i = \left\{\left(a_{11} - a_{22} \pm \sqrt{a_{11}^2 + 4a_{12}a_{21} - 2a_{11}a_{22} + a_{22}^2}\right), 2a_{12}\right\},$$

$$\hat{e}_i = \vec{e}_i/\|\vec{e}_i\|.$$

These expressions are in terms of the surface parameters u and v. Using the chain rule, it is simple to compute the required parametric derivatives. Note that in computing the parametric derivatives of e, f, and g, it is helpful to have a simple way of computing $\hat{n}_u$ and $\hat{n}_v$, the vector-valued derivatives of the unit surface normal with respect to the surface parameters u and v:

$$\hat{n}_u = \kappa_1(\vec{S}_u \cdot \hat{e}_1)\hat{e}_1 + \kappa_2(\vec{S}_u \cdot \hat{e}_2)\hat{e}_2,$$
$$\hat{n}_v = \kappa_1(\vec{S}_v \cdot \hat{e}_1)\hat{e}_1 + \kappa_2(\vec{S}_v \cdot \hat{e}_2)\hat{e}_2.$$

6.5.4. Continuity by penalty. We have described how our functional is evaluated. To complete the objective function that is being minimized we add to it a penalty function P for lack of continuity. This penalty function evaluates to zero when surfaces meet continuously and increases sharply with the magnitude of discontinuity. Because of the nature of gradient-based optimization, it is important for this penalty function to have zero gradient ∇P at its minimum point. In §6.5.4.1 we discuss the formulation of a penalty function for tangent continuity and in §6.5.4.2 we present a similar penalty function for G^2 continuity.

6.5.4.1. Tangent continuity. In this section we present a penalty function for incorporating tangent continuity into a surface optimization; the method is based on the numerical integration of a cross-boundary discontinuity function. We may form such a penalty function from parameterization-independent measures by computing the surface normal along either side of the patch–patch boundary. One such function integrates a function of the cosine of the angle formed by the normals

$$P_{G^1} = (1 - (\hat{n}_a \cdot \hat{n}_b))^2. \tag{6.12}$$

Note that as $(\hat{n}_a \cdot \hat{n}_b)$ approaches one, ∇P_{G^1} approaches zero.

This equation is satisfactory except that it is not strongly positive for large discontinuities; in fact its magnitude is limited to 2.0. Equation (6.13) computes a function of (6.12) such that its values range from 0 to ∞;

$$P_{G^1} = \left(\frac{1 - (\hat{n}_a \cdot \hat{n}_b)}{1 + (\hat{n}_a \cdot \hat{n}_b)}\right)^2, \tag{6.13}$$

as in equation (6.12), as $(\hat{n}_a \cdot \hat{n}_b)$ approaches one, ∇P_{G^1} approaches zero. Similarly, as $(\hat{n}_a \cdot \hat{n}_b)$ approaches minus one, ∇P_{G^1} grows without bound.

Since the normal vector is computed as part of evaluating the fairness functional, and the fairness functional is integrated using the Lobatto quadrature, we use the same quadrature rule to integrate (6.13). To determine the number of quadrature points required in the evaluation of $\int P_{G^1}$ we consider that a polynomial of order n is identically zero if it is equal to zero at n distinct locations. In this case (6.13) is a degenerate rational polynomial whose roots correspond to the roots of the determinant of the partial derivatives of the incident patches. This determinant is an order 15 polynomial, thus if (6.13) is sampled at least 15 times, we are guaranteed that if the result is zero then the function itself is also zero.

6.5.4.2. Curvature continuity. The problem of constructing a G^2 continuous network of polynomial patches is extremely difficult to solve exactly, due to the second-order analogue of the twist compatibility problem [22]. Our penalty based method avoids this problem by using methods similar to the G^1 penalty function to push the optimization toward a G^2 continuous solution. We construct the penalty function from the geometric measures of curvature, κ_1, κ_2, $\hat{e}_1$, $\hat{e}_2$. Pegna and Wolter [20] show that given a pair of patches that meet with G^1 continuity, patches need only have equal normal curvatures in a single transverse direction in order to ensure G^2 continuity. From this observation we derive a penalty function measuring the difference in normal curvature in the direction perpendicular to the patch–patch boundary. The normal curvature κ_n in any surface tangent direction $\hat{t}$ is $\kappa_n = \kappa_1(\hat{e}_1 \cdot \hat{t})^2 + \kappa_2(\hat{e}_2 \cdot \hat{t})^2$; therefore, because $\hat{e}_1 \perp \hat{e}_2$, the normal curvature in the direction $\perp \hat{t}$ is $\perp\kappa_n = \kappa_2(\hat{e}_1 \cdot \hat{t})^2 + \kappa_1(\hat{e}_2 \cdot \hat{t})^2$. If we refer to $\perp\kappa_n$ on adjacent patches as κ_a, κ_b, our G^2 penalty function is $P_{G^2} = (\kappa_a - \kappa_b)^2$. Note that this function has the necessary property that the gradient is zero when $\kappa_a - \kappa_b = 0$. In order to determine the number of samples required to accurately integrate this function, we could use a similar argument based on the order of the polynomials involved in its evaluation. Since the polynomial order is prohibitively high, we sample this function with the same density as the G^1 penalty and the MVS functional. We have found this to significantly improve the quality of problematic surfaces. For example, the suitcase corner blend computed with only G^1 penalties has large curvature discontinuities as shown in Fig. 6.13(2). In Fig. 6.13(3) we see the results of applying our penalty for curvature discontinuity; the resulting surface is nearly curvature continuous; Fig. 6.13(1) is a shaded rendering of the approximately G^2 surface.

6.5.4.3. Continuum methods. The G^1 (§6.5.4.1) and G^2 (§6.5.4.2) penalty functions are combined with the MVS functional to form the objective function

$$\int \left(\frac{d\kappa_n}{d\hat{e}_1}\right)^2 + \left(\frac{d\kappa_n}{d\hat{e}_2}\right)^2 dA + \int P_{G^1} ds + \int P_{G^2} ds. \tag{6.14}$$

Minimizing this objective function alone does not guarantee G^2 or even G^1 continuity. This is because using a single quintic patch in each network opening may not provide sufficient degrees of freedom to accurately approximate the theoretical minimum variation surface. Two solutions to this problem are possible. First, each patch of the network may be subdivided into multiple

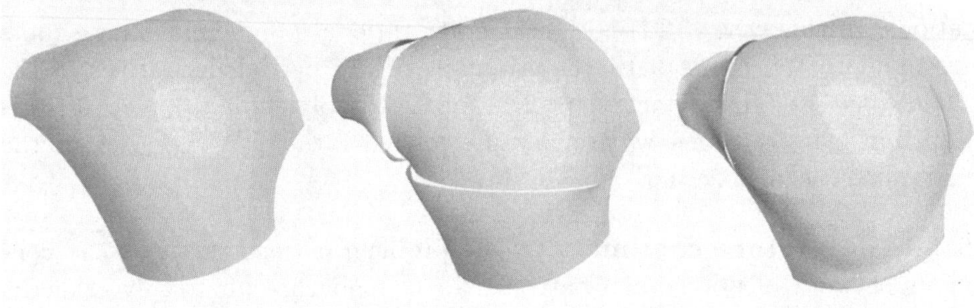

FIG. 6.13. *A curvature continuous suitcase corner.* (1) *A shaded rendering of the G^2 surface.* (2) *and* (3) *are displayed as functional offset surfaces, with the offset proportional to Gaussian curvature.* (2) *illustrates a G^1 suitcase corner with curvature discontinuities appearing as large gaps in the offset surface.* (3) *illustrates a nearly G^2 suitcase corner achieved using P_{G^2} as a penalty function.*

patches adding flexible arcs to the MVN, and augmented with the resulting additional degrees of freedom, the optimization may proceed. Second, we may use continuum methods to achieve final continuity. A continuum method computes a series of solutions to a problem by varying a parameter of the problem. In our case we modify (6.14) to allow the influence of the MVS functional to be slowly phased out:

$$(6.15)\quad w\int \left(\frac{d\kappa_n}{d\hat{e}_1}\right)^2 + \left(\frac{d\kappa_n}{d\hat{e}_2}\right)^2 dA + \int P_{G^1} ds + \int P_{G^2} ds, \qquad w \to 0.$$

Our approach iteratively minimizes (6.15), reducing w by a factor of two before each iteration. The iteration continues until the desired continuity is achieved. Peters has shown that quintic polynomial patches are sufficient for forming a G^1 continuous surface out of a network of polynomial patches. The degree requirements for patches forming a G^2 continuous surface are as yet unknown. If, after reducing w to zero, we find that G^1 continuity has not been achieved, we may either reapply the continuum method without any penalty for G^2 discontinuity or subdivide the network of patches and reapply the method retaining the penalty for G^2 discontinuity.

6.5.5. Initialization. The gradient descent scheme that is used starts with an initial surface and iteratively refines that surface until the optimal surface is reached. In this section we discuss ways of finding a suitable initial surface. In terms of the desired optimization, the goal is to find an initial point in the proper "valley" of the solution space, where the desired surface is found, as the minimal point. The optimization requires that initial values be provided for any parameters not explicitly set. The use of an optimized curve network initializes the control points on the perimeter of each patch. The interior control points are set so as to (1) achieve approximate G^1 continuity and (2) cancel the high-order derivatives at the patch corners. The first step initializes the twelve control points adjacent to the perimeter (cone, cylinder), and the second step initializes the four points in the center of the patch (cube) (Fig. 6.14). The heuristic used to position the control points adjacent to the perimeter uses linear interpolation of the normal vectors and magnitudes of difference vectors. Figure 6.14 and equation (6.16) demonstrate the approach, with the calculation of $\vec{p}_{13}$ (note that $\|\hat{H}'_i\| = 1$):

$$
\begin{aligned}
\vec{p}_{13} &= \vec{p}_{03} + \vec{F}'_3, \\
\vec{F}'_3 &= \|\vec{F}'_3\| \hat{F}'_3, \\
\|\vec{F}'_3\| &= \frac{2}{5}\|\vec{F}'_0\| + \frac{3}{5}\|\vec{F}'_5\|, \\
\hat{F}'_3 &= \frac{\hat{H}'_2 \times \hat{n}_{03} + \hat{H}'_3 \times \hat{n}_{03}}{\|\hat{H}'_2 \times \hat{n}_{03} + \hat{H}'_3 \times \hat{n}_{03}\|}, \\
\hat{n}_{03} &= \frac{\frac{2}{5}\hat{n}_{00} + \frac{3}{5}\hat{n}_{05}}{\left\|\frac{2}{5}\hat{n}_{00} + \frac{3}{5}\hat{n}_{05}\right\|}.
\end{aligned}
\tag{6.16}
$$

Alternatively, one could also use the construction due to Peters [21].

Finally, we position all the interior control points so that the twist vector and the third- and fourth-order derivatives are zero at the corners of the patch:

$$
\begin{aligned}
\vec{S}_{uv}(u,v) &= 0, \qquad \vec{S}_{uuv}(u,v) = 0, \\
\vec{S}_{uvv}(u,v) &= 0, \qquad \vec{S}_{uuvv}(u,v) = 0, \\
u, v &= \{0, 1\}.
\end{aligned}
$$

We have found this final technique to be simple and robust for a large variety of problems.

6.5.6. G^2 vertices. As an alternative to the G^2 penalty functions described in §6.5.4.2 we may construct the network of patches to meet with G^2 continuity at the vertices only, and maintain this continuity by construction during the minimization process. Because of the nature of the fairness functional, the G^2 continuity tends to "propagate" along the patch–patch boundaries. A comparison of surfaces with and without G^2 vertex continuity shows those with G^2 continuity to have superior overall curvature distribution.

The construction used to maintain G^2 vertex continuity of the surface is a simple extension of the construction used to maintain G^2 compatibility of the MVN (§6.4). An additional step is carried out after the principal curvatures at the vertices of the network have been established. This extra step of the

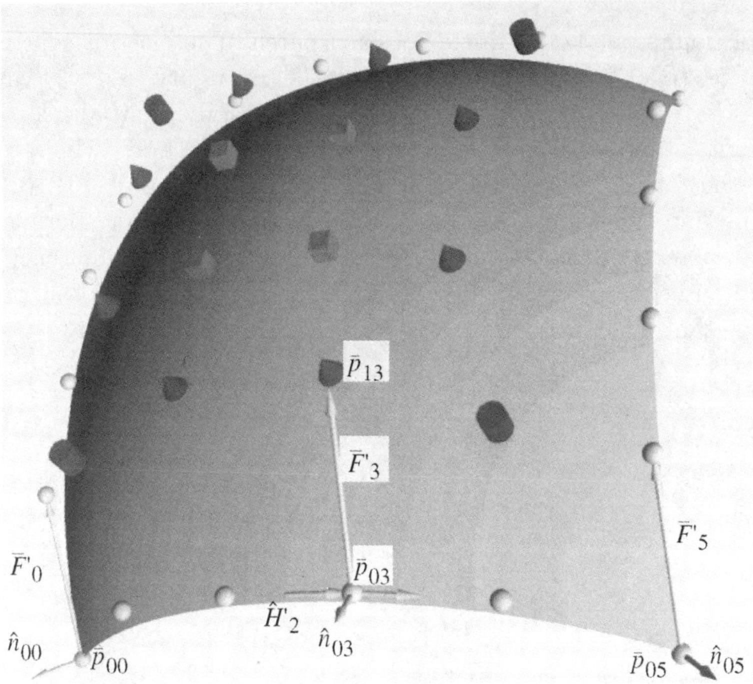

FIG. 6.14. *The control points of a biquintic Bézier patch are grouped as the* 20 *perimeter control points (sphere), the* 12 *adjacent control points (cone, cylinder), and the four central control points (cube). As an example of initialization,* $\overline{p}_{13}$ *is computed by linearly interpolating the surface normal vectors at the corners and the magnitudes of the corresponding difference vectors.*

construction requires that the twist vectors at the incident patch corners be compatible with the established curvatures, and thus $f = \hat{n} \cdot \vec{S}_{uv}$. The value of f is computed from the formulas for mean and Gaussian curvature:

$$\text{Gaussian } = \kappa_1 \kappa_2 = \frac{eg - f^2}{EG - F^2}$$

and

$$\text{mean } = \kappa_1 + \kappa_2 = \frac{gE - 2fF + eG}{EG - F^2},$$

where, as in (6.10) and (6.11),

$$\begin{aligned} E &= \vec{S}_u^2, & e &= \hat{n} \cdot \vec{S}_{uu}, \\ F &= \vec{S}_u \cdot \vec{S}_v, & f &= \hat{n} \cdot \vec{S}_{uv}. \\ G &= \vec{S}_v^2, & g &= \hat{n} \cdot \vec{S}_{vv}. \end{aligned}$$

The twist vector is adjusted to satisfy $f = \hat{n} \cdot \vec{S}_{uv}$. This is accomplished by forcing the tip of $\vec{S}_{uv}$ to lie in the plane perpendicular to $\hat{n}$ offset by distance f (Fig. 6.15).

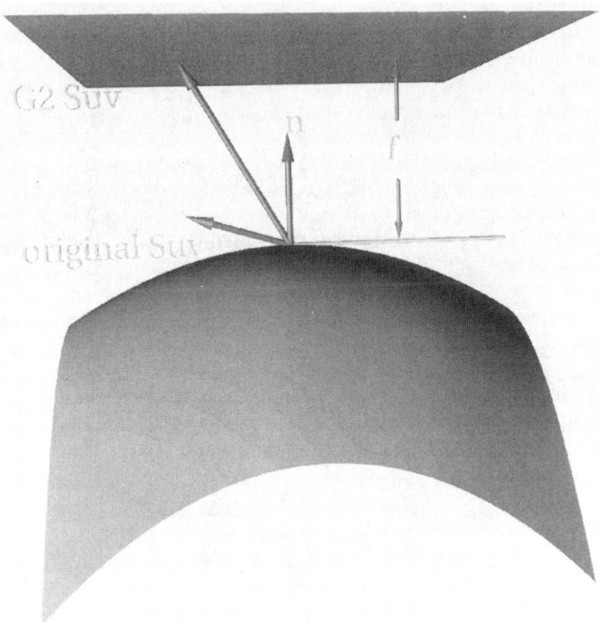

FIG. 6.15. *Construction of a G^2 vertex.*

6.6. Examples, Comparisons, and Applications

6.6.1. Minimum variation curves. In order to evaluate the utility of the MVC, we discuss a simple demonstration problem and compare the MVC with other curve representations on two test examples.

6.6.1.1. A sample problem: Corner blending. In this example, eight positional constraints are set symmetrically near the corners of a two-dimensional box; the goal is to round the corners of the box. Two possible solutions to the blending problem are presented. Figure 6.16(1) illustrates the case where tangent and curvature ($\kappa = 0$) constraints are specified at the eight points. These constraints result in a continuous curvature plot, with value zero along the sides of the box. The second solution (Fig. 6.16(2)) permits curvature discontinuities at the blend boundaries, which in turn allow the MVC to form circular blends at the corners of the square. In comparison, the first blend alternative is curvature continuous at the expense of a maximum curvature that is about 1.5 times the curvature of the circular arcs in the second solution.

6.6.1.2. A comparison of MVC, MEC, and natural splines. Figure 6.17 highlights the differences in fairness exhibited by a natural spline, an MEC, and an MVC constrained to pass through a given set of positional constraints [37]. Though the three resulting curves are nearly indistinguishable on paper, the curvature plots emphasize the differences between these three curve generation techniques. The natural spline has a "spiky" curvature plot

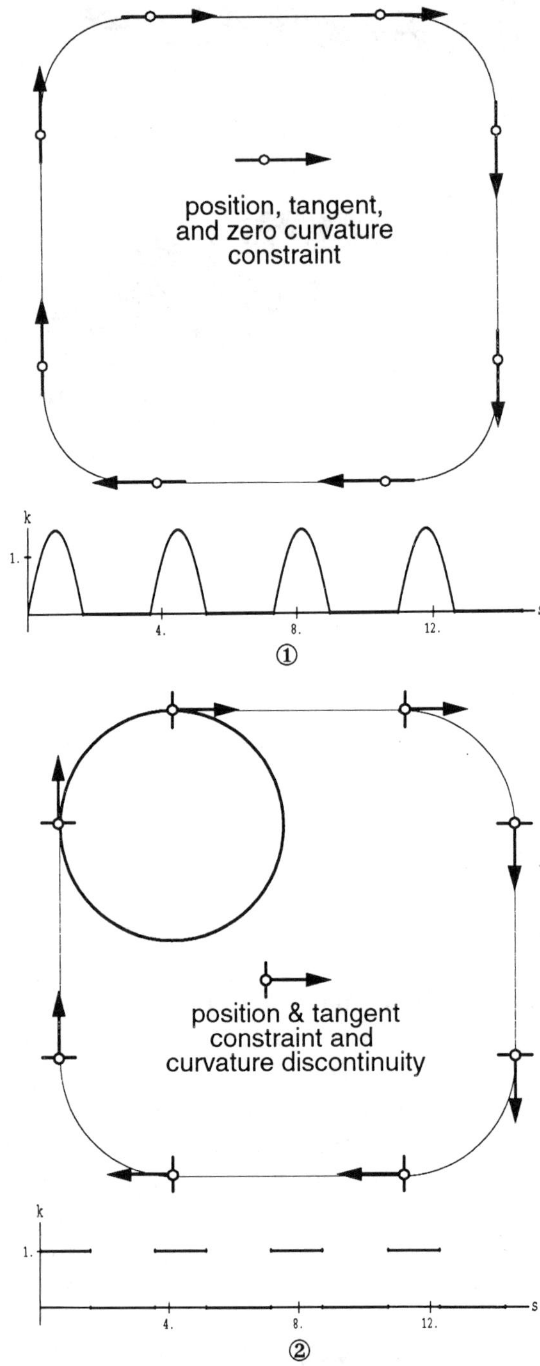

FIG. 6.16. (1) *A curvature continuous blend specified with position, tangent, and zero curvature constraints.* (2) *A circular blend specified with position and tangent constraints, and curvature discontinuities.*

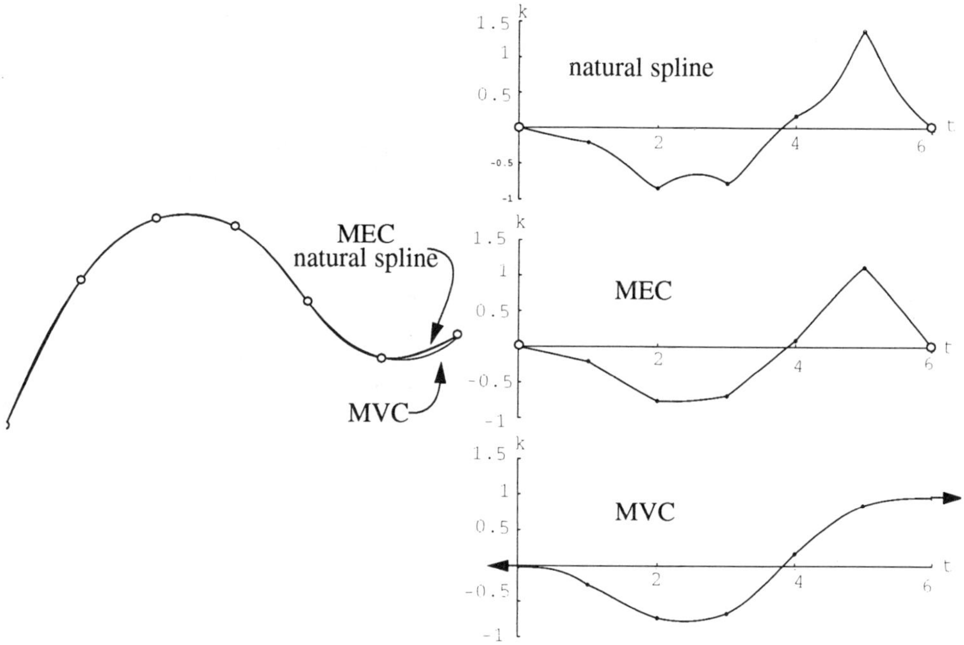

FIG. 6.17. *A comparison of a chord length parameterized natural spline, MEC, and MVC. Note that the MVC has a much smoother curvature distribution. Also, while the natural spline and MEC have zero curvature at their endpoints, the MVC has constant curvature at its endpoints, i.e., the derivative of curvature is zero.*

with peaks at several of the constraint points, and zero curvature at the endpoints. The MEC exhibits a considerably smoother curvature plot with slope discontinuities at the interpolated points, and zero curvature at the endpoints. Finally, the MVC has a very smooth curvature plot, free of sharp peaks and corners, and has constant curvature at the endpoints. Note also that the MVC has a lower maximum curvature.

6.6.1.3. MVC versus MEC space curves. In this example we examine the characteristics of MVC and MEC *space curves*. Figure 6.18 is a schematic view of the constraints used in this example. Figure 6.19 shows the MVC to be smoother and of slightly lower maximum curvature. Note, however, that the MVC, unlike the MEC, is torsion continuous.

6.6.2. Minimum variation surfaces. In order to evaluate the quality and usefulness of MVS, we examine a few interpolation and design problems. Special rendering techniques are used to assist in the evaluation of the quality of these surfaces. Functional shading is used to examine the distribution of curvature. In this case, Gaussian curvature is used to control the intensity of the surface. *Lines of reflection* are used to demonstrate G^1 and G^2 continuity [23]. They are generated by assuming that the surface is highly reflective and

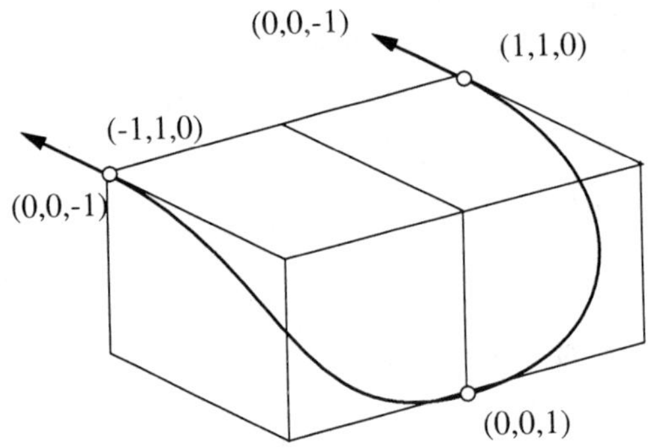

FIG. 6.18. *A simple set of positional constraints for a space curve.*

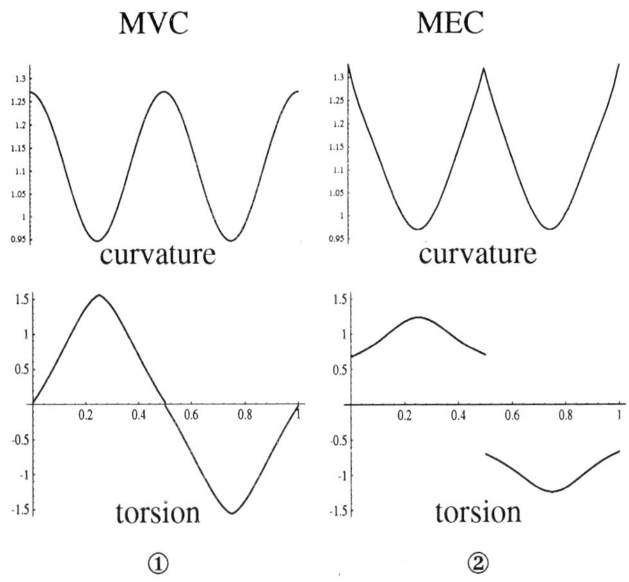

FIG. 6.19. *Curvature and torsion plots for the* MVC (1) *and the* MEC (2) *resulting from the constraint set shown in Fig.* 6.18.

is placed inside a large box with lines drawn on its walls. The surface is then rendered using environment mapping [1], [8]. Generally, smooth surfaces have smooth lines of reflection. G^2 surfaces discontinuities appear as kinks or G^1 discontinuities in the lines of reflection. G^1 surface discontinuities cause the lines of reflection to break up or become completely discontinuous. Finally, functional offset surfaces are used to illustrate the distribution of Gaussian curvature over the surface. The offset surfaces are computed by an offset from the original surface proportional to Gaussian curvature. Discontinuities of curvature appear as gaps in the resulting offset surface; discontinuities of the gradient of curvature appear as creases in the surface.

6.6.2.1. Spheres. The test task is to fit a surface through the eight corners of a cube. In Fig. 6.20 we compare the MVS functional with four other functionals; shading and offset are proportional to Gaussian curvature. In Fig. 6.20(2), only the G^1 penalty function is minimized; in 6.20(3), only the G^1 and G^2 penalty functions are minimized, in 6.20(4) a linearized approximation to strain energy is used resulting in improved curvature distribution; in 6.20(5) true strain energy is minimized, producing a surface with fairly uniform curvature. Finally, in 6.20(6) an MVS surface fitted to the corners of a cube produces a very close approximation to a sphere.

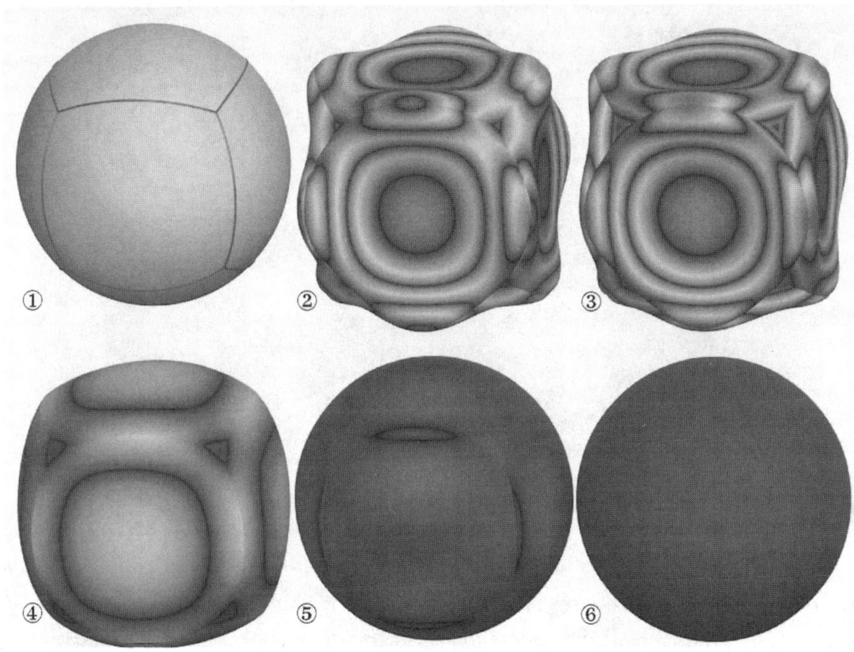

FIG. 6.20. *Surfaces interpolating the eight corners of a cube.* (1) *The MVN and MVS interpolating the cube. Pseudocoloring combined with functional offset surfaces illustrate the differences among the different objective functions.* (2) G^1 *penalty alone.* (3) $G^1 + G^2$ *penalties.* (4) *Linearized strain energy.* (5) *Strain energy.* (6) *MVS.*

6.6.2.2. Three handles. Figure 6.21 illustrates the application of the MVS to a more complicated example. In Fig. 6.21(1) we provide the MVN interpolated to create the G^1-MVS. Figure 6.21(2) illustrates the surface rendered with lines of reflection. In Fig. 6.21, strain energy (6.21(3) and 6.21(4)) and the MVS functional (6.21(5) and 6.21(6)) are compared. The differences are subtle; curvature varies more smoothly and is distributed more evenly over the MVS.

6.6.2.3. Tetrahedral frame. Figure 6.22 demonstrates how the MVS can handle a difficult blending problem in the shape of a tetrahedral frame.

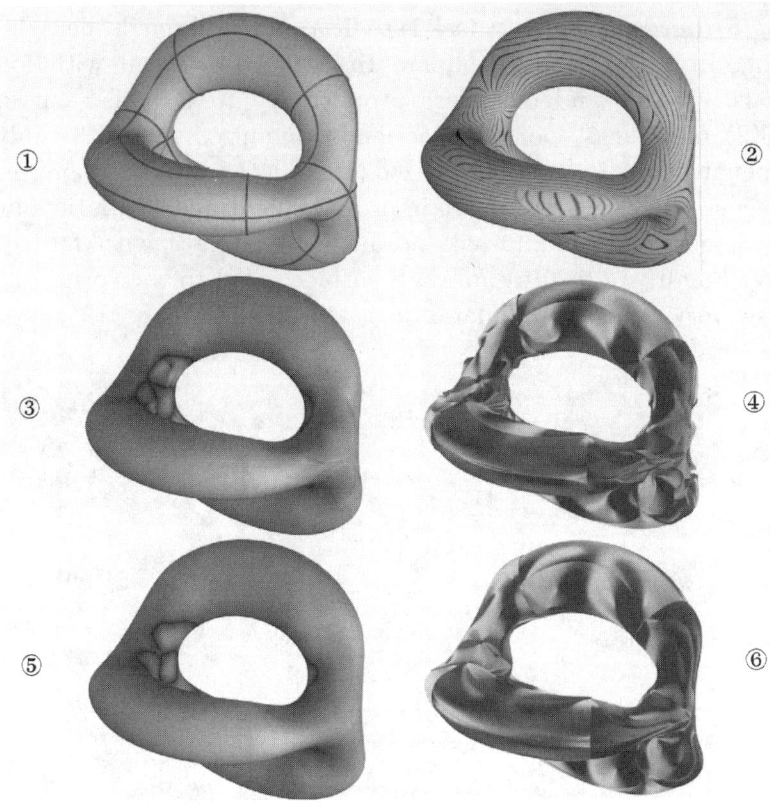

FIG. 6.21. *The MVN used in the computation of the surfaces is shown in* (1). (2) *is a rendering of the MVS with lines of reflection, demonstrating* G^1 *and* $\sim G^2$ *continuity.* (3) *and* (4) *show the Gaussian pseudocolored and Gaussian offset surface resulting from the surface computed using strain energy as an objective function.* (5) *and* (6) *illustrate these same rendering styles for the surface computed using curvature variation as an objective function. Curvature is more evenly distributed across the MVS.*

Figure 6.22(1) shows the minimum variation network (MVN); 6.22(2) shows the openings of the network filled, with the individual patches demarcated by black borders; 6.22(3) exhibits lines of reflection, and 6.22(4) is a simple lighted rendering. This example has caused considerable headaches for other surface generation systems.

6.7. Efficiency

In this section we discuss the cost of computing the MVS, what part of the optimization process dominates the expense, how to estimate relative computation times, and how computation times might be reduced. It is logical to discuss computational expense in terms of the number of times that the functional must be evaluated at a point on the surface.

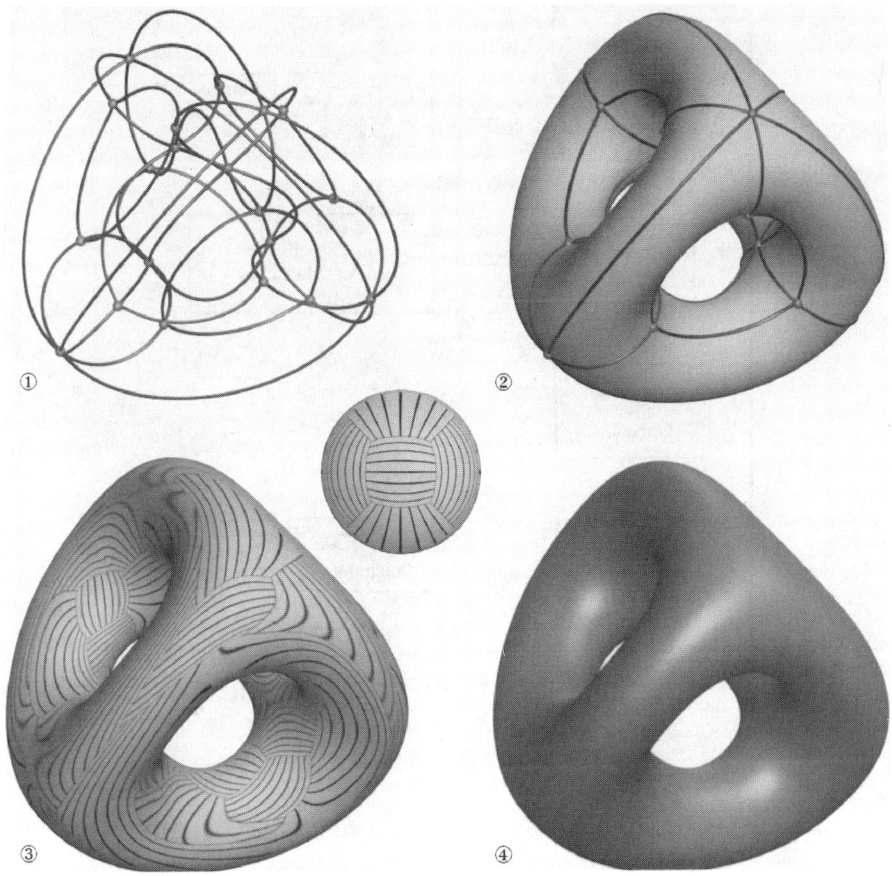

FIG. 6.22. *Tetra Thing.* (1) *The MVC network.* (2) *Individual patches demarcated by black boundaries.* (3) *Lines of reflection indicating the* G^1 *continuity of a sphere and of the tetrahedral shape.* (4) *A simple shaded rendering.*

First we examine the optimization procedure at a high level (Fig. 6.23). Each iteration of the optimization begins by computing the current value of the objective function; each patch in the quilt is integrated; this requires 200 functional evaluations for every triangular patch, and 400 evaluations for every tensor product patch. The computation of the penalty functions is of negligible cost, since we compute all of the required quantities during the integration phase.

The computation of the current gradient completely dominates the computation time. For each degree of freedom, the objective function of each affected patch must be recalculated twice. Depending on the type of variable, the number of patches affected ranges from one to several. If the degree of freedom (DOF) is interior to a single patch then only that patch is affected. If the DOF is part of a boundary curve definition then two patches are affected. In the most costly case, if the DOF is part of a vertex definition, then all patches incident to the vertex must be taken into account. From this it should be apparent

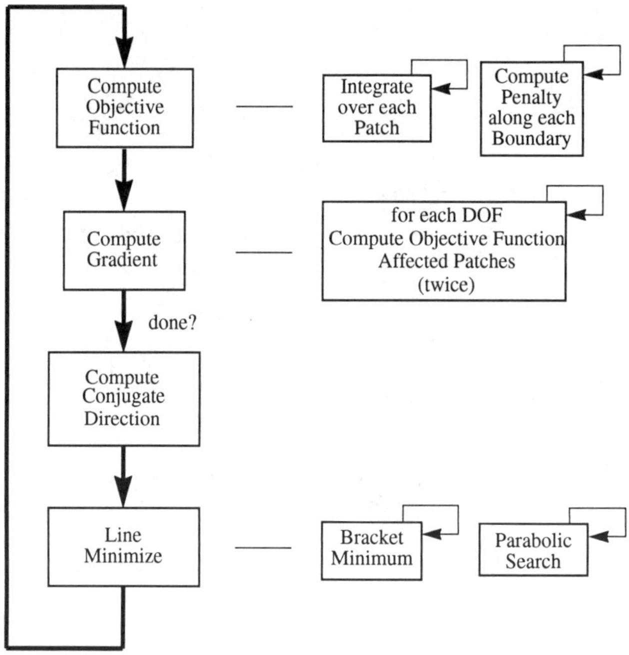

FIG. 6.23. *Optimization overview.*

that it is possible to compute an MVS with fixed curve network much more quickly than a surface with only point constraints. In the case of the flexible MVN, not only is the number of the DOF increased, but the cost of computing the associated partial derivatives is disproportionately higher because the number of patches associated with each DOF is higher. The cost of computing the conjugate direction and performing the line minimization is comparatively low.

We now consider a concrete example, the computation of the MVS sphere in Fig. 6.20. Because of its symmetry, this problem has only twelve DOF, and each of these is limited to the interior of a patch. In Fig. 6.24 we display a plot of the number of iterations versus the log of the normalized objective function. The optimization, including problem input, initialization, optimization, and result output, took approximately 91 seconds to complete and a total of 510 tensor product patch integrations were performed. Assuming no overhead whatsoever, each integration took ~ 0.18 seconds. Given the total number of iterations, we may also compute what portion of the 510 integrations were devoted to gradient computation. This optimization problem took 17 iterations, or ~ 5.3 seconds per iteration. Each iteration required a partial derivative computation for each of 12 DOF; each partial derivative requires two integrations. This accounts for a total of 408 integrations, which is roughly 80% of the computation. This percentage is conservative, since as the number of DOF grows, the number of integrations required for gradient computation increases much more quickly

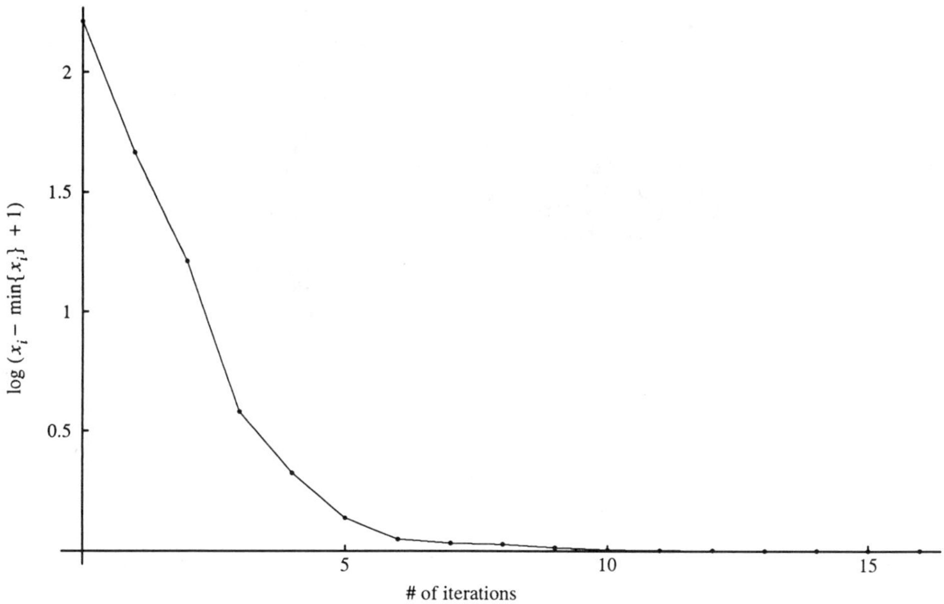

FIG. 6.24. *Sphere—number of iterations versus log (functional). Here we show the convergence of the optimization of the* MVS *sphere in Fig.* 6.20. *The optimization took approximately* 91 *seconds to complete. Taking advantage of symmetry, the number of* DOF *was reduced to* 12.

than the itegrations required for line minimization and objective function calculation. To understand why this is the case, consider two optimization problems, one of which requires one more tensor product patch than the other. In the larger problem, we gain at least 48 degrees of freedom corresponding to the 16 control points on the interior of the additional patch. These 48 additional DOF require 96 integrations to the gradient computation, while the additional patch introduces only one additional integration in the evaluation of the objective function and one additional integration in the line minimization loop.

This particular optimization problem converges quickly. This is due to the fact that the solution has zero "energy." The computation of more general surfaces, such as the one defined by two position constraints (Fig 6.25), converges more slowly (Fig. 6.26). In this case the optimization has 29 degrees of freedom; these DOF are from the curvature at the corners of the patch, the curves of the network, and the interior of the patch, all reduced by symmetry. The optimization took about 30 minutes to complete, and used a total of 8646 integrations distributed over 136 iterations. To compute the fraction of this total that is devoted to derivative calculation, we multiply twice the number of DOF by the number of iterations, yielding a total of 7888 integrations, or about 91% of the total.

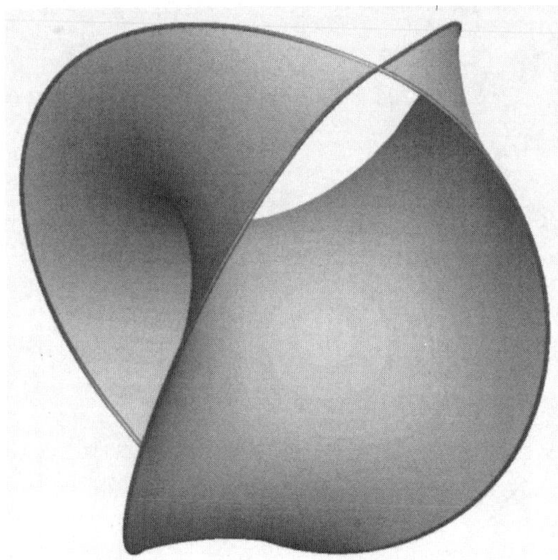

FIG. 6.25. *OnePatch—an isometric view of a surface defined by two point constraints, all other degrees of freedom were used to optimize the surface.*

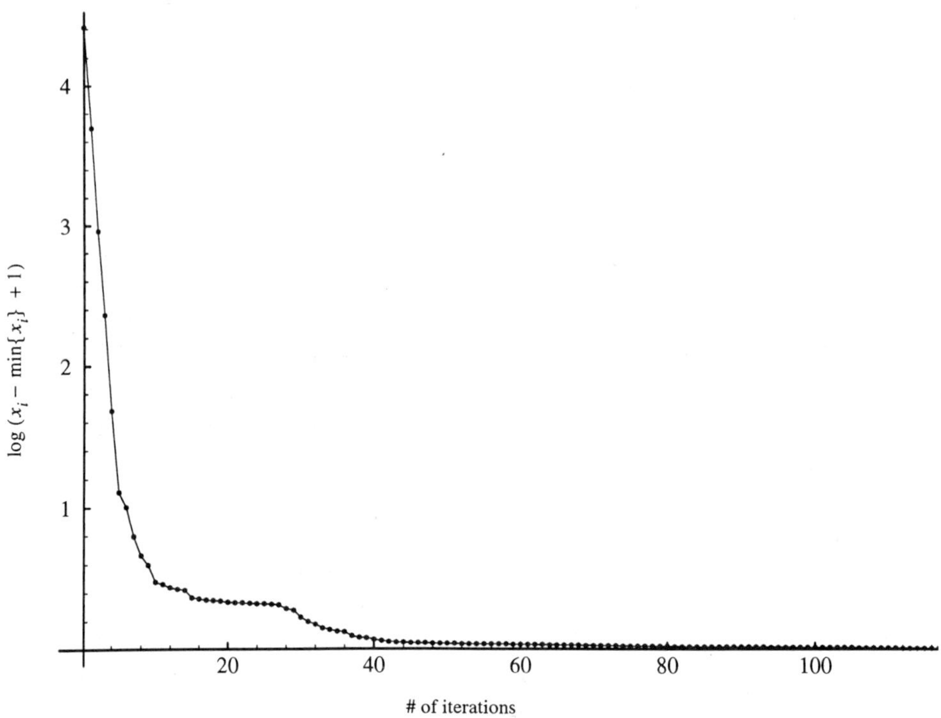

FIG. 6.26. *OnePatch—number of iterations versus log (functional). This plot illustrates the convergence of the optimization of the MVS sculpture in Fig. 6.25. The optimization took approximately* 30 *minutes to complete. Taking advantage of symmetry, the number of DOF was reduced to* 29.

From these examples, we see that improvement in the efficiency of the integration of the functional will have a direct impact on the efficiency of the algorithm. Counterintuitively, we have found that reducing the number of integration points actually slows the optimization process. This is because the accuracy of the gradient computation suffers and the descent scheme spends lots of time "wandering around." In order to improve performance, work needs to be aimed at restructuring the functional so that it may be integrated analytically.

6.8. Conclusions

Throughout our work we have observed that minimizing the *variation* of curvature rather than the magnitude of curvature leads to smoother and more pleasing curves or surfaces. In many cases the differences are small, and curves computed with different functionals are almost indistinguishable. However, an analysis of their curvature distribution reveals significant differences. For the MVC the extremal magnitudes of curvature are smaller, and they exhibit smoothly varying and uniformly distributed curvature.

For surfaces, the differences can be more obvious, particularly when we try to fit a smooth surface of nontrivial topology through a sparse set of constraint points. The tetrahedral frame, and to a lesser degree even the torus, are cases in point. Previous attempts at modeling these surfaces from just 20 or 16 points, respectively, have readily resulted in G^1 continuous surfaces, but these surfaces typically have shown undesirable bulges or ripples.

Admittedly, the presented optimization-based approaches to creating smooth interpolating curves and surfaces are much slower than the constructive approaches widely discussed in the literature on computer-aided geometric design. However, for interactive design, the optimization procedure can be terminated earlier, resulting in faster but cruder approximations of the ideal MVS in the early stages of the design. Only in the final refinement stages of a design would the process be run to a very fine approximation; however, in that phase, run-times of several minutes or even hours may well be acceptable. In any case, this issue will probably become less important as faster multiprocessor workstations become the norm rather than the exception. For applications where the quality of curved forms really counts, the MVC and the MVS provide solutions that are superior to other procedures.

Acknowledgments

The authors thank Silicon Graphics, Inc. for its generous support.

References

[1] J. F. Blinn, *Texture and reflection in computer generated images*, Comm. ACM, 10 (1976), pp. 542–547.

[2] W. Böhm, *On cyclides in geometric modeling*, Comput. Aided Geom. Design, 7 (1990), pp. 243–255.

[3] C. R. Calladine, *Gaussian curvature and shell structures*, The Mathematics of Surfaces, Clarendon Press, Oxford, England, 1986, pp. 179–196.

[4] G. Celniker and D. Gossard, *Deformable curve and surface finite-elements for free-form shape design*, Computer Graphics 25, 4 (1991), pp. 257–266.

[5] M. P. Do Carmo, *Differential Geometry of Curves and Surfaces*, Prentice Hall, Englewood Cliffs, NJ, 1976, p. 145.

[6] G. Farin, *Curves and Surfaces for Computer Aided Geometric Design, A Practical Guide*, second ed., Academic Press, San Diego, CA, 1990.

[7] S. D. Fisher and J. W. Jerome, *Stable and unstable elastica equilibrium and the problem of minimum curvature*, J. Math. Anal. Appl., 53 (1976), pp. 367–376.

[8] N. Greene, *Environment mapping and other applications of world projections*, IEEE Computer Graphics and Applications, 11 (1986), pp. 21–29.

[9] H. Hagen and G. Schulze, *Automatic smoothing with geometric surface patches*, Computer Aided Geometric Design, 4 (1987), pp. 231–236.

[10] J. W. Jerome, *Smooth interpolating curves of prescribed length and minimum curvature*, Proceedings of the American Mathematical Society, 51 (1975), pp. 62–66.

[11] A. K. Jones, *Nonrectangular surface patches with curvature continuity*, Comput. Aided Design, 6 (1988), pp. 325–335.

[12] M. Kallay and B. Ravani, *Optimal twist vectors as a tool for interpolating a network of curves with a minimum energy surface*, Comput. Aided Geom. Design, 6 (1990), pp. 465–473.

[13] M. A. Malcolm, *On the computation of nonlinear spline functions*, SIAM J. Numer. Anal., 15 (1977), pp. 254–282.

[14] S. Mann, C. Loop, M. Lounsbery, D. Meyers, J. Painter, T. DeRose, and K. Sloan, *A survey of parametric scattered data fitting using triangular interpolants* in Curve and Surface Modeling, H. Hagen, ed., Society for Industrial and Applied Mathematics, Philadelphia, PA, 1992.

[15] E. Mehlum, *Nonlinear splines*, in Computer Aided Geometric Design, R. E. Barnhill and R. F. Riesenfeld, eds., Academic Press, Orlando, FL, 1974.

[16] G. M. Nielson, *A method for interpolating scattered data based upon a minimum norm network*, Mathematics of Computation, 161 (1983), pp. 253–271.

[17] G. M. Nielson and R. Franke, *Surface construction based upon triangulations*, in Surfaces in CAGD, R. E. Barnhill and W. Böhm, eds., North-Holland, Amsterdam, 1983.

[18] S. C. Ohlin, *Splines for Engineers, Eurographics '87*, G. Marechal, ed., North-Holland, Amsterdam, pp. 555–565.

[19] ——, *2-D and 3-D curve interpolation by consistent splines*, IBM Nederland NV, CAD/CAM Systems Support Group, Amsterdam, 1985.

[20] J. Pegna and F. E. Wolter, *A simple practical criterion to guarantee second order smoothness of blend surfaces*, in Advances in Design Automation—1989—Volume One—Computer-Aided and Computational Design, B. Ravani, ed., ASME, Montreal, Quebec, Canada, September 17–21, 1989.

[21] J. Peters, *Smooth interpolation of a mesh of curves*, Constr. Approx., 7 (1991), pp. 221–247.

[22] ——, *Local smooth surface interpolation : a classification*, Comput. Aided Geom. Design, 7 (1990), pp. 191–195.

[23] T. Poeschl, *Detecting surface irregularities using isophotes*, Comput. Aided Geom. Design, 1 (1984), pp. 163–168.

[24] H. Pottmann, *Scattered data interpolation based upon generalized minimum norm networks*, Constr. Approx. 7(1991) pp. 247–256.

[25] A. Pramila, *Ship hull surface using finite elements*, International Shipbuilding Progress, 25 (1987), pp. 97–107.

[26] M. J. Pratt, *Cyclides in computer aided geometric design*, Comput. Aided Geom. Design, 7 (1990), pp. 221–242.

[27] W. H. Press, B. P. Flannery, S. A. Teukolsky, and W. T. Vetterling, *Numerical Recipes in C*, Cambridge University Press, 1988.

[28] T. Rando and J. A. Roulier, *Designing faired parametric surfaces*, Comput. Aided Design, 23 (1991), pp. 492–497.

[29] J. Roulier, T. Rando, and B. Piper, *Fairness and monotone curvature*, Approximation Theory and Functional Analysis, C. K. Chui, ed., Academic Press, New York, 1991.

[30] R. F. Sarraga, *G^1 interpolation of generally unrestricted cubic Bézier curves*, Computer Aided Geometric Design, 6 (1987), pp. 23–40.

[31] L. A. Shirman and C. H. Séquin, *Procedural construction of patch-boundary curves*, in Curves and Surfaces, L. L. Schumaker, ed., Academic Press, Boston, 1991.

[32] ——, *Curve interpolation with geometrically continuous splines*, Comput. Aided Design, 5 (1992), pp. 267–277.

[33] ——, *Curve interpolation with curvature continuous splines*, Comput. Aided Design, 5 (1992), pp. 278–286.

[34] ——, *Local surface interpolation with Bézier patches*, Comput. Aided Geom. Design, 24 (1987), pp. 279–295.

[35] P. H. Todd and R. J. Y. McLeod, *Numerical estimation of the curvature of surfaces*, Computer-Aided Design, 18 (1986), pp. 33–37.

[36] C. J. K. Williams, *Use of structural analogy in generation of smooth surfaces for engineering purposes*, Computer-Aided Design, 19 (1987), pp. 310–322.

[37] C. H. Woodford, *Smooth curve interpolation*, BIT, 9 (1969), pp. 69–77.

CHAPTER

7

Convexity Preserving Surface Interpolation

Tim Gallagher and Bruce Piper

7.1. Introduction

Given a discrete set of points in $\mathcal{R}^3$, we desire an algorithm to interpolate the points with a smooth, convex surface. Furthermore, we want the final surface to reflect any shape characteristics present in the initial data, while maintaining a visual simplicity that directly conveys these characteristics to a designer.

As designers demand more efficient, spatially intuitive algorithms for the computer-aided design and processing of surfaces, it becomes essential to understand geometric shape characteristics. The convexity, smoothness, or curvature of a surface is often more important than the choice of functions used to represent the surface. In this chapter we consider a surface design algorithm which emphasizes particular shape characteristics by composing simple surfaces which already possess the qualities we desire. In fact, "simpler" surface descriptions will often lead to faster and better solutions precisely because they allow the designers to work in the spatially intuitive terms they understand [9].

There are, at present, effective methods to find convex surfaces which approximate given data, including one by Andersson et al. [1]. There it is noted that not only is convexity the basic and most frequently required shape characteristic, but also that designing a smooth convex approximating surface to given data is very difficult because it is a nonlinear, constrained optimization problem with many variables and constraints. This complexity is in part a reflection of the representation of the surface. It would be advantageous if the choice of representation facilitated both a formal statement of the convexity of the final surface and the designer's intuition of convexity.

We begin with the convex hull of a certain set of equal spheres in §7.2 and generalize characteristics of this hull in terms of planes, cones, cylinders, spheres, tori, and cyclides. In an attempt to build up intuition we start with the simpler cases, leaving full analyses of the surfaces until we arrive at our most general case. The parametrization of the surfaces will be of secondary importance, but in any case there are direct techniques to get rational parametrizations; see [3] and [12]. Proceeding from the simplest case of the convex hull of a set of spheres, each of the same radius, we replace the planar subsets of the surface with spherical components in §7.3. This replacement is not done haphazardly, but by a continuous deformation of the surface under constraints to be specified later. This requires replacing the cylindrical subsets with toroidal components in §§7.3.1 and 7.3.2. With further generalizing, we allow the radii of the initial spheres to be unequal in §7.4, requiring the toroidal components to be replaced with cyclidal ones. A particular cyclide is conveniently represented as the image of a simpler surface under an inversion in a sphere which is dependent on the spheres to be fit by the cyclide. Sections 7.2 and 7.3 are intended more to develop intuition, while §§7.4 and 7.5 focus on calculations and rigorous constraints. Proof that the derived surfaces do satisfy many fairness properties is presented in §7.6.

Examples of the three types of components for a specific surface may be seen unassembled in Figs. 7.1(a)–7.1(c), and assembled in Fig. 7.1(d). Figure 7.1(a) consists of spherical patches which interpolate the initial data points in the interior of the patches, Fig. 7.1(b) consists of spherical patches which provide smooth surfaces between these interpolating patches, and Fig. 7.1(c) consists of the cyclidal patches which fill in the remaining edge gaps.

7.1.1. Definitions. For any subset of $\mathcal{R}^3$, a plane *supports* the set if and only if the set lies entirely on one side of the plane (including in the plane itself). A *half space* is the portion of space on one side of a plane, always taken to be closed unless otherwise specified. The *positive* support of a plane is that half space defined by the plane and containing some reference (herein assumed to be the given points unless otherwise noted). Thus if f is a plane (or a facet of a polyhedron) then H_f^+ is the closed half space bounded by the plane of f and containing the reference. The following three statements can serve as definitions of a *convex* set: (1) the line segment between any two points of the set is contained in the set, or (2) for each point on the boundary of the set there exists a (hyper)plane containing the point and supporting the set, or (3) the set can be represented as the intersection of half spaces. A *convex surface* is then the boundary of a convex set. We will use the *convex closure* of a set X to mean the intersection of all closed convex supersets of the set, written $cc(X)$, and the *convex hull* of X to mean the boundary of the convex closure of X, written $ch(X)$.

The algorithm accepts as input a convex set of discrete data P and a set of tangent planes $\{T_p\}_{p\in P}$. Here P *convex* means that the data lie on their convex hull. It is assumed that each tangent plane T_p has already been chosen such that it is the desired/required tangent plane to the final surface at the point of interpolation p, and furthermore that T_p is consistent with convexity, i.e., that it supports the data. A vector originating in a plane is *inward pointing* if it lies in the positive support of the plane. Each tangent plane T_p can be represented by the point p and the inward pointing unit normal, n_p, to the plane. The set of tangent planes defines a convex polyhedron which must contain the convex

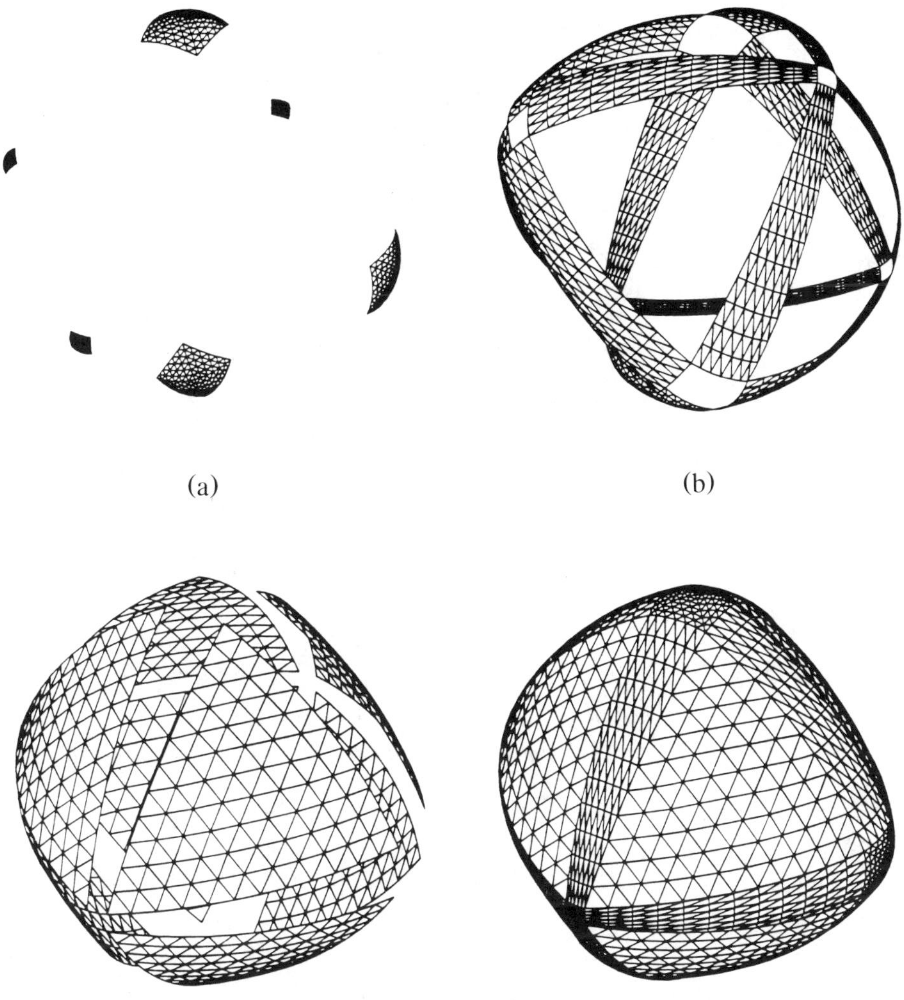

(a) (b) (c) (d)

FIG. 7.1. (a) *Spherical "vertex" patches for the example surface. The original data points are on the interiors of these patches.* (b) *Cyclidal "edge" patches for the example surface.* (c) *Spherical "facet" patches for the example surface.* (d) *A composite surface assembled from the three types of patches shown in* (a)–(c).

closure of the original data set because the convex closure is the smallest convex set containing the data and the tangent plane polyhedron is *one* convex set containing the data; see Fig. 7.2. The tangent plane polyhedron can also be represented explicitly as $T_P := \bigcap_{p\in P} H^+_{T_p}$, where the positive support of any T_p is taken to be the half space containing the set P.

7.2. The Initial Surface—The Convex Hull of Spheres

More details on the contents of this section will appear shortly and will also be included in [10].

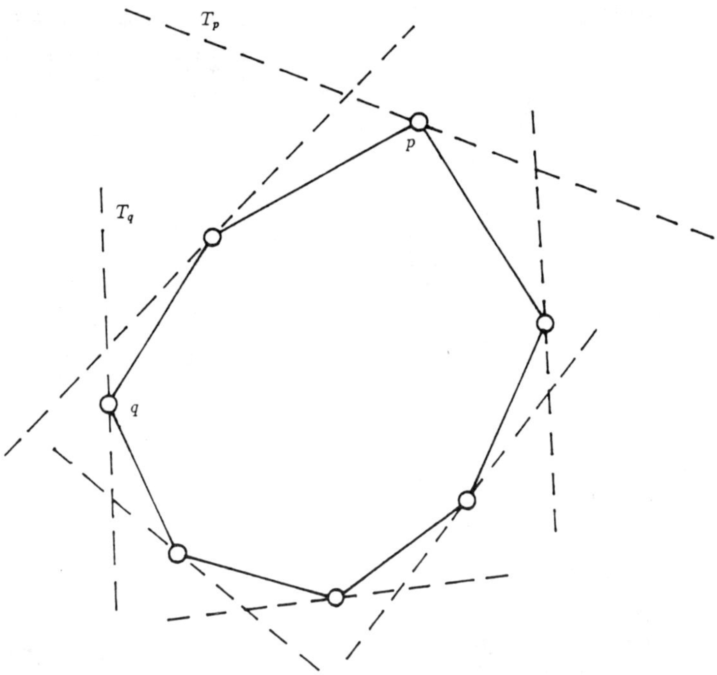

FIG. 7.2. *Relationship of the tangent plane polyhedron to the convex hull of the data points. The chosen tangent planes are represented by the broken lines.*

We begin by considering each vertex p, the tangent plane T_p, and the tangent plane polyhedron, and then by constructing a sphere tangent to T_p at p, in the positive support of T_p. Thus if the radius of the sphere B_p is ϵ_p, the center of the sphere is $c_p = p + \epsilon_p n_p$. This *vertex sphere*, B_p, interpolates p and is convex. We impose two restrictions on the radii of the vertex spheres. If the radius of a vertex sphere is zero, then the final surface will have a corner at p, i.e., it will not be smooth. It may be the case that the tangent plane polyhedron is such that the radius of some vertex sphere must be zero. For instance, this may happen if certain data points are linear combinations of the others. In such a case *any* convex interpolating surface must have a corner at the point. Since we are herein primarily concerned with smooth surfaces, for the remainder of this discussion it will be assumed that the data are such that none of the vertex spheres is so constrained. We do not entirely discount such cases because they may prove useful in the design process, but they do not require specific mention (cf. [10]).

Because any convex interpolating surface with the T_p as tangent planes must lie in the tangent plane polyhedron, we will not allow the radius to be so large that some B_p is not wholly contained in this polyhedron. Thus the allowable range for this radius is $\epsilon_p > 0$ and ϵ_p small enough such that every vertex sphere is contained in the tangent plane polyhedron. Now define $\mathcal{B}$ to be the convex hull of the union of the set of vertex spheres, thus $\mathcal{B} := ch(\bigcup_{p \in P} B_p)$. This surface, we claim, is an interpolating, convex, and tangent plane continuous surface with piecewise constant curvature. To generalize this construction we will consider a decomposition of the surface.

7.2.1. The surface tessellation. The *tessellation* of a surface is a decomposition of the surface into components (generally images of subsets of $\mathcal{R}^2$ under a bicontinuous, one-to-one map) and a description of how the components meet. The tessellation of the convex hull of a discrete set of points (e.g., $ch(P)$) is the usual collection of facets (convex polygons) and degenerate facets, i.e., vertices (points or corners) and edges (lines) along with a description of how they meet. The tessellation of the convex hull of a collection of spheres can be similarly described as a decomposition into patches which are intuitively like vertices, edges, and facets, but even when every vertex sphere has a common radius ϵ_p, the tessellation of the surface $\mathcal{B}$ may be different from that of $ch(P)$ or that of the convex hull of the centers of the vertex spheres, $ch(\cup_{p\in P} c_p)$. We associate a *vertex sphere* of the tessellation with every maximal connected subset of a sphere in the convex hull of the spheres. Such a subset occurs wherever one of the given spheres is not wholly contained in the interior of the convex closure of the spheres. We will see examples where this subset need not be simply connected. If the intersection of a vertex sphere and the convex hull of all of the vertex spheres is not connected, then, in terms of the tessellation, we will refer to each of the components as different vertex spheres, both of which are defined by the same radius and center.

The convex hull of two arbitrary (nonnested) spheres is a subset of a cone together with two pseudohemispheres. An *edge cone* is associated with any connected subset of such a cone on the surface $\mathcal{B}$. If the two spheres are equal (i.e., have equal radii) then the cone is an *edge cylinder*. The convex hull of three general position spheres is bounded by planar surfaces. A *facet plane* is associated with any connected planar subset of $\mathcal{B}$. The facet plane is the entire hyperplane containing the planar subset and must be tangent to at least three of the vertex spheres and support all of the vertex spheres on the same side. The description of how these three types of components (or patches cut from the associated surfaces) meet and cover $\mathcal{B}$ is the tessellation of $\mathcal{B}$. The following examples are intended to make the reader wary of assuming that the convex hull of spheres has too much in common with the convex hull of discrete points.

7.2.1.1. Example. Let f and f_1 be two facets of the much larger polyhedron $ch(P)$ with vertices p, q, r and p, q, s, respectively, and joined along edge e (obviously with vertices p, q), as in Fig. 7.3(a). While the convex hull of the vertex spheres may possess the same tessellation as the data (see Fig. 7.3(b)) with the appropriate choice of tangent planes at p, q, r, and s and one $\epsilon(= \epsilon_p = \epsilon_q = \epsilon_r = \epsilon_s)$, the convex hull of the vertex spheres may have an edge cylinder between the vertex spheres B_r and B_s; see Fig. 7.3(c). That is, the tessellation of the surface has two facet planes B_f and B_{f_1} with vertex

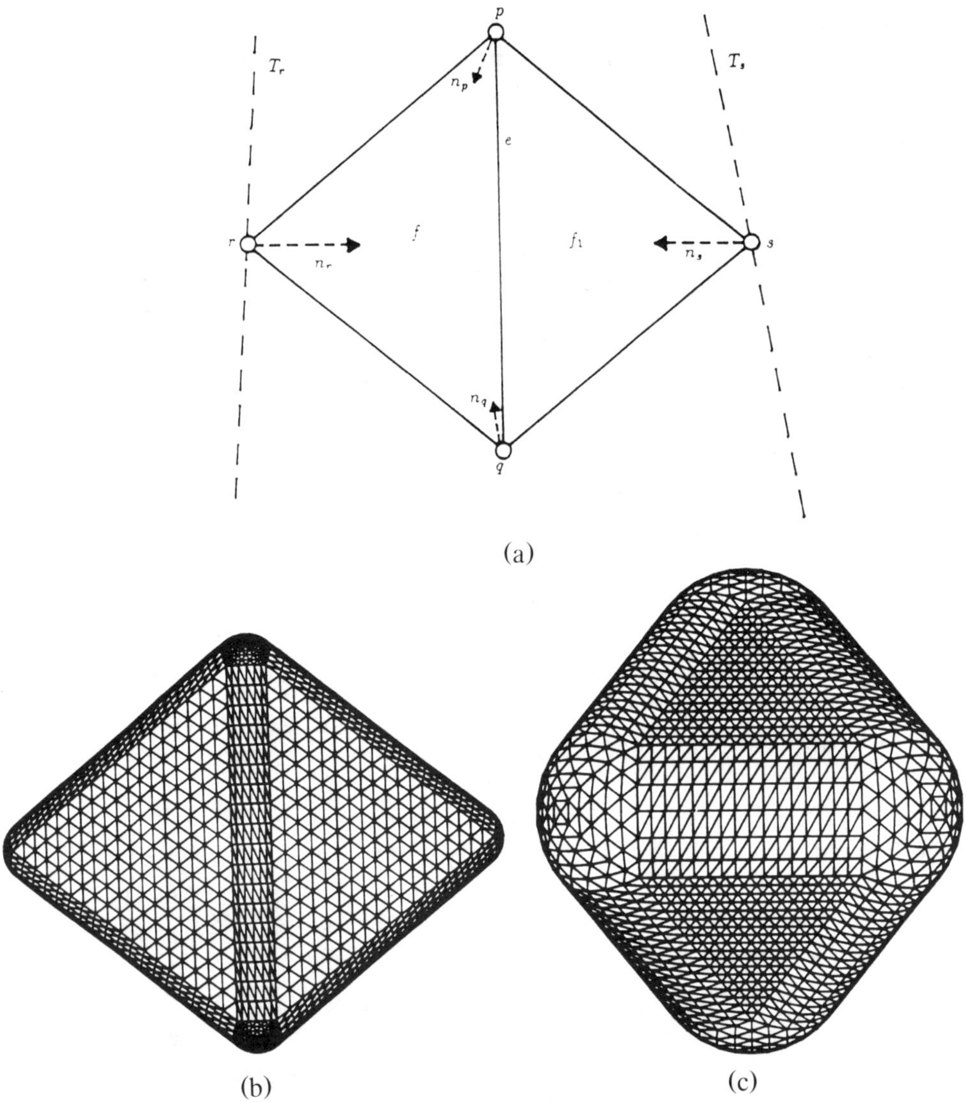

FIG. 7.3. (a) *Subset of convex hull of data with normals to the tangent planes at the data points. Note that* n_p *and* n_q *have a larger into-the-page component than* n_r *and* n_s. (b) *With small enough values for vertex sphere radii the convex hull of the vertex spheres has a similar tesellation to the convex hull of the data shown in* (a). (c) *For larger vertex sphere radii, the tesellation of the convex hull of the vertex spheres may become structurally different from that of the initial data shown in* (a).

spheres B_p, B_s, B_r, and B_q, B_r, B_s, respectively, joined along an edge cylinder between the vertex spheres B_r and B_s. Thus, within the bound on the radii of vertex spheres from the tangent planes, it is simple to generate an example where the tessellation of $ch(P)$ is different from that of $\mathcal{B}$, and actually varies as a function of ϵ.

7.2.1.2. Example. Consider four noncoplanar points p, q, r, s, forming a very flat tetrahedron (i.e., the lengths of its legs pq, pr, and ps are all much greater than its height, which is the distance from p to the base qrs); see Fig. 7.4(a). Suppose we choose tangent planes such that the normal to each tangent plane at each vertex p, q, r, and s, when extended, passes through the opposite facet of the tetrahedron. Then we can choose radii for the vertex spheres such that the center c_p of B_p (the top sphere of the tetrahedron) lies in the triangle c_q, c_r, c_s and B_q, B_r, and B_s are slightly smaller than B_p.

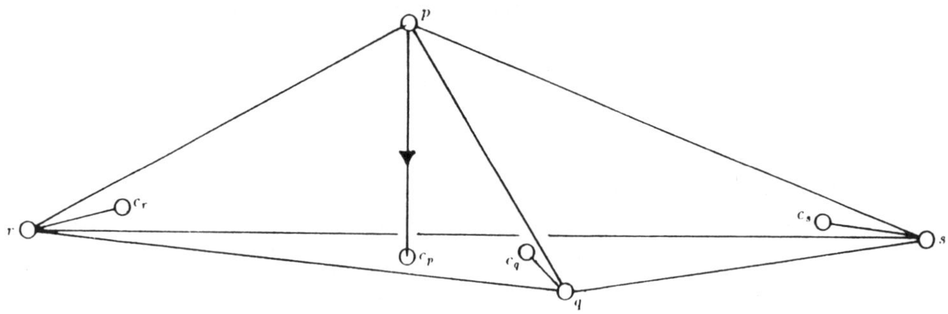

(a)

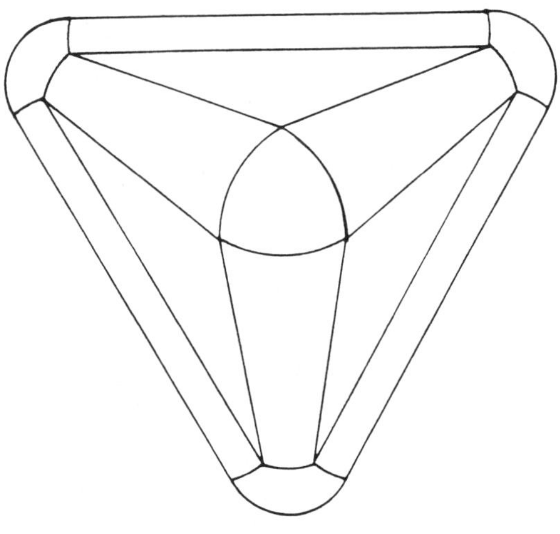

(b)

FIG. 7.4. (a) *Tetrahedron of data with the altitude from p to the opposite face much less than the lengths of the three edges meeting p. The tangent planes at the data points are chosen so that their normals, if extended, would intersect the opposite face of the tetrahedron.* (b) *The tesellation of convex hull of the vertex spheres may be different from that of the convex hull of the centers of the vertex spheres. This is both a top and bottom view of the surface because it is symmetric in the plane of centers of the spheres.*

The tessellation of $ch(B_p, B_q, B_r, B_s)$ is that of two tetrahedrons sharing a common face, i.e., six facet planes, nine edge cylinders, and five vertex spheres, which is symmetrical in the plane of c_p, c_q, and c_r; see Fig. 7.4(b). Thus the structure of $ch(P)$ (a tetrahedron) and $ch(\bigcup_{p \in P} c_p)$ (a triangle) are different from each other and from that of the surface $\mathcal{B}$. Note that the tessellation of $\mathcal{B}$ includes the sphere B_p twice, i.e., as a different vertex sphere for each place it bulges out from $\mathcal{B}$, but while the point p must be interpolated by one such bulge there is no point that must be interpolated by the other bulge. (This difference will be pertinent later in §7.5.5.) Before beginning the decomposition of $\mathcal{B}$, it is necessary to understand its tessellation. Hence we require an algorithm to generate the convex hull of an arbitrary collection of spheres. The authors are preparing a paper on such an algorithm which will also be included in [10].

7.2.2. Generating the convex hull of spheres. Such an algorithm is not as straightforward as one for the convex hull of a finite set of points. The algorithm is essentially a Jarvis march [13, p. 104] on the collection of centers of spheres with an ϵ_p offset at each step; however, special cases arise. Although in the case of a finite set of points, a point that is in the positive support of every facet of a polyhedron is in the interior of the body, it is possible for a sphere to be in the positive support of every facet plane but still be outside the body, i.e., to contribute to the convex hull. There are now two separate cases to consider. In the first case the sphere lies in the positive support of two facet planes but outside of the edge cone between them, and in the second case the sphere lies in the positive support of every facet plane of a vertex sphere, and inside every edge cone of the vertex sphere, but outside of the vertex sphere.

In the first case the tessellation is normal. The new vertex sphere B_q is connected by two new edge cones to the old vertex spheres, and by two facet planes which each connect to the old edge cone. The analogy with the tessellation of the convex hull of a discrete set of points is strained because such a point never has only two edges and two facets, but it holds because the sphere is joined by four other surfaces.

In the second case the tessellation changes radically. The new vertex sphere B_q is connected only to the old vertex sphere by a conical edge cone; see Fig. 7.5. This new edge cone is not bounded by any facets at all. A new sphere B_r may be chosen such that it can "see" part of the vertex sphere B_q and no other part of the surface, including the new *link edge cone* created by the addition of B_q. This creates a second link edge cone between B_q and B_r and we can repeat this process indefinitely as long as each successive sphere shrinks to fit outside the body, but it "sees" only a portion of a vertex sphere. (It is in this way that link edge cones are responsible for spherical subsets of the convex hull which may not be simply connected.) Note that link edge cones do not touch facet planes. For the remainder of this paper *vertex*, *edge*, and *facet*

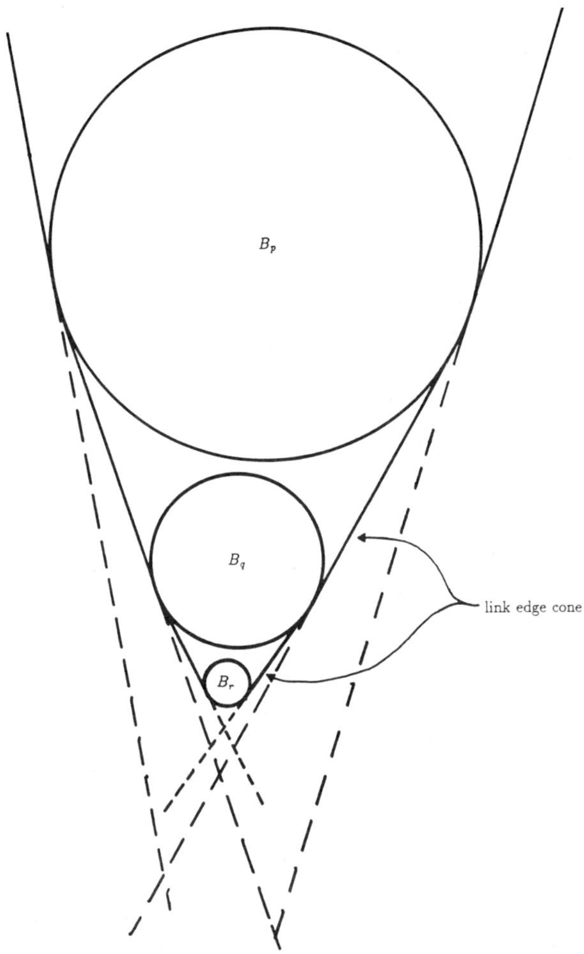

FIG. 7.5. *A two-dimensional cross section of an example where link edge cones can exist in the convex hull of spheres.*

will refer to tessellation components of the surface being constructed, i.e., *not* of the convex hull of the original set P.

7.2.3. Calculation of tessellation components. Again, initially we restrict every vertex sphere radius to be equal ($\epsilon = \epsilon_p = \epsilon_q = \epsilon_r = \ldots$ etc.). Recall the surface $\mathcal{B} := ch(\bigcup_{p \in P} B_p)$ and visualize the components as follows. Consider any connected planar subset of this convex surface. The corresponding facet plane must touch at least three vertex spheres B_p, B_q, and B_r. Define B_f to be the *facet plane* tangent to the vertex spheres B_p, B_q, and B_r of that facet, which positively supports every vertex sphere, where the positive support of a facet plane is taken to be the half space containing the set P. The facet plane will be parallel with the plane of centers of the vertex spheres because the centers lie on one side of the plane and are all the distance ϵ from it. Then the *facet patch* $\hat{B}_f$, the trimmed surface cut from B_f, is the convex hull of the points of tangency of the plane B_f and the vertex spheres of that facet. That is, the facet patch is a triangle (in general) with

$$x_{fp} := B_f \cap B_p, \qquad x_{fq} := B_f \cap B_q, \qquad x_{fr} := B_f \cap B_r$$

as vertices. Data storage should be such that the vertices in every facet are ordered consistently, so that

$$n_f = \frac{(c_q - c_p) \times (c_r - c_p)}{\|(c_q - c_p) \times (c_r - c_p)\|}$$

is the inward pointing unit normal to the facet patch. The tangent points are then

$$x_{fp} = c_p - \epsilon_p n_f, \qquad x_{fq} = c_q - \epsilon_q n_f, \qquad x_{fr} = c_r - \epsilon_r n_f.$$

Next consider any two vertex spheres B_p and B_q sharing facet planes B_f and B_{f_1} and define $\mathcal{T}_{pq}$ to be the edge cylinder uniquely determined by the two spheres. This cylinder is tangent to the vertex spheres in great circles of the vertex spheres. The *edge patch* $\hat{\mathcal{T}}_{pq}$, the trimmed surface cut from $\mathcal{T}_{pq}$, is the subset of $\mathcal{T}_{pq}$ bounded between these two great circles and the two facet patches of the edge; see Fig. 7.6. This is a four-sided patch with corners x_{fp}, x_{f_1p}, x_{fq}, and x_{f_1q}. The side of the patch between x_{fp} and x_{f_1p} is the arc of the great circle of B_p in the plane of c_p, x_{fp}, and x_{f_1p}. The side between x_{fp} and x_{fq} is the straight line between the two points, and the cylinder $\mathcal{T}_{pq}$ and the plane B_f meet tangentially along this line. Note that some of the special cases (cf. §7.2.2) that arise in the convex hull of arbitrary spheres cannot occur when the radii of the spheres are equal except for the extremely degenerate case when the centers of such (three or more) spheres are colinear. In this case the intermediate spheres meet the surface (a cylinder) in circles only, i.e., in subsets of empty area. Accordingly we ignore such an extreme case.

Finally, for any vertex sphere B_p of $\mathcal{B}$ the *vertex patch* $\hat{B}_p$ is the spherical m-gon on B_p with the tangent points of B_p and its m adjacent facet planes as vertices, and the great circle arcs between B_p and its m adjacent edge cylinders as edges. Examples of these patches may be seen in Figs. 7.1(a)–7.1(c) with further examples of vertex patches in Figs. 7.14(a)–7.14(f). We can now describe the surface $\mathcal{B}$ as the union of these three types of patches.

7.2.4. Patches as maps of $\mathcal{R}^2$. Because the surface $\mathcal{B}$ is convex, tessellation divides $\mathcal{B}$ into a family of cusp free, nonintersecting patches bounded by straight lines and great circles on planes, cylinders/cones, and spheres. Representations for such patches may be found in [3] and [12]. Planes, cylinders/cones, and spheres are regular (cusp-free) surfaces [4] and thus a neighborhood of each point on each surface is the image under some homeomorphism (bicontinuous and one to one) $\mathbf{x} : D \subset \mathcal{R}^2 \to \mathcal{R}^3$ of a domain D. Each patch is

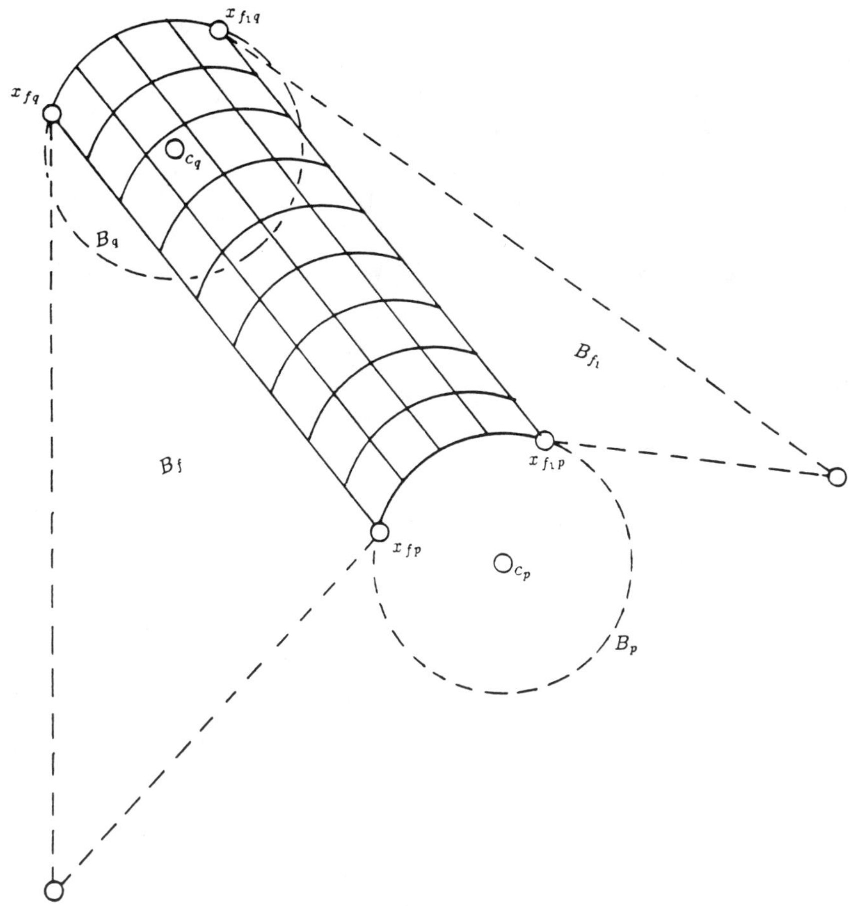

FIG. 7.6. *An edge patch when vertex spheres are equal and facet spheres have infinite radius, i.e., a subset of a cylinder.*

the image of a subset of the inverse image of the surface from which the patch is trimmed. Because the boundary of each patch is composed of line segments or circular arcs which intersect with each other only at their common tangent points, i.e., the x_{fp}, the domain of each patch is a simply connected subset of D bounded by a simple closed curve. Later (in §7.6) we will utilize a different domain for our patches, but for now this shall suffice.

7.2.5. Summary. We specifically chose the vertex sphere radii ϵ_p such that the surface $\mathcal{B}$ would possess the tangent plane T_p at p for every $p \in P$ (cf. §7.2, ϵ_p restrictions), hence the surface interpolates the data. Furthermore, this surface is clearly convex by the definition of the convex hull.

This surface is also tangent plane continuous. This should be intuitive because we are taking the convex hull of the vertex spheres, bodies which are smooth initially. Over each patch this is obvious because spheres, planes, and cylinders are smooth. There are then three types of boundaries to consider: vertex–facet, vertex–edge and facet–edge. Each consists of points at which the tangent plane of the patch on either side of the boundary is identical because the surfaces from which the patches are cut are tangent along this boundary cut. Thus every supporting plane which touches the convex hull of the vertex spheres must be tangent to the convex hull because it is tangent to the sphere, cylinder (or cone), or plane at the points where it touches the surface. A more detailed proof of tangent plane continuity may be found in [10].

Although the constructed surface possesses most of the desired characteristics, it is poor from a design standpoint because the distribution of curvature is very uneven and rigid. While a greater curvature near vertices seems intuitive, it is aesthetically unpleasant that the curvature should be zero except at the vertices, or that a designer should be limited to planar patches. To modify this, we will now generalize the types of surfaces from which we cut the patches.

7.3. Varying Facet Sphere Radii (Equal Vertex Spheres)

Replacing the flat facet planes with more general surfaces will require different surfaces to fit the edges where facet planes meet. In particular, the torus becomes an essential tool for surface fitting.

7.3.1. The torus. A torus may be defined as the surface of revolution of a circle about some line in the plane of the circle, or the envelope of a sphere revolved about some line. This line is the *axis* of the torus. The radius of the revolving circle (or sphere) is the *tube radius* of the torus. The center of this revolving circle (sphere) sweeps out the *center circle* of the torus, and the radius of the center circle is the *radius of the torus*. The *plane of the torus* is the plane of the center circle of the torus and intersects the axis of the torus perpendicularly at the *center* of the torus.

Remark. The envelope of a revolved sphere is the same as the envelope of the sphere being rolled along a circular path.

7.3.2. Facet spheres, facet tori, and tangent points. Instead of considering each B_f as a plane tangent to the vertex spheres of a facet, think of it as a sphere of infinite radius, $r_f = \infty$. Let us examine what happens as r_f decreases from infinity through some finite values and the shrinking sphere remains tangent to the same vertex spheres. As we vary these radii in a continuous fashion we are *continuously deforming* the surface in a smooth manner. The most important constraint on such a deformation is that we maintain the initial tessellation of the surface because only then will our intuition be free to examine the surface behavior locally, knowing how the different patches meet. To preserve convexity we only consider the case where the *facet sphere* contains the vertex spheres of that facet.

Remark. There is a very special case which occurs when tangent planes to four or more data points and vertex sphere radii are chosen so that (1) these vertex spheres share a facet plane, and (2) their tangent points do not lie on a circle. In this case the facet sphere is constrained to remain a plane because in general a varying sphere cannot be kept tangent to four spheres.

Otherwise, in the generic case, there exists a unique torus, $\mathcal{T}_f$, defined by the three vertex spheres of the facet f. This torus is defined by a center circle and a tube radius. The circle is uniquely defined by the centers of the vertex spheres, say c_p, c_q, and c_r, and the tube radius is $\epsilon(=\epsilon_p=\epsilon_q=\epsilon_r)$.

What happens to the tangent point of a facet sphere and one of its vertex spheres as the radius of the facet sphere changes? *If the facet sphere remains tangent to the three vertex spheres then it must remain tangent to this torus.* The tangent point of B_p and B_f can be calculated as the unique point on the line of centers of the vertex and facet sphere, a distance ϵ_p from the center of the vertex sphere and r_f from the center of the facet sphere. This point is also the intersection of the facet sphere and the ray from the center of the facet sphere through the center of the vertex sphere. Thus,

$$x_{fp} = r_f \frac{(c_p - c_f)}{\|(c_p - c_f)\|} + c_f = \frac{\epsilon_p}{r_f - \epsilon_p}(c_p - c_f) + c_p. \tag{7.1}$$

From the one extreme of $r_f = \infty$ and B_f being a plane atop the torus as before, to the other extreme of $r_f = r_{f_{\min}}$ (the radius of the torus added to the radius of the vertex spheres) and B_f being the sphere of minimum radius containing the three vertex spheres, the point of tangency between a facet and vertex sphere must travel along a quarter arc of a great circle of the vertex sphere because this point is always on the surface of the torus. During this transition the center of the facet sphere, c_f, travels from infinity to the center of the torus. The point c_f may be calculated as follows. Because the facet sphere is symmetric with respect to the vertex spheres (since they are equal), the center must lie on the axis of the torus. If $c_{\mathcal{T}}$ is the center of the torus, there exist only two points on the line $\{c_{\mathcal{T}} + \lambda n_f | -\infty \le \lambda \le \infty\}$ (n_f defined in §7.2.3) and the distance $r_f - \epsilon$ away from all three centers c_p, c_q, and c_r. Since n_f is inward pointing we take the value of $\lambda \ge 0$. Observe that the tangent points vary continuously with the radius of a facet sphere. The trace of the point of tangency between a facet sphere B_f and a vertex sphere B_p as the radius r_f of the facet sphere varies is called the *path* of x_{fp}; explicitly $\{x_{fp}(r_f) | r_{f_{\min}} \le r_f \le \infty\}$.

Now we have a one parameter family of spheres, $r_{f_{\min}} \le r_f \le \infty$, from which to cut facet patches, but a stricter lower bound on the radius of each may be required to preserve the tessellation. We shall postpone a rigorous development of bounds to preserve interpolation and convexity until §7.5.

7.3.3. Patches and their boundaries. Suppose we choose some fixed facet spheres. For the edge patches, consider vertex spheres B_p and B_q sharing facet spheres B_f and B_{f_1}. These facet spheres will generally intersect in a circle, and cannot be fitted with a cylinder, but we can roll one vertex sphere to the other while maintaining tangency with the two facet spheres. Because the two facet spheres meet in a circle, the envelope of the rolling sphere is an arc of a torus, $\mathcal{T}_{pq}$; see Fig. 7.7. The axis of this torus is the line of centers of the two facet spheres $\{(1-\lambda)c_f + \lambda c_{f_1} | -\infty \leq \lambda \leq \infty\}$. The four sides of the edge patch $\hat{\mathcal{T}}_{pq}$ are circular arcs which are subsets of lines of curvature of the torus. Two are great circles of its vertex spheres as before. The other two are traced out as the tangent point between one facet and vertex sphere x_{fp} (or x_{f_1p}) is rotated to the tangent point between that facet sphere and the edge's other vertex sphere x_{fq} (or x_{f_1q}) about the axis of $\mathcal{T}_{pq}$.

The three-sided facet patch $\hat{B}_f$ has three such arcs; note that these arcs are not from great circles of the facet sphere B_f. To calculate such an arc one needs two endpoints and a center. (Generally two endpoints and a center determine *two* arcs of a circle, but later we will place constraints on the patches so that we always know which arc we mean.) The endpoints we have

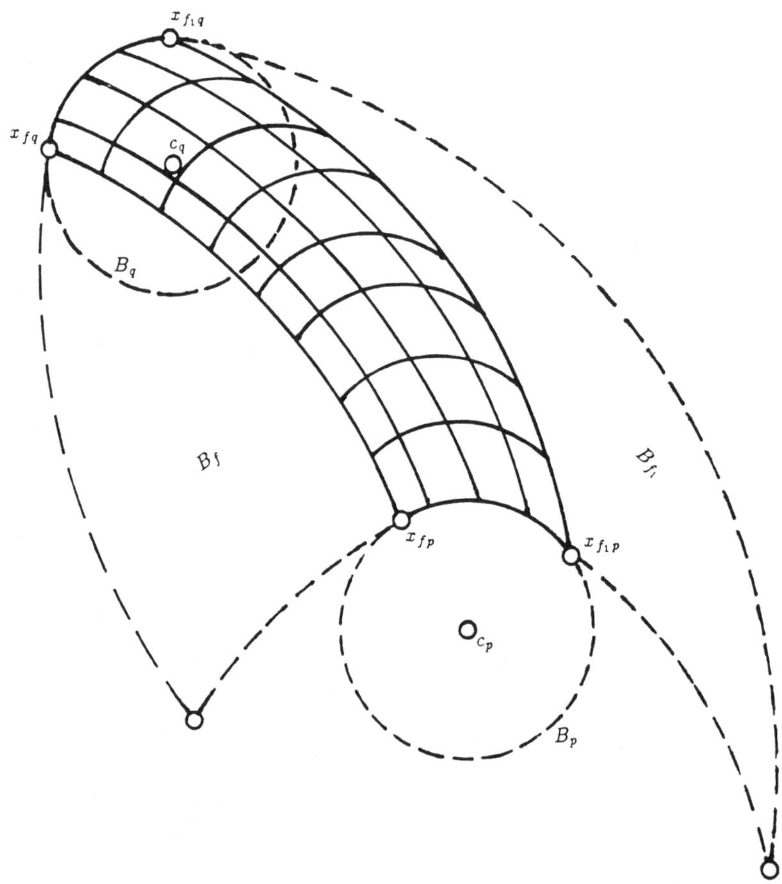

FIG. 7.7. *An edge patch with equal vertex spheres and finite radii facet spheres. Note that the edge torus is the rotation of either of the vertex spheres about the line of centers of the facet spheres.*

already (x_{fp} and x_{fq}), and the center is simply the closest point on the line of centers of the facet spheres to the line of the endpoints of the arc because the plane which contains the boundary arc is perpendicular to the axis of the edge torus $\mathcal{T}_{pq}$.

The vertex patch $\hat{B}_p$ is still a spherical m-gon with the tangent points with its facet spheres as vertices and great circles where it meets its edge tori for boundary arcs.

As we vary the radius of a facet patch the tangent points and boundaries continuously deform through a series of circular arcs. Provided certain conditions are met, these arcs deform "nicely," i.e., roughly so that the boundary of the inverse image of each patch (cf. §7.2.4) remains a simple closed curve. If this is preserved then our patches can still be consistently defined as the image of the simply connected region bounded by the curve.

7.3.4. Bounds on facet sphere radii. The primary reason to restrict the radii of the facet spheres is to insure that the surface tessellation remains the same, allowing us to speak of bounds (which are possibly more strict) such that the surface remains cusp free and locally non-self-intersecting, and to guarantee interpolation of the data points. Each of these considerations can be more clearly addressed in §7.5 after further generalization.

7.3.5. Summary. We take the union of the three types of patches in §7.3.3 for our generalization of $\mathcal{B}$. Instead of spheres, planes, and cylinders we have spheres and tori, with the original surface being the special case with every facet sphere of infinite radius. With certain restrictions on facet sphere radii, which will be explicit in §7.5.5, the new surface still possesses the tangent plane T_p at p for every $p \in P$ and thus it interpolates the data.

While the new surface is a better design tool than the original, it requires each vertex sphere to be of the same radius ϵ. Next we further generalize the descriptions of patch sources to allow each vertex sphere to have a different radius.

7.4. Unequal Vertex Sphere Radii

We begin, as before, by defining a vertex sphere B_p for each $p \in P$, but now each sphere is of a different radius $\epsilon_p > 0$, where each ϵ_p is chosen such that the vertex sphere lies within the tangent plane polyhedron. The convex hull of the vertex spheres is now composed of subsets of spheres and planes as in the initial case, and subsets of cones as mentioned in §7.2.1. It has a similar tessellation except for conical instead of cylindrical edge components and it is possible to have link edge cones as well. These unequal vertex sphere radii require generalized surfaces for the facet and edge tori which we now explore.

7.4.1. The Dupin cyclide. In §7.3.2 we considered the torus $\mathcal{T}_f$ (uniquely defined by the three vertex spheres of the facet f) to visualize the paths of tangent points between the B_f and a vertex sphere B_p. Now that ϵ_p, ϵ_q, and ϵ_r are not all equal; such a torus no longer exists. However, examination of the figure of the vertex spheres in intersection with the plane of centers of the vertex spheres suggests that there exists a canonical body containing the spheres and tangent to them along the same tangent paths that a tangent sphere would follow; see Fig. 7.8. It is basically a "bulging torus" (with the possibility of a pinch point or "pinch" curve). In fact, this surface satisfies the classical definition of a *Dupin cyclide* [6, p. 265–266], [12, p. 220].

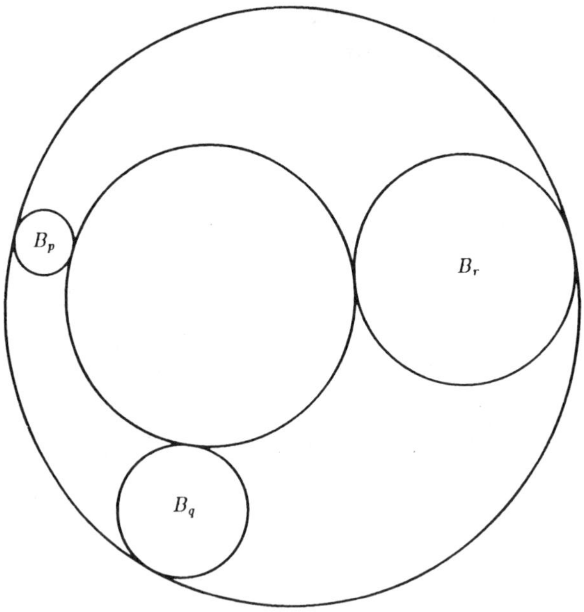

FIG. 7.8. *The intersection of three vertex spheres with their plane of centers. There is a natural generalization of the torus which fits them.*

A cyclide is the envelope of a variable sphere that touches three fixed spheres in a continuous manner.

In the case $\epsilon = \epsilon_p = \epsilon_q = \epsilon_r$ (as before) the torus $\mathcal{T}_f$ may by thought of as such a cyclide in two ways. There are two families of spheres whose common envelope is the torus. The first and more obvious family of spheres is the one whose members are all of the same radius ($\epsilon =$ the tube radius of the torus) and whose centers lie on the center circle of the torus, i.e., all those spheres "inside" the torus. The second family consists of spheres whose centers all lie on the axis of the torus. Every member of each family is tangent to every member of the other family. Furthermore, every member of each family is tangent to the torus along a circle which is a line of curvature of the torus.

If we fix three members of the first family (our three vertex spheres for example), then the second family represents the variable sphere (our facet sphere) in the definition of a cyclide. This variable sphere then becomes all balls containing all three vertex spheres and tangent to them, all balls not containing any of these three vertex spheres and tangent to them, and both planes tangent to the three spheres and not separating them. Thus the torus is the envelope of the second family and is a cyclide. Or we may take three members of the second family as our fixed spheres, and then the first family represents the variable sphere. This variable sphere becomes every sphere "inside" the torus. The torus is then the envelope of the first family and is a cyclide. The choice of three fixed spheres from either family results in the same torus.

If the radii of the vertex spheres are unequal then the envelope $\mathcal{T}_f$ is no longer a torus, but instead is a more general cyclide; see Figs. 7.9(a)–7.9(c). (For a given cyclide of this type both families of variable spheres are *Dupin series* of spheres for which there is a great deal of classical theory [6].) The facet sphere B_f is tangent to B_p, B_q, and B_r as before, but the plane of the tangent points x_{fp}, x_{fq}, and x_{fr} is no longer parallel to the plane of the centers of the vertex spheres. Thus, calculating these tangent points is no different from before, but calculating the paths on which they travel (as the radius of the facet sphere varies) and finding the center of the facet sphere is. There is a simple way to represent this surface $\mathcal{T}_f$, but first let us consider the other torus in the last construction.

For any edge (with vertex spheres B_p, B_q and facet spheres B_f, B_{f_1}) we defined $\mathcal{T}_{pq}$ to be the torus generated by rolling B_p to B_q, maintaining tangency with B_f and B_{f_1}. Since $\epsilon_p \neq \epsilon_q$, a torus will not suffice. A rolling ball whose radius varies from ϵ_p to ϵ_q is required. There are many cyclides (of the type mentioned above) which would fit the two vertex spheres because it takes three vertex spheres to uniquely determine such a cyclide. The additional restriction that the surface be tangent to both facet spheres determines a set of canonical variations for the radius of this rolling ball, and we will choose one of these. Exactly what bodies to use as generalizations of the edge and facet tori will require a detour through the land of inversive geometry.

7.4.2. Inversion. Excellent expositions of the following definitions and concepts may be found in [2], [6]–[8], and [11]. The discussion by Johnsson [11] is particularly direct.

Two circles are said to be *externally* tangent when they lie on opposite sides of their tangent line at their point of contact and *internally* tangent when they lie on the same side. If a circle is tangent to two others we say that it has *like* contact with them if it is either externally tangent or internally tangent to both; it has *unlike* contact if it is externally tangent to one and internally tangent to the other.

Given a point p and a circle C (sphere S) with center O and radius r, and any line through p intersecting C (S), the *power* of the point p with respect to C (S) is the product of the distances from the point to the points of intersection and is independent of the line. Thus we can always take the line through the center of the circle (sphere) and calculate the power of p as

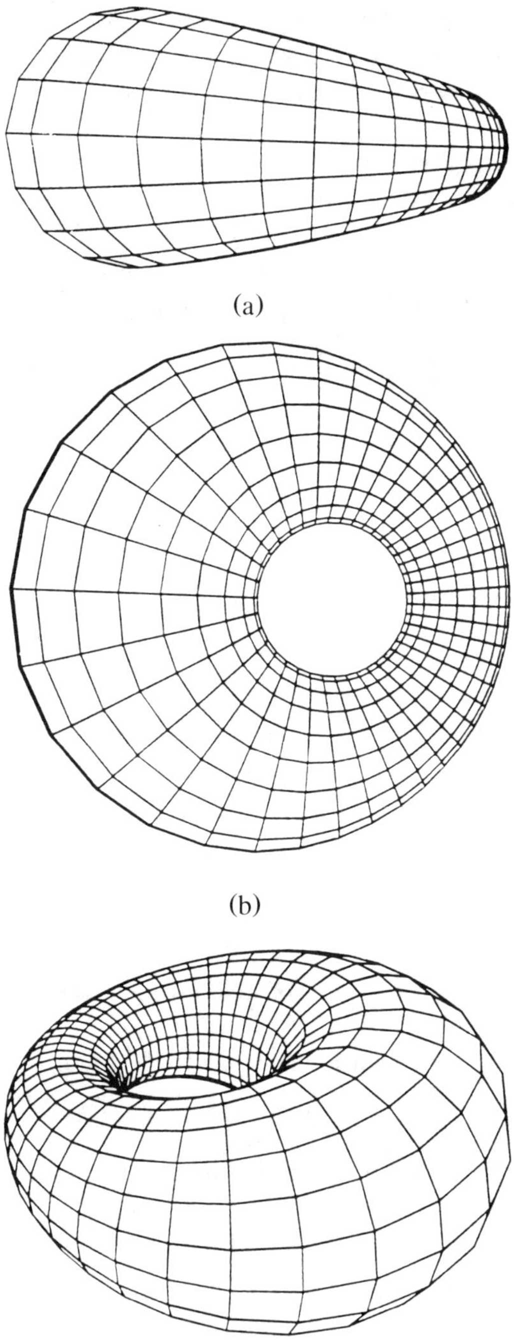

(a)

(b)

(c)

FIG. 7.9. (a) *Cyclide seen from a side view.* (b) *Cyclide seen from a top view.* (c) *Generic cyclide containing three unequal vertex spheres.*

$(\|O-p\|-r)\cdot(\|O-p\|+r)$. The *radical axis* (*plane*) of two circles (spheres) is the line (plane) of points having equal power with respect to the circles (spheres). The radical axes (planes) of pairs of three fixed circles (spheres) meet in a point (line) called the *radical center* (*axis*) of the three circles (spheres). The point of intersection of two such axes has equal power with respect to all three circles and thus lies on the third axis. (Of course, if the centers of the three circles or spheres are colinear then this point is the ideal point at infinity.) The point on the line of centers of two spheres and on their radical plane is the *radical center* of the two spheres.

Given a point p and a circle C (sphere S) with center O and radius r, the *inverse* of p with respect to C (S) is the point p' on the ray from O through p such that $\|O-p\|\cdot\|O-p'\|=r^2$. This is equivalent to

$$p' = O + r^2\frac{(p-O)}{\|p-O\|^2}.$$

Clearly this is an involution, i.e., p' is the inverse of p if and only if p is the inverse of p'. If we define the image of the center of inversion to be the ideal point and extend R^3 to R^3_∞ with this point, then this inversion is a conformal mapping [6, p. 228] of R^3_∞ onto itself, $I_S : R^3_\infty \to R^3_\infty$, and maps circles to circles and spheres to spheres (where lines and planes are circles and spheres containing the ideal point and line, respectively). However, the center of a circle (sphere) does not map to the center of its image circle (sphere) because the map is not linear. If K is some subset of R^3_∞, we write $K' = I_S(K)$, where the prime notation will always refer to the image of an inversion in the sphere S or the circle C. Given any two circles (spheres), the intersection of their line of centers with a line through the endpoints of parallel radii (line segments from the center to the boundaries) is their (*external*) *center of similitude*, and is independent of the angle the radii make with the line of centers. This is the center of a dilatation of one into the other. Clearly this point cannot lie between the two centers on the line through them. For any two circles (spheres) there always exists at least one *midcircle* (*midsphere*) which inverts the two given circles (spheres) into each other. In the older literature this is also referred to as the circle (sphere) of *antisimilitude*. The newer term is due to Coxeter [7, p. 121]. The center of a midcircle (midsphere) is the center of similitude of the two given circles (spheres). Because the two points nearest each other on the two circles (spheres) are mutually inverse, p and p', and we know the center of inversion O, we can calculate the radius from $\|O-p\|\cdot\|O-p'\|=r^2$. Clearly the midcircle (midsphere) of two equal circles (spheres) is the midline (midplane) of the two, which is partial justification for calling inversion in a circle or sphere *reflection* in a circle or sphere.

THEOREM 7.4.1. *The image of a cyclide under inversion in a sphere is another cyclide.*

Proof. Fix a sphere of inversion. Since spheres invert into spheres the three fixed spheres and one variable sphere in the definition of a cyclide are mapped to another three fixed spheres and one variable sphere. And because the map preserves coincidence, the image of the variable sphere will touch the images of the three fixed spheres in a continuous manner. Thus the envelope of the image of the variable sphere satisfies the definition of a cyclide.

Before we invert any of the cyclides let us consider some useful theorems.

THEOREM 7.4.2 [8, p. 119]. *Two nonintersecting [planar] circles can always be inverted into two concentric [planar] circles.*

Thus two nonintersecting spheres can always be inverted into two concentric spheres.

THEOREM 7.4.3 [11, p. 97]. *Any two [planar] circles can be transformed by inversion into equal [planar] circles.*

Thus any two spheres can be transformed by inversion into equal spheres.

This is accomplished by inversion in a circle (sphere) centered on the midcircle (midsphere) of the two circles (spheres). Intuitively, the midcircle (midsphere) is mapped to a line (plane) through which the two circles (spheres) are each reflections of the other.

THEOREM 7.4.4 [11, p. 56]. *The inverse of the center of a given circle is the same as the inverse of the center of inversion with regard to the circle inverse to the given circle.*

Thus (in terms of spheres) if spheres K and K' with centers O_K and $O_{K'}$, respectively, are mutually inverse with respect to the sphere S with center O, then

$$I_S(O_K) = I_{K'}(O). \tag{7.2}$$

COROLLARY 7.4.1 [11, p. 56]. *To construct the inverse of a given circle by locating its center, find the inverse of the center of inversion with regard to it, and then invert this point with regard to the circle of inversion.*

Again in terms of spheres, we have

$$O_{K'} = I_S(I_S(O_{K'})) = I_S(I_K(O)). \tag{7.3}$$

7.4.3. Circles of tangency (boundary arcs). Every boundary curve of the tessellation is a subset of the intersection where two surfaces (associated with connected subsets of the surface $\mathcal{B}$) meet tangentially, either facet–edge or vertex–edge, both of which are the intersection of a sphere and a cyclide. Likewise the path traced on a vertex sphere by its tangent point with a facet sphere (as the radius of the facet sphere varies) is the intersection of a sphere with a cyclide. Thus it is often necessary to calculate this locus of points. This section explains that such a locus is indeed a subset of a circle by demonstrating that it is the intersection of two spheres. The center of this circle can be calculated as the radical center of the two spheres.

Consider again the facet sphere B_f with vertex spheres B_p, B_q, and B_r of three unequal radii as the fixed spheres of a cyclide, $\mathcal{T}_f$, which is the envelope of a variable sphere tangent to all three in a continuous manner. If $\mathcal{T}_f$ can be inverted into a torus $\mathcal{T}_f'$, this is optimal. In such a case, as we saw in §7.3.2, since B_p', B_q', and B_r' are equal, the image of the variable facet sphere B_f' is tangent to the images of the vertex spheres along great circles of these images. Since circles are mapped to circles (because every circle is the intersection of two spheres which are mapped to two intersecting spheres), the inverse of the great circle $\mathcal{T}_f' \cap B_p'$ is a circle on B_p with a center other than c_p. If $\mathcal{T}_f$ cannot be inverted into a torus, an inversion is still useful, and one can examine characteristics of the cyclide through those of its image.

Let us examine the intersection of the cyclide with the plane of centers of the three vertex spheres; recall Fig. 7.8. Intersecting the vertex spheres with this plane yields three great circles: C_p, C_q, and C_r. Because the convex hull of the set of all vertex spheres has a facet plane tangent to B_p, B_q, and B_r, we know that *none of the three circles can lie such that every line segment between it and one of the other two intersects the third circle.* This is a greater restriction than that none of the three lies in the convex closure of the other two. In particular we have that none of the three circles can contain either of the others. So, because none is in the convex closure of the other two, there always exists a circle of minimum radius containing and touching all three circles; call this C_1. This circle is internally tangent to the given three and is a great circle of the sphere, S_1, of minimum radius which contains all three vertex spheres, and this sphere is a member of the variable family of the cyclide. Being internally tangent to C_p, C_q, and C_r, C_1 has like contact with all three. According to Coolidge [6, p. 168] the condition that none of the circles contains either of the others ensures that there exists a second real circle having like contact with the circles C_p, C_q, and C_r. This is a subset of the larger problem of Apollonius; see [6] and [11]. If the discs bounded by these circles have empty intersection, then this second circle is external to the three given circles. If the discs have a nonempty intersection, then the second circle is internal to the given three. If the intersection of the discs is a single point, the point can be considered an internally or externally tangent circle of zero radius. See Fig. 7.10 for examples of these cases. In any case, this circle, C_2, is a great circle of a sphere, S_2, which is a member of the variable family of the cyclide and this sphere may be thought of as the smallest such member. The sphere S_1 is the smallest sphere containing the cyclide, and S_2 is the sphere fitting the "hole" in its "center." Depending on how C_1 and C_2 intersect we can greatly simplify the cyclide with an inversion. Either they are nested and separated, nested and tangent, or intersect at two points; see the left column of Fig. 7.11. (Because C_1 contains C_p, C_q, and C_r, and because C_2 touches C_p, C_q, and C_r, C_1 must contain at least three points of C_2. Thus the only cases are the three mentioned.)

In the first case we invert S_1 and S_2 into concentric spheres in a sphere of inversion S with center O at the external center of similitude of the two (cf. Theorem 7.4.2), see the top row of Fig. 7.11. Observe that the image

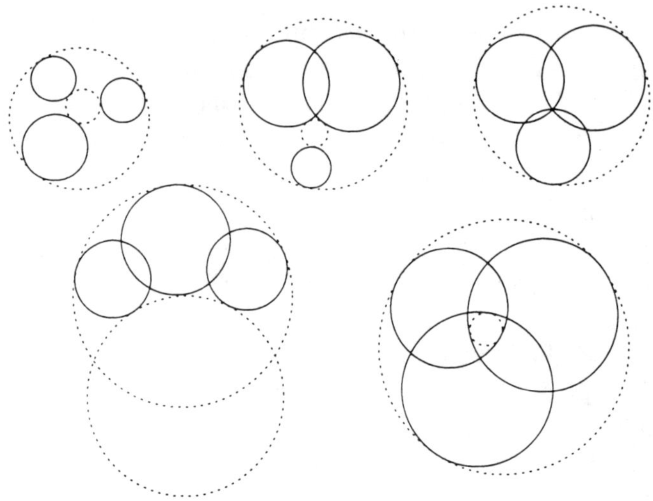

FIG. 7.10. *Various ways that two circles (dashed lines) can have like tangency with three given circles (solid lines) provided that none of the three contains either of the other two given circles.*

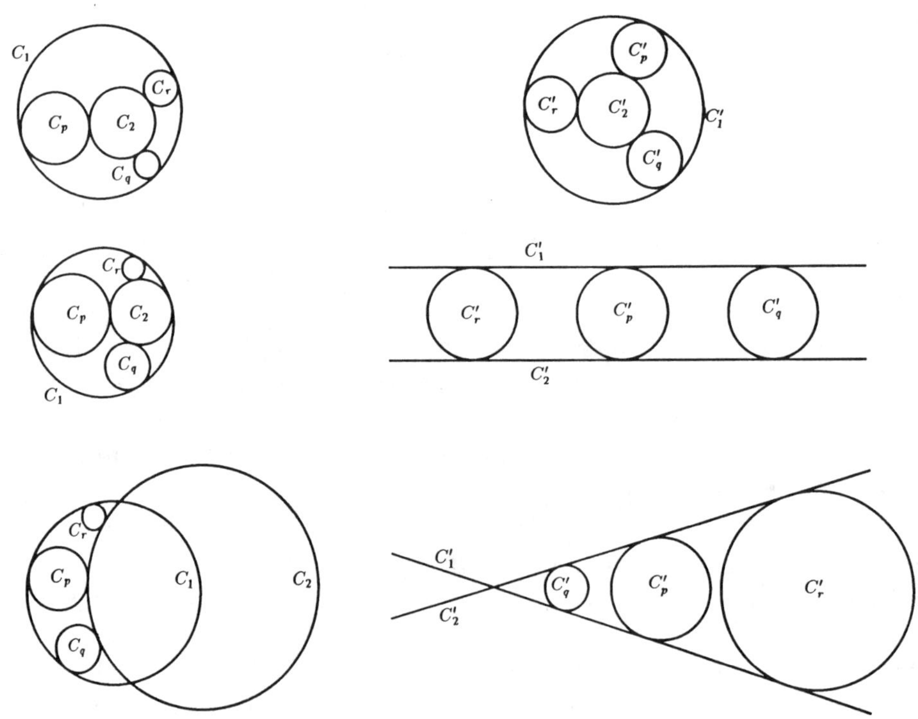

FIG. 7.11. *The left column represents the three possible types of cyclides which fit any three given spheres sharing a planar subset of the convex hull. The right column is the respective images of these three cyclides under a special inversion meant to simplify them.*

of the cyclide is now a torus and thus the image of the circle of tangency (between $\mathcal{T}_f$ and B_p) is a great circle of the image B'_p of the vertex sphere B_p (because the torus is the envelope of the revolving sphere). This great circle is the intersection of the image vertex sphere and H, the plane determined by the four points $x_{fp}(\infty)'$, $x_{fp}(r_f)'$, $O_{B'_p}$, and ∞. Thus the circle of tangency is the intersection of B_p and $I_S(H)$, where $I_S(H)$ is the sphere determined by $x_{fp}(\infty), x_{fp}(r_f), I_{B_p}(O) = I_S(I_S(I_{B_p}(O))) = I_S(O_{B'_p})$ (from (7.3)), and $O = I_S(\infty)$. Observe that these last two points are fixed, i.e., they are not functions of the facet sphere radii.

In the second case we invert the cyclide in a sphere S with center O at the point of tangency of S_1 and S_2; see the center row of Fig. 7.11. The images of C_1 and C_2 are two parallel lines with the three equal images of the circles C_p, C_q, and C_r between them. The images of S_1 and S_2 are planes, and the image of the cyclide is a cylinder with the three image spheres inside. Thus the image of the circle of tangency is a great circle of one image sphere. As in the first case, we choose the image of the plane of this great circle as the second sphere from which to generate the radical center of the circle of tangency.

For the last case the intersection of the cyclide with the plane of centers of the vertex spheres is the two intersecting circles C_1 and C_2; see the bottom row of Fig. 7.11. Call these two points of intersection a and b, and invert in sphere S with center $O = a$. The images of C_1 and C_2 are straight lines intersecting at the image of b with the images of C_p, C_q, and C_r in one sector. The image of the cyclide is a double semi-infinite cone with vertex $I_S(b) = b'$, and the images of the three spheres are spheres in (one half of) the cone, and their centers all lie on the axis of the cone. The image of a circle of tangency is not a great circle, but it is the intersection of B'_p and a sphere K' (which we are now defining in the image space) centered at b' and orthogonal to the cone and B'_p where they meet. Thus the pre-image K of this sphere K' will intersect B_p in its circle of tangency with the cyclide. The center of this image (i.e., the sphere we are looking for) is $I_S(I_{K'}(O))$ (from equation (7.3)) and its radius is the distance between this center and x_{fp}.

Hence in all cases the locus of the points of tangency between a sphere and cyclide is a circle and one can calculate its center via one of the three previous constructions.

Computationally it is far simpler to compute the circle of tangency between a vertex sphere and a facet cyclide by computing $x_{fp}(r_f)$ for three different values of r_f. Every other circle of tangency on the surface will be the intersection of a sphere with a cyclide which is the inversion of a torus, i.e., the second and third cases above are not strictly necessary to consider for computational purposes.

7.4.4. Centers of facet spheres and tangent points. Finding the centers of the facet spheres is simple. If there exists an inversion such that the image of the cyclide $\mathcal{T}_f$ is a torus $\mathcal{T}'_f$ (i.e., as required by Theorem 7.4.2, $C_1 \cap C_2 = \emptyset$), then the center of the image of the facet sphere is on the perpendicular to the plane of the torus through the center of the torus. Applying (7.3) to the center of the image of the facet sphere gives the center of B_f, $O_{B_f} = I_C(I_{B'_f}(O))$. However, there is a simpler way to calculate this center. It must satisfy the set of equations

$$(7.4)\quad \|O_{B_f} - c_p\| = r_f - \epsilon_p, \quad \|O_{B_f} - c_q\| = r_f - \epsilon_q, \quad \|O_{B_f} - c_r\| = r_f - \epsilon_r,$$

which is equivalent to O_{B_f} being a point of intersection of three spheres centered at c_p, c_q, and c_r of radii $r_f - \epsilon_p$, $r_f - \epsilon_q$, and $r_f - \epsilon_r$, respectively. These three spheres intersect pairwise in their radical planes, which intersect in the radical axis of the spheres. This line contains the radical center of the great circles (of the vertex spheres intersected with the plane of their centers) and intersects each sphere twice (once if $r_f = r_{f_{\min}}$). Solving for where this line intersects the sphere of radius $r_f - \epsilon_p$ centered at c_p yields two points. We want the one in the positive support of the plane of the centers of the spheres. With the vertices ordered so that n_f is inward pointing, this is easy. Thus, to reiterate, O_{B_f} is the point on the ray at the radical center of the three circles (intersections of vertex spheres with the plane of their centers) in the positive support of the plane of centers, perpendicular to the plane of centers, and $r_f - \epsilon_p$ from c_p. In practice this is much simpler than performing the double inversion of (7.3), although it may be less intuitive.

It may be the case that the cyclide $\mathcal{T}_f$ cannot be inverted into a torus (i.e., $C_1 \cap C_2 \neq \emptyset$). This torus was just a device to facilitate visualization of the tangent paths and the centers of facet spheres. The calculation (from (7.4)) of the centers of a facet sphere is clearly independent of being able to invert the cyclide into a torus because there always exists a minimum sphere containing the three vertex spheres of the facet. Calculating the center of the arc of the tangent path $x_{fp}(r_f)$ in these other cases was covered in §7.4.3.

Together with (7.1) we can now solve for the tangent points which are still the endpoints of the boundary arcs.

7.4.5. Edge cyclides. Recall the edge cyclide $\mathcal{T}_{pq}$ with unequal vertex spheres B_p, B_q and facet spheres B_f, B_{f_1}. We can invert B_p and B_q into two equal spheres (in a point on their midsphere B_{pq} by Theorem 7.4.3) which can then be fit with some torus called the *image torus* because it is the image under inversion of the desired cyclide; inverting this image torus yields an edge cyclide. The term *image torus* will be reserved for the inversion in a sphere of an edge cyclide. The choice of this torus is not unique because the two equal spheres (the images of the two vertex spheres) can be fit with many tori. In the previous case ($\epsilon = \epsilon_p = \epsilon_q$) the torus was the envelope of an ϵ-ball as it rolled from B_p to B_q (tangent to B_f and B_{f_1}), where this rolling was equivalent to revolving B_p about the line of centers of B_f and B_{f_1}. Thus (because tangency is preserved) the image torus must be tangent to B'_f and B'_{f_1}, i.e., its axis is the line of the centers of B'_f and B'_{f_1}. This is enough to uniquely determine a torus once we have fixed an inversion, but we have many choices for an inversion and will choose it to satisfy two more conditions.

7.4.5.3. Radial monotonicity. The first condition on the edge patch is that the arc between B_p and B_q possess *radial monotonicity*, i.e., the maximum and minimum radii of the ϵ-ball as it rolls from B_p to B_q are ϵ_p and ϵ_q. This is a desirable characteristic because we would like to not introduce curvature variations into the edge patch that are not explicit in the two vertex and two facet spheres, i.e., radial monotonicity gives a minimal variation of the radius of the rolling ball. A sufficient condition to insure radial monotonicity of the patch is that the centers of B'_p and B'_q are not separated by the plane through the center, O, of the sphere of inversion, S, and the axis of the image torus, because the centers of the maximum and minimum interior spheres of the cyclide are on this plane. The axis of the image torus is the line of centers of the images (under inversion) of the facet spheres sharing the vertex spheres. Thus the plane of O and the axis of the image torus is the plane of O and the centers of the images of the facet spheres, which is also the plane of O and the centers of the facet spheres. So radial monotonicity is guaranteed if and only if the plane of O, c_f, and c_{f_1} does not separate c_p and c_q.

The midsphere, B_{pq}, of B_p and B_q separates their centers c_p and c_q so that B_{pq} and the line through c_p and c_q intersect in two points, one between c_p and c_q, and one which is *not* between them and is on the side of the line of the smaller vertex sphere. If O is chosen as this second point, then the plane of O, c_f, and c_{f_1} intersects the line through c_p and c_q in O which is not between c_p and c_q, and thus there exists a radially monotonic arc of the edge cyclide between the two vertex spheres.

Of course, there are two arcs of the cyclide "between" B_p and B_q (refer to Fig. 7.8), and only when these spheres are the smallest and largest interior spheres of the cyclide will both arcs be radially monotonic. When this is not the case, we never want to choose the radially nonmonotonic arc. Initially when every facet sphere is a plane and the edge cyclide is a cone, it is clear that our patch is from the shorter "arc." As we decrease the facet sphere radii and continuously deform an edge cyclide we continue to have the shorter, radially monotonic arc until the line through c_f and c_{f_1} crosses the line through c_p and c_q. When these two lines intersect, then the vertex spheres are the extreme interior spheres of the cyclide. If either of the facet sphere radii decrease further, then we no longer obtain the radially monotonic arc. If the tessellation of the convex hull of the vertex spheres is the same as the tessellation of the convex hull of the *centers* of the vertex spheres, then the previous bounds of the $r_{f_{\min}}$ are sufficient to guarantee that the line of centers of the facet spheres does not cross the line of centers of the vertex spheres. However, if the tessellations are different (cf. examples of §7.2) then the radii of the facet spheres may require further restriction (specified in §7.5.1.3). Even when the edge cyclide is a torus and radial monotonicity is automatic it is sensible to impose such a bound because then it is symmetric with the generic edge cyclide case, and we are guaranteed that the arcs of the cyclide (torus) belonging to facet patches are less than π and thus we can consistently choose the correct patch from the surface of the cyclide (cf. §7.6.3 for precise bounds).

7.4.5.4. Continuity of the map. Continuity of the map means that as ϵ_p and ϵ_q approach some common ϵ, the generated edge cyclide approaches a torus. Radial monotonicity is sufficient to ensure this continuity because as the two vertex spheres become equal, their midsphere approaches their midplane, and the second intersection of their line of centers with this midsphere approaches the ideal point. As the center of inversion approaches the ideal point the inversion approaches the identity map and thus the inverse image of the image torus of the cyclide approaches a torus.

7.4.6. Patches. Again, the edge patch $\hat{\mathcal{T}}_{pq}$ to be cut from the edge cyclide $\mathcal{T}_{pq}$ is a four-sided patch; see Fig. 7.12. These four sides are arcs of circular lines of curvature of the cyclide. Examine first the sides bounding the vertex spheres. Since the edge patches are always lifted from cyclides which have been inverted into tori, we can use the first case of §7.4.4 to calculate the circles of tangency, i.e., the boundary arcs. Because the same cyclide is determined if we roll an ϵ-ball from B_p to B_q while tangent to and *inside* B_f and B_{f_1}, or if we roll an ϵ-ball from B_f to B_{f_1} while tangent to and *outside* B_p and B_q

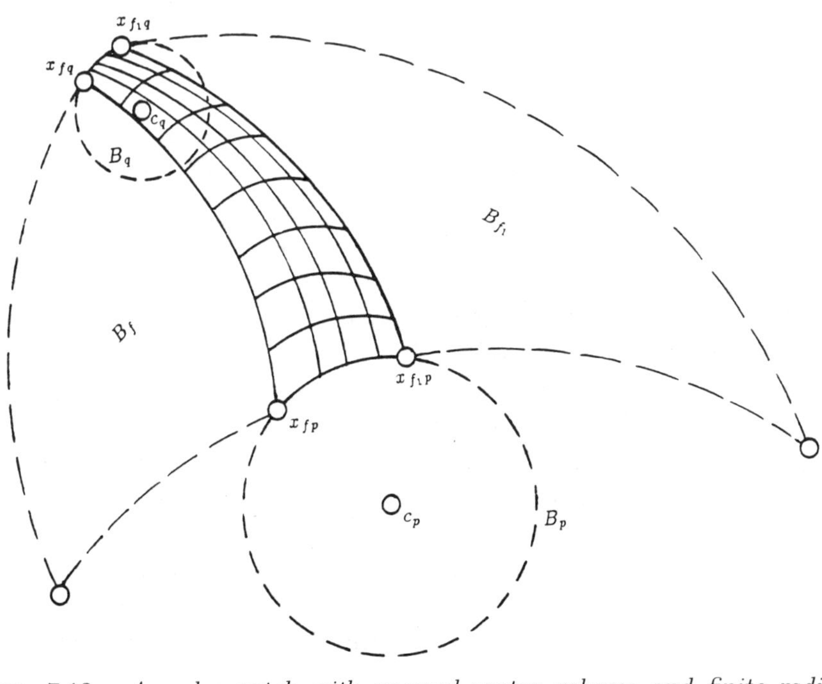

FIG. 7.12. *An edge patch with unequal vertex spheres and finite radii facet spheres.*

(while the radius of the rolling ϵ-ball varies appropriately for each case), that construction is symmetric in its treatment of the vertex spheres and the facet spheres. That is, to calculate the remaining two sides of the edge patch which are shared with facet patches, where we have the endpoints x_{fp} and x_{fq} and need the center of the arc, we can apply the same technique by fixing a sphere of inversion S, with center O, mapping B_f and B_{f_1} to equal spheres. We can then consider that the image torus $\mathcal{T}'_{pq}$ is formed by "rolling" B'_f to B'_{f_1} tangent to B'_p and B'_q.

Until we varied the radii of the vertex spheres it was unnecessary to consider link edge patches because they did not exist. Now, however, they are possible. A link edge patch, by definition, is the conical subset of the convex hull of its two vertex spheres. The boundary where a link edge patch meets either of its vertex patches is a circle on the surface of the vertex sphere and is straightforward to calculate.

Facet patches are determined as before, i.e., their boundaries are the arcs they share with the edge patches adjacent to them. After specifying the facet and edge patches, what are we left with for vertex patches? The m points of intersection of a vertex sphere and its m facet spheres are still the vertices of a spherical m-gon. The edges of this m-gon are no longer arcs of great circles, but arcs of circles specified above where the edge cyclide and vertex sphere are tangent.

There is a special case of vertex patches which can occur. If a vertex sphere has a link edge, then the vertex patch is *not* simply connected, i.e., it has a "hole" where the link edge meets it, and the domain of the patch will likewise have a "hole." Thus it will have *two* boundaries, one exterior composed of tangent arcs with edge cyclide patches as usual, and one interior which is the circle where the link edge meets the vertex sphere. Note that a vertex sphere may have many such "holes" if it has many link edges.

7.4.7. Summary. Again, take the union of these facet, edge, and vertex patches for the generalization of $\mathcal{B}$. Once the radii of the facet spheres are fixed, the surface is uniquely determined. The new surface is an interpolant because it possesses the tangent plane T_p at p for every $p \in P$ (cf. §7.5.5). Tangent plane continuity follows because the surfaces from which the patches are cut are tangent along patch boundaries. This surface possesses a better distribution of curvature than its predecessors, and it is far more flexible, allowing unequal vertex spheres and variable facet spheres.

7.5. Bounds on the Radii of Facet Spheres

Two things should be intuitively clear. First, when every facet sphere radius is very large (i.e., its curvature is within some suitably small neighborhood of zero) the surface behaves nicely. Second, if any of the radii become too small things may go wrong. We would like to characterize what can go wrong and what "too small" is, and then restrict the facet sphere radii from such cases. The rest of this section is concerned with specifying terminology relating the facet sphere radii to the degeneracy of all three types of patches (§7.5.1), and outlining a geometric algorithm for continuously decreasing the radii in a manner that preserves the tessellation of our surface (§7.5.4) and the interpolation of the initial data (§7.5.5). We then argue that the algorithm is sufficient to prevent the degeneracy of the patches under the types of continuous deformations associated with varying the facet sphere radii (§7.5.6).

Until now, this chapter has focused on establishing the reader's intuition. The rest of §7.5 establishes a technical description of this intuition. Section 7.6 applies this technicality to defend shape characteristics of the surface.

7.5.1. Degenerate patches. Because maintaining the tessellation of the surface allows us to describe the deformed surface in terms of the original convex hull of spheres, we require a technical description of how this tessellation can collapse. We shall say that a patch is *nondegenerate* if it is the image under a bicontinuous, one to one map of a simply connected domain of $\mathcal{R}^2$ bounded by a simple closed curve which maps the boundary of the domain to the boundary of the patch (cf. §7.2.4). Otherwise the patch is *degenerate*. Since, for our surface, our patches are cut from portions of surfaces that do not self-intersect, this is equivalent to saying that a patch is degenerate if any of two boundary arcs intersect except pairwise at the vertex–facet tangent points which are the corners of the patch, or equivalently that its inverse image (under the map from $\mathcal{R}^2$ as in §7.2.4) has a nonsimple boundary.

We need to be sure that the patches can always be defined as being on one side of their boundary arcs, i.e., that the "inside" of the domain is bounded by a simple closed curve and thus clearly defined. As long as the patches remain nondegenerate, this definition of the patches is clear and will allow us to make certain statements about the convexity of the surface. Before we describe how patches can become degenerate we must develop necessary terminology.

7.5.1.5. Interior angles of patches. Let be given a facet sphere B_f with vertex spheres B_p, B_q, and B_r, and facet spheres B_{f_1} sharing B_p and B_q, and B_{f_2} sharing B_p and B_r, and let $\mathcal{T}_{pq}$ and $\mathcal{T}_{pr}$ be the appropriate edge cyclides; see Fig. 7.13. Consider the point x_{fp} of contact of B_f, B_p, $\mathcal{T}_{pq}$, and $\mathcal{T}_{pr}$, and the tangent plane $T_{x_{fp}}$ to all four surfaces at that point. We know that the boundary arcs between the pairs of adjacent patches are circles. The intersection of the planes of these circles with $T_{x_{fp}}$ yields four vectors in $T_{x_{fp}}$ with origins at x_{fp} in the directions of the boundaries away from x_{fp}. Calling these vectors t_{fpq}, t_{fpr}, t_{fpf_1}, and t_{fpf_2} (refer to Fig. 7.13), we see that they are tangent to the four circular arcs at the point x_{fp}. Denote the angle between t_{fpq} and t_{fpr} by α_{fp} and the angle between t_{fpf_1} and t_{fpf_2} by γ_{fp}. We assign sign to these angles by taking the cross-product of the two vectors and dotting the result with the vector from x_{fp} to the center of the vertex sphere, where

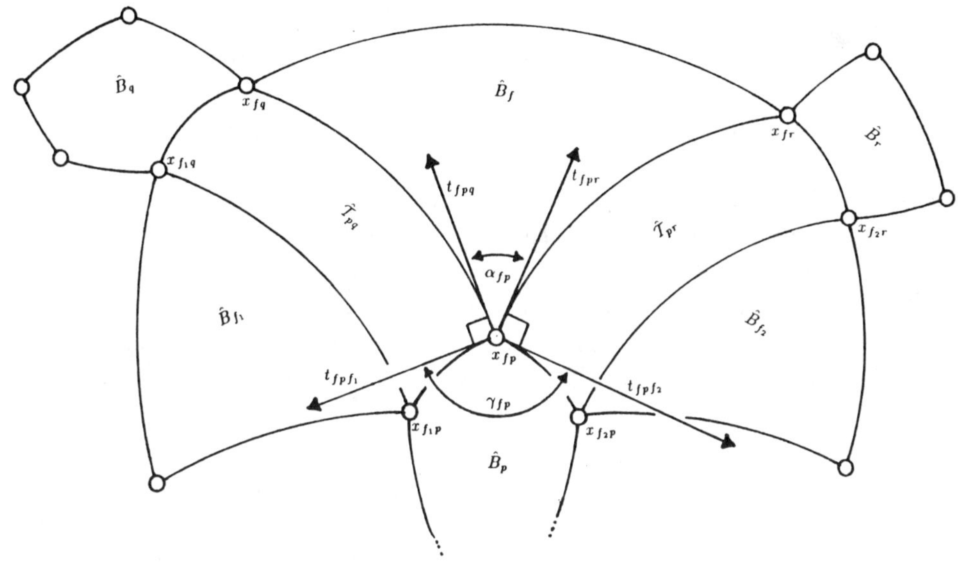

FIG. 7.13. *The relation of the angles and tangent vectors at a tangent point x_{fp} to the surrounding patches.*

the order of the cross-product is taken so as to yield a positive angle when every facet sphere radius is infinite.

Before we specify restrictions on the facet sphere radii, note that if $\epsilon_p = \epsilon_q$, then $\mathcal{T}_{pq}$ is a torus and thus t_{fpq} is perpendicular to t_{fpf_1}. If $\epsilon_p \neq \epsilon_q$, then $\mathcal{T}_{pq}$ is the image of a torus under inversion in a sphere. On this torus the images of the circular boundaries between the of B_p, $\mathcal{T}_{pq}$, and B_f are perpendicular at the image of x_{fp}. Because inversion in a sphere is a conformal map, the circular boundaries are perpendicular at x_{fp} and thus t_{fpq} is perpendicular to t_{fpf_1}. Hence $\alpha_{fp} = \pi - \gamma_{fp}$, that is, $0 < \alpha_{fp} < \pi$ if and only if $0 < \gamma_{fp} < \pi$. Accordingly, we need only restrict the facet sphere radii in terms of α_{fp} or γ_{fp}, but not both. To keep the adjacent boundary arcs of a patch from intersecting, and thus to prevent degeneracy, a necessary restriction is $\alpha_{fp} > 0$ and $\gamma_{fp} > 0$. The angle γ_{fp} turns out to be the simpler one to work with because it lies on a vertex sphere which does not vary with the facet radii.

7.5.1.6. Half angles of patches. Recall that the cyclide $\mathcal{T}_f$ and the vertex sphere B_p intersect along a circle which contains the path of x_{fp} because the varying facet sphere B_f must meet both the cyclide and the vertex sphere. Let us examine the variation of the angles γ_{fp} in the corners of the vertex patches. Initially, when every facet sphere radius is infinite, the path of x_{fp} lies between the two edges of $\hat{B}_p$ intersecting neither except at x_{fp} [10]. Thus there are positive angles which we can label as follows. Denote the angle between the edge fpf_1 and the path of x_{fp} by γ_{fpf_1} and the angle between the edge fpf_2 and the path of x_{fp} by γ_{fpf_2} and observe that $\gamma_{fp} = \gamma_{fpf_1} + \gamma_{fpf_2}$. Bounding each half angle between 0 and $\frac{\pi}{2}$ will not only bound γ_{fp} between 0 and π (and α_{fp} between 0 and π), but will restrict them symmetrically and more independently of each other. Thus instead of considering γ_{fp} as a function of r_f, r_{f_1}, and r_{f_2}, we consider γ_{fpf_1} as a function of r_f and r_{f_1}.

7.5.1.7. Monotonicity of angles. The claim is that each of these half angles varies monotonically with the radii of the facet spheres. The algorithm will make use of this fact to guarantee behavior of these angles as facet sphere radii are varied. Let $\mathcal{T}_f$ and $\mathcal{T}_{f_1}$ be the cyclides associated with the vertex spheres of the facets B_f and B_{f_1}, respectively. Then we define the circles $C_p = \mathcal{T}_{pq} \cap B_p$, $C_f = \mathcal{T}_f \cap B_p$, and $C_{f_1} = \mathcal{T}_{f_1} \cap B_p$. Note that C_f contains the path of x_{fp}. The angle γ_{fpf_1} is between C_p and C_f at x_{fp}, and the angle γ_{f_1pf} is between C_p and C_{f_1} at x_{f_1p}. If we fix r_f and decrease r_{f_1}, how do these angles vary? Recall in §7.4.5 we saw how to generate an edge cyclide with an inversion in a sphere, S, with center, O, on the midsphere of two vertex spheres; see Fig. 7.12. Because this inversion is conformal, we can examine the angles between the images (under inversion) of these circles on the image of the vertex sphere. Because $\mathcal{T}'_{pq}$ is a torus, C'_p is a great circle of B'_p and x'_{fp} is fixed since r_f is fixed, and as we vary r_{f_1}, C'_p rotates about the line through x'_{fp} and the center $O_{B'_p}$ of the image of the vertex sphere. Thus the second intersection of C'_p and C'_f travels in a fixed direction around the circle C'_f, and the angle between the two circles must therefore vary monotonically. Hence the angle γ_{fpf_1} decreases *monotonically* and γ_{f_1pf} increases *monotonically.* Similarly, if we fix r_{f_1} and decrease r_f, then γ_{f_1pf} monotonically decreases and γ_{fpf_1} monotonically increases.

7.5.1.8. Angles and path cross: Nondegeneracy. In addition to keeping the angle γ_{fp} nonzero, it is also necessary to prevent the edges of patches from crossing. In this section we examine how the adjacent edges of a vertex patch can cross.

Now consider the second intersections of C_p with C_f or C_{f_1} (other than x_{fp} or x_{f_1p}, respectively) as we vary r_{f_1} or r_f, respectively. Fix r_f and decrease r_{f_1} and examine the image of the inversion where C'_p is a great circle of B'_p. We wish to avoid the edge of the vertex patch (which is a subset of C_p) crossing the tangent path of x_{fp} (a subset of C_f) as we vary the facet sphere radii. Since decreasing r_{f_1} is a rotation of C'_p about the axis of B'_p through x'_{fp} and $O_{B'_p}$, there are three cases to consider.

If C'_f is a great circle of B'_p then both points of intersection of C'_p and C'_f are fixed. Thus the second intersection of the two circles cannot move, and the edge fpf_1 cannot cross the path of x_{fp} until $\gamma_{fpf_1} = 0$ and $C_p \equiv C_f$ and x_{f_1p} is on the circle C_f of the path of x_{fp}.

If C'_f is *not* a great circle of B'_p then the second intersection is not fixed and it moves along the fixed circle C'_f as r_{f_1} decreases. If the side of B'_p bounded by C'_f and containing fpf_1 is larger than a hemisphere of B'_p, then the second intersection moves along C'_f away from $x'_{fp}(r_{f_{\min}})$ and toward $x'_{fp}(\infty)$. Thus the second intersection of C_p and C_f encounters the path of x_{fp} when $\gamma_{fpf_1} = 0$ and C_p is tangent to C_f at x_{fp} and *before* x_{f_1p} crosses path of x_{fp}.

If the side of B'_p bounded by C'_f and containing fpf_1 is less than a hemisphere of B'_p, then the second intersection moves along C'_f toward $x'_{fp}(r_{f_{\min}})$ and away from $x'_{fp}(\infty)$. Thus the second intersection of C_p and C_f encounters the path of x_{fp} when x_{f_1p} is on the path of x_{fp} (x_{f_1p} is then the second intersection) and $\gamma_{fpf_1} > 0$.

Similarly fixing r_{f_1} and decreasing r_f has an analogous effect, so that if we do not let any of the facet sphere radii decrease so much that (1) any tangent point crosses the plane of the path of an adjacent tangent point, and (2) any angle γ_{fpf_1} becomes less than zero or greater than $\frac{\pi}{2}$, we are always assured of the following: that for every facet vertex pair f, p, $0 \leq \alpha_{fp} \leq \pi$ and $0 \leq \gamma_{fp} \leq \pi$, and that two neighboring edges of a vertex patch are nonintersecting, and moreover that we maintain a symmetry between the facet sphere radii such that we can decrease them one at a time in any order. To test for these conditions we can check the following: (1) The tangent point x_{f_1p} is on the path of x_{fp} if and only if the facet sphere B_{f_1} is tangent to the other vertex sphere, B_r, of the facet sphere B_f, i.e., $\|c_{f_1} - c_r\| = r_{f_1} - \epsilon_r$. (2) The angle γ_{fpf_1} is zero if and only if C_p and C_r are tangent at x_{fp}, or, equivalently, if the distance between the two intersections of these circles becomes zero. The angle is a right angle if and only if the two circles are perpendicular at x_{fp}. This is equivalent to their image circles, C'_p and C'_r, under the inversion being perpendicular at the image, x'_{fp}, of the tangent point. In the image C'_p is a great circle of the sphere upon which both image circles lie, thus the condition is equivalent to the center of the image circle C'_f being in the plane of the image circle C'_p.

7.5.2. Restricting the arc length of an edge cyclide. As mentioned in §7.4.5.1, for a given edge cyclide it may be necessary to restrict some facet sphere radii such that the line of centers of the two facet spheres does not cross the line of centers of the vertex spheres during the continuous deformation of the surface. This section establishes a symmetrical bound on the radii r_f and r_{f_1} of the two facet spheres. It is sufficient to prevent the four points c_f, c_{f_1}, c_p, and c_q from becoming coplanar, which is equivalent to the four points x_{fp}, x_{f_1p}, c_p, and c_q being coplanar. On the surface of the vertex sphere B_p this is equivalent to the (plane of the) great circle through x_{fp} and x_{f_1p} (which varies with r_f and r_{f_1}) not crossing the two points of intersection of the line through c_p and c_q with B_p (which are fixed). Any plane through c_p and c_q which supports $x_{fp}(\infty)$ and $x_{f_1p}(\infty)$ on the same side will suffice as a bound on the points x_{fp} and x_{f_1p}, but we require one irrespective of the order in which we consider r_f and r_{f_1}. The two points $x_{fp}(\infty)$ and $x_{f_1p}(\infty)$ each determine one plane with the pair of points c_p and c_q thus defining two great circles on B_p through the line of c_p and c_q. We choose the plane of the great circle on B_p

which bisects the angle between these two and supports both points $x_{fp}(\infty)$ and $x_{f_1p}(\infty)$. This plane is then used to bound the paths of the tangent points x_{fp} and x_{f_1p}, which in turn bounds the radii of the facet spheres.

7.5.3. Separation of edge patches. It is useful, as we will see in §7.6.3, to be able to say that two edge patches $\hat{\mathcal{T}}_{pq}$ and $\hat{\mathcal{T}}_{pr}$ which meet at x_{fp} cannot otherwise intersect. It is sufficient (as we will see later) to show that the boundary arcs of $\hat{\mathcal{T}}_{pq}$ from x_{fp} to x_{fq} and of $\hat{\mathcal{T}}_{pr}$ from x_{fp} to x_{fr} are separated by some planes. Maintaining the bounds on the half angles defined above, these boundaries are separated if neither x_{fq} nor x_{fr} cross the plane of the path of x_{fp}.

7.5.4. A geometric algorithm for nondegenerate patches. Visual tweaking of the facet sphere radii is easy because the geometric constraints are obvious to the eye. However, one would like a simple automated algorithm to arrive at some fair solution of finite values for the facet sphere radii.

Geometrically, maintaining the nondegeneracy of patches requires maintaining the following conditions on the facet sphere radii and the associated tangent points:

(7.5a) x_{fp} must lie on the original arc of the tangent path between $x_{fp}(\infty)$ and $x_{fp}(r_{f_{\min}})$. This is equivalent to requiring B_f to always contain its vertex spheres;

(7.5b) x_{fp} is restricted to remain on the original sides (the sides they were on when all facet spheres were of infinite radius) of the paths of x_{f_1p} and x_{f_2p} on B_p (cf. §7.5.1.4). This is equivalent to requiring B_f to always contain every vertex sphere of any facet sphere with which it shares two vertex spheres;

(7.5c) x_{fq} and x_{fr} are restricted to remain on their original sides of the plane of the path of x_{fp} (cf. §7.5.3);

(7.5d) the line through c_f and c_{f_1} is restricted to not cross the line through c_p and c_q (cf. §§7.4.5.1 and 7.5.2);

(7.5e) every half angle at every tangent point remains properly bounded, i.e., $0 < \gamma_{fpf_1} < \frac{\pi}{2}$ (cf. §7.5.1.4).

With one further constraint to keep the vertex patches nondegenerate (which we shall mention in §7.5.5), these conditions suffice to maintain nondegeneracy of all of the patches. The planes determining conditions (7.5a)–(7.5d) are fixed before, and regardless of, any choice of the facet sphere radii. Condition (7.5e) varies with the radii. We rely on sufficient (instead of strictly necessary) bounds to maintain the symmetry of the algorithm, thus plucking a sequence of particular solutions from the nonlinear haze. First we solve for each r_f such that (7.5a)–(7.5c) are satisfied and call the most restrictive (largest value) over each facet sphere for all of its vertex spheres the new $r_{f_{\min}}$. This preprocessing is a symmetric restriction, i.e., the order in which we find these bounds on the radii of the facet spheres is irrelevant. In some cases (cf.

§7.4.5.1) it is never necessary to further restrict the radii to preserve (7.5d), but in any event a symmetric bound can be applied to the radii (cf. §7.5.2) if necessary and $r_{f_{\min}}$ is replaced with the new value if it is larger than the old value.

In the iterative portion of the algorithm we hold every facet sphere radii but one, say r_f, fixed then we solve for the minimum value of r_f which preserves condition (7.5e) for each vertex sphere tangent to the facet sphere. Choose the largest of these bounds over all of the vertex spheres of the facet sphere and call this value r_{f*}. Perform this operation for each facet sphere radius, holding every other one at the original value. The monotonic variation of the angles (cf. §7.5.1.3) is crucial to the patches remaining well behaved. Again note that this method of restricting the facet sphere radii is symmetrical as mentioned above. Fixing each $r_f = r_{f*}$ (the new value) we see that condition (7.5e) is altered, possibly allowing further reduction of some of the facet sphere radii, but the first four conditions do not vary with the choice of the facet sphere radii. We can perform several iterations, but any choices found must satisfy interpolation constraints (covered in the next section) *before* we proceed to the next iteration.

There are many redundant calculations in such an algorithm which are worth considering to improve efficiency. For instance, x_{fp} can cross the plane of the path of x_{f_1p} if and only if x_{fq} crosses the plane of the path of x_{f_1q}.

7.5.5. Interpolation constraints. There are multiple types of interpolation constraints on the algorithm. The most obvious are those to ensure that the surface interpolates the original data points. These are treated first. Then there are constraints to guarantee that the link edges are preserved. There are also constraints on the algorithm to maintain nondegeneracy of patches which are most easily implemented as interpolation constraints.

Consider the spherical m-gon of the vertex patch $\hat{B}_p$ with vertices $x_{f_ip}(r_{f_i})$ for $i = 0, 1, 2, \ldots m-1$; see Fig. 7.14a. Each edge of this m-gon is the circular arc of the intersection between the vertex sphere B_p and the the edge torus $\mathcal{T}_{pq}$ bounded by the points $x_{f_ip}(r_{f_i})$ and $x_{f_{i+1}p}(r_{f_{i+1}})$. When $r_{f_i} = \infty$ for $i = 0, 1, 2, \ldots m-1$, $\mathcal{B}$ is the convex hull of spheres and the point p must be in this m-gon (referred to as the *maximal* m-gon) because we chose the vertex sphere radii such that the tangent plane T_p would meet the surface $\mathcal{B}$ at the point p, thus preserving interpolation. If $r_{f_i} = r_{f_i*}$ for $i = 0, 1, 2, \ldots m-1$, where r_{f_i*} is determined in §7.5.4, then it may be that p is no longer in the m-gon, which we will refer to as the *minimal* m-gon (for each iteration of the geometric algorithm!). There are five simple cases to consider depending on where the point p is located: (1) in the interior of the minimal m-gon (Fig. 7.14b), (2) on an edge of the minimal m-gon (Fig. 7.14c), (3) on the path of a tangent point, between the maximal and minimal m-gon (Fig. 7.14d), (4) on the edge of the maximal m-gon (Fig. 7.14e), or (5) in one of m spherical quadrilaterals bounded by two tangent arcs and an edge each of the maximal

(a)

(b)

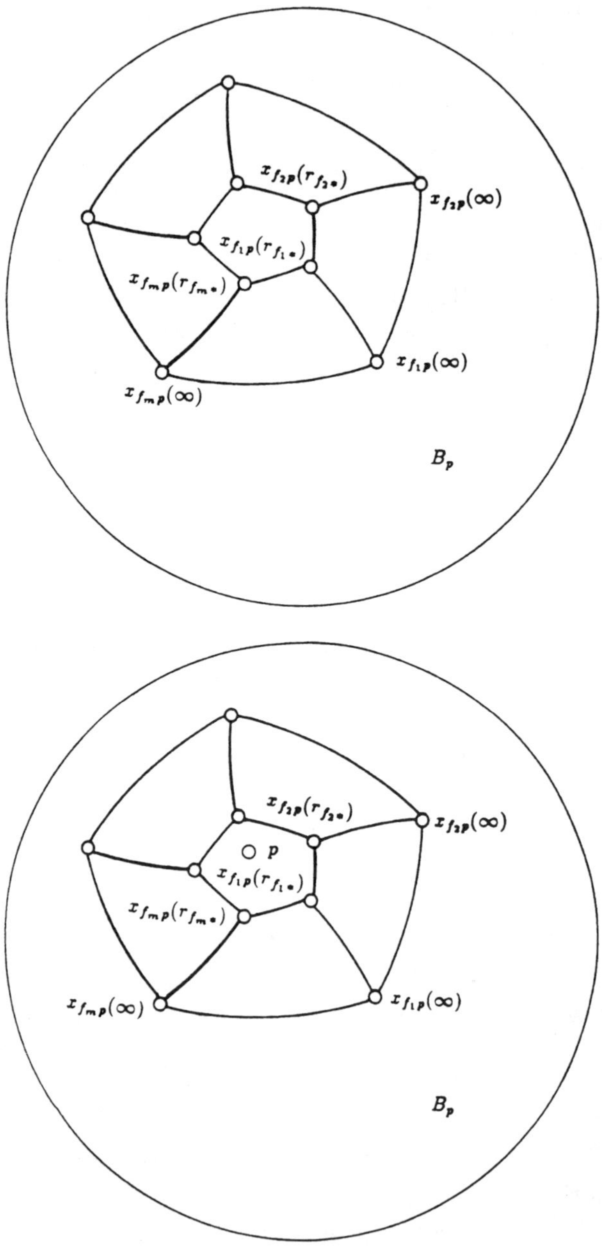

FIG. 7.14. (a) *The vertex sphere* B_p *and its maximal* m-gon *are compared to a minimal* m-gon *after one iteration of the quick algorithm.* (b) *The point* p *lies in the interior of the vertex patch and thus is interpolated by these choices of facet sphere radii. No further restriction need be placed on the* $r_{f_{i*}}$.

(c)

(d)

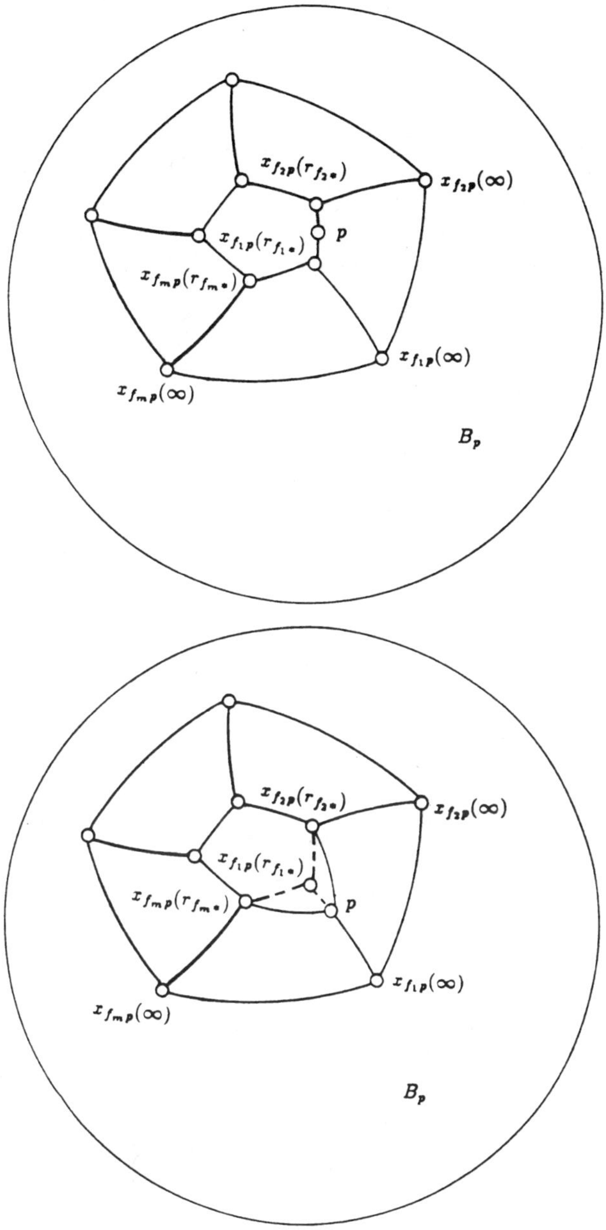

FIG. 7.14. (c) *The point* p *lies on the boundary of the vertex patch and thus is interpolated by these choices of facet sphere radii. No further restriction need be placed on the* $r_{f_{i*}}$. (d) *The point* p *lies in the path of a tangent point of the vertex sphere and a facet sphere. The new value of the radius of this facet sphere cannot be any smaller than the value which makes* p *and* $x_{f_1}(r_{f_{1*}})$ *identical.*

(e)

(f)

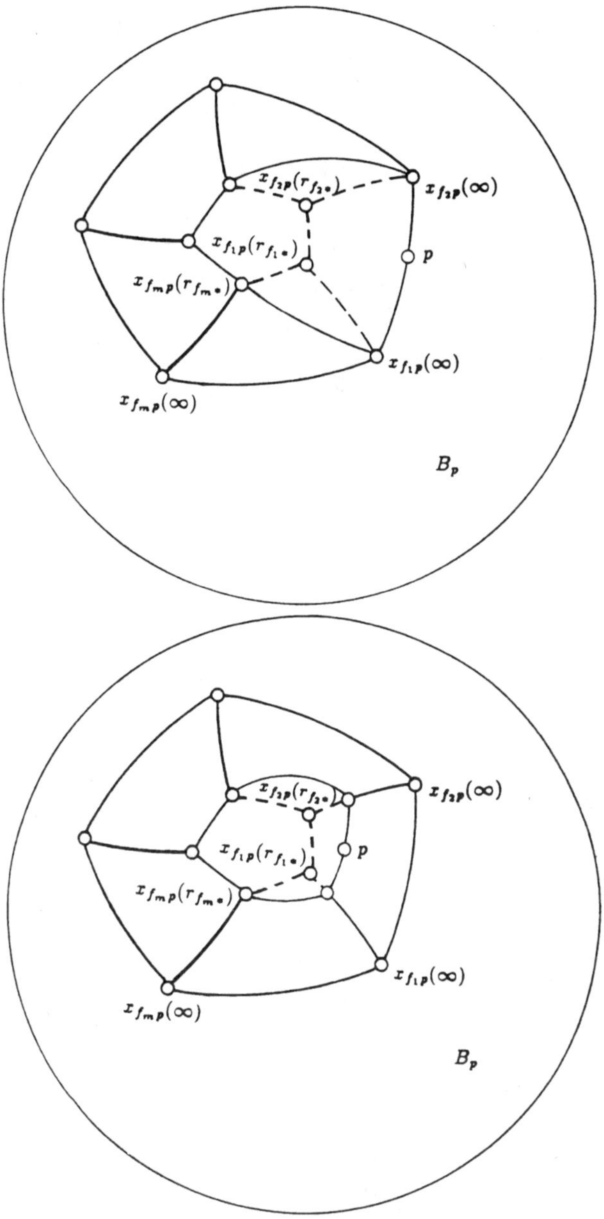

FIG. 7.14. (e) *The point* p *lies on the boundary of the maximal* m-gon. *Both adjacent facet spheres must have their radii fixed at* ∞. (f) *The point* p *lies in the interior of one of the quadrilaterals bounded by the maximal and minimal* m-gons *and the tangent paths. This requires the two neighboring facet spheres to have their new radii changed such that the boundary arc passing between them has* p *on the proper side.*

m-gon and the minimal m-gon (Fig. 7.14f). The maximal m-gon is decomposed into $m+1$ regions: m such quadrilaterals and the minimal m-gon. If p is in this minimal m-gon (or on an edge of it), then interpolation puts no further restrictions on the r_{f_i*}. If p is on a tangent path of x_{f_ip}, i.e., there exists a finite $r_{f_i} > r_{f_i*}$ such that $p = x_{f_ip}(r_{f_i})$, then this r_{f_i} becomes the new r_{f_i*}, and the other facet spheres tangent to B_p are not further restricted. If p is on an edge of the m-gon between $x_{f_ip}(\infty)$ and $x_{f_{i+1}p}(\infty)$, then r_{f_i*} and $r_{f_{i+1}*}$ are fixed at ∞, but the other facets are not further restricted. If p lies in the interior of one of these quadrilaterals then we choose new values for r_{f_i*} and $r_{f_{i+1}*}$ such that p lies on the circular boundary arc on the vertex sphere between $x_{f_ip}(r_{f_i*})$ and $x_{f_{i+1}p}(r_{f_{i+1}*})$. There are many such choices and any one will suffice, provided that it is chosen symmetrically with respect to the two radii. Any choice of the radii must ensure that the boundary arc between the tangent points does not cross the point p as it deforms from the initial boundary arc between $x_{f_ip}(\infty)$ and $x_{f_{i+1}p}(\infty)$. The deformation of this arc is not clear because it is not necessarily a great circle of B_p. However, if we consider the inverse of the diagram in the sphere used to generate the edge cyclide which shares this boundary, then the image of the boundary is a great circle of the image of the vertex sphere. As in the method outlined in §7.5.2, take as a bounding plane on the paths of the (images of the) tangent points x_{f_ip} and $x_{f_{i+1}p}$ the plane of the great circle on the image of the vertex sphere, which supports $x_{f_ip}(\infty)$ and $x_{f_{i+1}p}(\infty)$ and bisects the angle of the great circles through each of them and the image of the point p. This then provides a symmetric bound in the original space. For each facet sphere we choose the most restrictive bound over all of its vertex spheres to be the bound on its radius.

There are some special cases to consider due to unusual tessellations of the surface. If we are considering decreasing the radius of a facet sphere which touches a vertex sphere which touches a link edge patch, then the link edge (or conical subset) meets the vertex sphere in a circle. This circle lies within the maximal m-gon, and no r_f will be made so small that x_{fp} lies within this circle. Thus a link edge places an additional restriction on the radius of every facet sphere tangent to a vertex sphere of a link edge. If a vertex sphere has multiple link edges then each must remain inside the m-gon as the facet sphere radii are decreased. The cases are similar to the cases outlined above for the data point although the circle of intersection may extend into more than one of the spherical quadrilaterals so that restriction would have to be made on multiple facet sphere radii to accommodate the link edge. In such a case the relevant facet sphere radii should be restricted in a manner which is symmetrical; this is because, for example, the link edge and the two adjacent tangent points of any tangent point put an absolute minimum bound on the radius of the facet sphere of the given tangent point, which might in turn restrict the radii of the facet sphere of the adjacent tangent points to ∞. Choosing a symmetrical bound in this case is slightly more complicated than for the interpolation of the point p. Performing the same inversion described in the previous paragraph, we have that the boundary arcs in the image space are great circles of the image sphere. Now let us restrict our attention to one particular quadrilateral, which the link edge circle intersects the interior of, and let it be the one bounded by the paths of x_{fp} and x_{f_1p}. Remember that we are in the image space now. Fixing r_f we decrease r_{f_1} until one of three things happens: x'_{f_1p} meets the (image of the) link edge circle, x'_{f_1p} meets the end of the (image of the) path arc, or the great circle boundary becomes tangent to the (image of the) link edge circle at a point z_{f_1} inside the (image) quadrilateral. Fixing r_{f_1} and decreasing r_f we have three similar cases, where in the last case we label the point z_f. Due to symmetry there are really only six combinations of occurrences to consider. Remember that we are trying to choose a plane of a great circle on the image of B_p to bound the tangent points and thus the facet sphere radii. For brevity we outline only one such case.

Suppose independently decreasing each radius results in the third occurrence mentioned, wherein the great circle boundary becomes tangent to the (image of the) link edge circle at a point z_{f_1} inside the (image) quadrilateral. Then, for a bound, we choose the (plane of the) great circle tangent to the (image of the) link edge circle at the midpoint of the arc between the tangent points z_f and z_{f_1} of the boundary arcs to the link edge circle. This plane bounds the paths of the tangent points which then provides bounds on the facet sphere radii. The other combinations of occurrences have similar solutions. Each case provides bounds on the radii of the facet spheres tangent to the vertex sphere once the bounding great circles are mapped back through the inversion. The most restrictive of the bounds over each facet sphere should be chosen as the interpolation bound. It should be noted that an arbitrary but finite number of link edges can possibly meet the vertex sphere and thus such interpolation bounds should be calculated as the most restrictive over all of the link edges of the vertex sphere, and then over all vertex spheres of a facet sphere.

Preserving interpolation of the data may thus require that some of the r_{f*} found in one iteration of the geometric algorithm in §7.5.4 not be used, but instead be increased as specified above, *before* proceeding to set $r_f = r_{f*}$ for every facet sphere and further iterating the algorithm. Again, the monotonic variation of the angles of the patches forces the patches to be well behaved enough for this to work. Of course, those facet sphere radii fixed by interpolation constraints will not be changed by further iterations.

Another special case can occur even when the vertex radii are all equal. As mentioned in the example of §7.2.1.2, it is possible for a vertex patch to have no data point to be interpolated. In this case the only restriction on the radii of the facet spheres tangent to this vertex patch is that its facet patches (and hence the vertex patch) do not degenerate.

Finally there is a special constraint which is not immediately one of interpolation, but which we treat in this manner for simplicity. For each pair of edge cyclides which share a vertex sphere but are not adjacent, we cannot allow their circular boundaries on the vertex sphere to intersect (the patch would be degenerate) (cf. §7.5.6). Because these boundaries can behave quite poorly, we separate them in a manner which is a balance between simplicity and symmetry with the case of adjacent edges. Consider each such pair of cyclides when every facet sphere radius is infinite. Their boundaries are two fixed circular arcs on the vertex sphere. Choose the great circle on the vertex sphere which bisects the angle between the (planes of the) two circles and separates them. Then prevent either of the boundaries from crossing this circle by exactly the same method that the boundaries were prevented from crossing the link edge circles above. This provides the final set of restrictions on the facet radii and again can be performed as a preprocessing step, not varying with the facet radii, because it is dependent on the initial surface only.

At each iteration *any* choice of facet sphere radii between the old choices from the previous iteration and the minimum new choices will ensure nondegenerate patches. This is so because of the symmetric nature of the geometric algorithm and because interpolation is ensured at every iteration. The algorithm is designed so that a radius is never increased to preserve interpolation, instead a radius is never decreased so far that this becomes necessary. Moreover, because facet sphere radii never need be increased, there is no danger of choosing some value that will invalidate the current choices for the other facet sphere radii.

7.5.6. Preventing degenerate patches via continuous deformations. In this section we prove that the given algorithm creates nondegenerate patches.

There are two ways for a facet patch to degenerate by a continuous deformation of its boundaries. Initially, when all of the facet sphere radii are infinite, each facet patch is a planar triangle (with interior angles between 0 and π). Likewise, the inverse under the map $\mathbf{x}$ of our patch (cf. §7.2.4) is a simply connected domain (some triangle) bounded by a simple closed curve. Then, because we only continuously deform the facet patch, $\hat{B}_f$ can become degenerate if and only if we can find a continuous deformation which makes the boundaries intersect one another. Because the tangent points which are the images of the corners of the domain are always distinct, the boundaries can intersect one another in only two cases: (1) Two boundary arcs meet in a second point other than their common tangent point. (2) One of the boundary arcs crosses both of the others. We can place algebraic conditions on both cases as follows.

If we fix the radii r_{f_1} and r_{f_2} and allow r_f to decrease, the angle α_{fp} decreases and can become zero or negative. If α_{fp} is negative, then the two boundary arcs between the facet patch and the two edge patches intersect as in case (1) and $\hat{B}_f$ is degenerate. This is prevented by (7.5e).

While decreasing r_f decreases α_{fp}, decreasing r_{f_1} or r_{f_2} *increases* α_{fp} which could thus become greater than π. There are two possible ways for the domain triangle to continuously deform as in case (2). It must be that one of the corners of the circular triangle has crossed one of the edges, and thus, for some intermediate value, must have been *on* the interior of that edge as in the two possible subcases: (2a) Both arcs sharing the corner lie on the same side of the crossed arc. (2b) The two arcs sharing the corner lie on opposite sides of the crossed arc. Case (2b) cannot occur because the corner on the boundary cannot be continuously moved off of the boundary without two boundary arcs intersecting, and thus (2b) could not be arrived at by a continuous deformation of the domain triangle with no intersecting boundary arcs. Case (2a) can be arrived at via a continuous deformation but the interior angle α_{fp} at the corner on the boundary arc must be greater than π and thus must have passed through the value π which we had disallowed in (7.5e).

Vertex patches are generally more complicated than facet patches simply because they have more sides. Vertex patches degenerate if they become non-simple and preventing α_{fp} (or γ_{fp}) from exceeding zero or π will guarantee that adjacent sides of a vertex patch do not intersect. This is prevented by (7.5b) and (7.5e). However, there may be many sides to a vertex patch which require further restrictions. The special interpolation constraint of the previous section will suffice to keep these other boundaries separated.

Because the four corner angles of an edge patch are right angles, an edge patch can degenerate (by our original definition) if and only if two opposite sides of the patch intersect. However, because every edge cyclide is the image of a torus, the radius of the rolling ϵ-ball which determines the cyclide can never be zero, and thus *two opposite boundaries of an edge cyclide must always be separated by the radical plane of the two spheres (which share those boundaries with it) and to which it is tangent.* As the radii of the two facet spheres B_f and B_{f_1} approach equal values and the facet spheres approach equal spheres, the "width," or the distance across the edge patch between them, approaches zero. When the radii are equal and the spheres are equal, the two boundary arcs of the edge patch are identical. Hence an edge patch is degenerate if the two facet spheres it is tangent to are the same sphere, in which case the edge patch can be considered as a patch of zero area and ignored. The radii r_f and r_{f_1} of the facet spheres can become equal if and only if the tangent points $x_{fp}(r_f)$ and $x_{f_1p}(r_{f_1})$ are identical (and the tangent points $x_{fq}(r_f)$ and $x_{f_1q}(r_{f_1})$ are identical, but this is the same condition). Thus it is clear how the "width" of the edge patch approaches zero, i.e., how the edge patch degenerates. By the above argument, any larger values for the facet sphere radii result in a nondegenerate edge patch. If we allow these radii to be even smaller than this, i.e., so that both of the tangent points have crossed the path of the other, then we have the ugly case where the interior of the patch $\hat{B}_f$ intersects the interior of the patch $\hat{B}_{f_1}$, the edge patch no longer supports either of the facets patches, and the edge patch meets each of the facet patches in a cusp. An edge patch is also considered degenerate in this case and we thus avoid the crossing of tangent paths with constraint (7.5b).

7.5.7. Summary. The minimum valid radii for the facet spheres are the minimum radii such that the tessellation remains the same, every patch in the tessellation is nondegenerate and interpolation of the data points is preserved. Some characteristics of the tessellation are preserved with interpolation constraints. To solve this rigorously requires a nonlinear optimization algorithm, because while the constraints on the tangent points in terms of the neighboring tangent points (or equivalently on the centers of the facet spheres in terms of the centers of the neighboring facet spheres) are linear, these points are not linear functions of the radii of the facet spheres, and the radii of each is dependent on the radii of every neighbor. We have constructed symmetric bounds on the radii to find particular solutions to this problem. Solving for the bounds of each iteration involves only algebraic conditions no worse than quadratic equations. Virtually every condition is based on intersecting certain circles on certain spheres.

7.6. Convexity

Proceeding along lines of argument suggested by do Carmo [4] we first show that the surface possesses certain local properties everywhere. Then, because the surface is a continuous deformation of a convex surface, we will see that our bounds on the facet sphere radii are sufficient to preserve convexity.

We will refer to the initial surface, the convex hull of the vertex spheres, as $\mathcal{B}^\infty$ to distinguish it from the surface $\mathcal{B}$ with generic facet sphere radii. We considered patch domains as subsets of R^2 in §7.2.4; in this section it will be convenient to treat them as subsets of a sphere. We can choose some point inside the convex closure of the surface as the center of a unit sphere, $\mathcal{S}$, and (because the surface is convex) the projection of the surface onto the sphere is a homeomorphism between the surface and the sphere. Using this unit sphere as the domain for each patch has the advantage that (because both the sphere and the surface are closed) the manner in which all of the patches meet and all of the patch domains meet is similar. Thus the surface can be considered as a homeomorphic map $\mathbf{x} : \mathcal{S} \to \mathcal{B}$. Note that the domains of the patches are fixed by the tessellation of the initial surface $\mathcal{B}^\infty$ and it is the parameterized map from the domain to the surface which varies as we deform the surface. Thus, $\mathbf{x}$ is a projection only initially for the surface $\mathcal{B}^\infty$ and, in general, as $\mathbf{x}$ varies with the facet sphere radii, it is not a projection. The nondegeneracy of a patch is equivalent to the map of its boundary being one to one.

A property holds *locally* near a point on the surface if and only if there is a nonempty domain neighborhood of the inverse of the point such that the property holds for the image of that neighborhood. This definition of *local* is a crucial distinction between a neighborhood of a point in 3-space, and a neighborhood *on* the surface. We shall say a surface is *locally convex* if there is a neighborhood on the surface of every point such that the neighborhood is on one side of the tangent plane to the surface at that point (cf. the second definition of a convex set in §7.1.1). (This is different from do Carmo's definition [4, p. 174] because his surfaces are defined to be non-self-intersecting and therefore his neighborhoods of points on the surface can be defined as "balls" in 3-space. We must begin with this weaker definition in terms of neighborhoods on the surface.) A surface is *locally non-self-intersecting* if the map from the domain is locally one to one.

At every point on the generic surface $\mathcal{B}$ there exists a unique and well-defined tangent plane. This is clear at the interior point of any patch. On any patch boundary, the patches sharing the boundary meet tangentially and so have a common tangent plane.

7.6.1. Local convexity. Every patch is convex and thus locally convex at any interior point. Because there is a well-defined tangent plane everywhere, the tangent plane to a point on the boundary of a patch is the same for all patches sharing that point. We want to show that there is a neighborhood of that point on the surface which is on one side of the tangent plane at that point.

If the point is a tangent point x_{fp}, then the tangent plane to all four patches sharing the point is the plane tangent to the facet sphere at x_{fp}. Since the vertex sphere and both edge cyclides are contained in the facet sphere and the tangent plane supports the facet sphere because it is convex, the tangent plane supports all four patches on the same side.

Similarly, along the boundary between a facet patch and an edge patch there is a unique tangent plane at each point. The edge cyclide is the envelope of a variable sphere internally tangent to the facet sphere. Thus the tangent plane supports the facet sphere and everything inside of it (especially the cyclide) on the same side.

At any point along the boundary between a vertex sphere and an edge cyclide there is also a unique tangent plane. It is tangent to the cyclide at a point on the convex hull of the cyclide and so it supports the convex hull of the cyclide, which in turn contains the vertex sphere. Thus both patches are supported on the same side of the tangent plane.

Hence we have shown that the surface is locally convex.

7.6.2. Cusp free. Initially $\mathcal{B}^{\infty}$ is convex and therefore cusp free everywhere, particularly at the point $x_{fp}(\infty)$. We first show that $\mathcal{B}$ is cusp free at the point $x_{fp}(r_f)$ if $r_f > r_{f*}$. Recall Fig. 7.13 and the notation of §7.5.1. If $r_f > r_{f*}$, then none of the four angles at $x_{fp}(r_f)$ in the tangent plane $T_{x_{fp}}$ can become zero or negative and so none of the four sectors defined by the tangent vectors can overlap.

Intersect $\mathcal{B}$ with any plane $H \neq T_{x_{fp}}$ containing $x_{fp}(r_f)$. Then $l = H \cap T_{x_{fp}}$ is a line which contains $x_{fp}(r_f)$ and lies in the tangent plane, and the curve $H \cap \mathcal{B}$ is composed of two arcs which are smooth and meet at $x_{fp}(r_f)$. The curve passes from one patch to another of the four at the tangent point with a well defined tangent line l. Either the curve has a cusp at $x_{fp}(r_f)$ or it has a continuous tangent vector. The curve has a cusp if and only if the tangent vectors to the two arcs away from $x_{fp}(r_f)$ coincide. These two tangent vectors lie on l in $T_{x_{fp}}$ through $x_{fp}(r_f)$, and so each of these two vectors to the curve at x_{fp} and away from x_{fp} is in one of these sectors, depending on which patches H intersects. Because none of these four sectors can overlap, the two tangent vectors cannot be in the same sector and cannot coincide. Thus the curve has no cusp at x_{fp}. Because the choice of H was arbitrary, the surface $\mathcal{B}$ has no cusp at x_{fp} if $r_f > r_{f*}$.

If a patch is nondegenerate then the patch must locally lie on one side of the plane of each of its boundary arcs. Thus two adjacent patches can meet in a cusp if and only if they meet in a cusp along the entire boundary arc, including the endpoints. We have just seen that this cannot happen, provided the bounds of the algorithm are maintained.

Hence the surface is cusp free everywhere, and therefore tangent plane continuous, if $r_f > r_{f*}$.

7.6.3. Locally non-self-intersecting. Again, at any point interior to a patch the desired property is clear. What about at a boundary point between patches? It is sufficient (and convenient for the proof in §7.6.4) to show that at any tangent point x_{fp} the four patches $(\hat{B}_f, \hat{\mathcal{T}}_{pq}, \hat{B}_p, \hat{\mathcal{T}}_{pr})$ sharing the tangent point intersect only along their common boundaries. Because two unequal tangent spheres can meet at only one point, it is obvious that $\hat{B}_f \cap \hat{B}_p = x_{fp}$.

Each of the four boundary arcs is shared by a sphere and a cyclide, thus it is sufficient to consider only one of the four possible cases; the rest are similar. The facet sphere B_f and an edge cyclide $\mathcal{T}_{pq}$ must intersect in a circle. If the patches $\hat{B}_f$ and $\hat{\mathcal{T}}_{pq}$ intersect at all, it must be in a subset of this circle. But by definition the only subset of the circle that is in $\hat{\mathcal{T}}_{pq}$ is the arc between x_{fp} and x_{fq} it shares with $\hat{B}_f$. Thus the two patches can intersect nowhere else.

The only case left is to show that $\hat{\mathcal{T}}_{pq} \cap \hat{\mathcal{T}}_{pr} = x_{fp}$. We proceed to show that $\hat{\mathcal{T}}_{pq}$ and $\hat{\mathcal{T}}_{pr}$ are supported on opposite sides of the plane of the path of x_{fp} on B_p; call it H_{fp}. The method is to show that there is a convex cone (a four-sided pyramid) containing each edge patch which is separated by the plane from the other cone.

First, we need to show that the plane of the four corner tangent points supports the patch (that these four points are coplanar is not difficult [12]). The edge cyclide $\mathcal{T}_{pq}$ (which can be visualized as a torus for this argument) is divided into four patches by the four lines of curvature (circles) C_f, C_{f_1}, C_p, and C_q, where the first two are intersections of the cyclide with its facet spheres and the second two are intersections of the cyclide with its vertex spheres. Observe that C_f and C_{f_1} can only lie in the portion of the cyclide which is convex, i.e., where the Gaussian curvature is greater than, or equal to, zero, or equivalently, that portion of the cyclide on its convex hull, $\mathcal{T}_{pq} \cap ch(\mathcal{T}_{pq})$.

These four circles define four patches and our observation implies that two of them are convex; call them patch 1 and patch 2, where patch 1 is the patch bounded by the arcs of C_f and C_{f_1} which are less than π. Call the other patch between C_f and C_{f_1} patch 3, and the other patch between C_p and C_q patch 4. We will refer to the plane of the four tangent points x_{fp}, x_{fq}, x_{f_1q}, and x_{f_1p} (which are the pairwise intersections of the lines of curvature on $\mathcal{T}_{pq}$) as the *base plane*. Observe that the base plane cuts each circle into two arcs and that either patches 1 and 4 or patches 2 and 3 have all four of their boundary arcs on the same side of the base plane. Specifically, either patch 1 or patch 2, but *not* both, has all four boundary arcs on the same side of the base plane. Finally, observe that one of the convex patches has all four of its arcs on one side of the base plane if and only if its arcs of C_f and C_{f_1} are less than π. Thus, if we specify that our patch be both *convex* and be bounded by arcs of C_f and C_{f_1}, which are less than π, then we can choose only the patch with all four arcs on one side of the base plane with arcs of C_f and C_{f_1} less than π, i.e., patch 1.

Remark. This argument is also justification for the claim made throughout this chapter that each boundary arc requires only two endpoints and a center to be uniquely determined. Because of restriction (7.5d) we always choose those arcs of C_f and C_{f_1} which are less than π, and because of the preceding argument we always choose those arcs of C_p and C_q which are on the same side of the base plane as the first two arcs. The interior of the patch is likewise uniquely defined.

Nutbourne and Martin [12, p. 260] derive an explicit formula for the directed distance of a point in the interior of the patch from the base plane. It can be shown [10] that this distance function is always nonnegative for our edge patch since the two vectors t_{fpq} and t_{fpr} (cf. §7.5.1) both lie on the same side of the base plane (this was partially the purpose of the preceding argument). Since the directed distance function is always of one sign the edge patch is indeed supported by this base plane.

Then, again recalling the notation of §7.5, our edge patch is bounded by the following: the base plane of x_{fp}, x_{f_1p}, x_{f_1q}, and x_{fq}; the plane of t_{fpq}, x_{fq}, and x_{fp}; the plane of t_{fpf_1}, x_{f_1p}, and x_{fp}; and the plane of t_{fpq} and t_{fpf_1}. These four planes are concurrent at x_{fp} and thus the edge patch is contained in the convex cone bounded by the four vectors t_{fpq}, t_{fpf_1}, $x_{fq} - x_{fp}$, and $x_{f_1p} - x_{fp}$. If we can show that these four vectors are on the same side of H_{fp} then we have shown that this edge patch is supported by the plane. Since x_{fp} is in the plane H_{fp}, restriction (7.5e) requires t_{fpq} and t_{fpf_1} to remain on the same side of H_{fp}, and (7.5b) and (7.5c) require x_{fq} and x_{f_1p} to remain on the same side of H_{fp}. Because t_{fpq} is $x_{fq} - x_{fp}$ initially, these four vectors are always on the same side of H_{fp}.

And because H_{fp} separates x_{f_2r} and x_{f_1p}, (7.5b) requires that $\hat{\mathcal{T}}_{pq}$ and $\hat{\mathcal{T}}_{pr}$ are supported and separated by H_{fp}, i.e., $\hat{\mathcal{T}}_{pq} \cap \hat{\mathcal{T}}_{pr} = x_{fp}$, as desired. Thus we have not only shown that the surface is locally non-self-intersecting, we have also shown that any two adjacent patches (i.e., two patches which shared boundary points in the initial convex hull of spheres) cannot intersect elsewhere.

7.6.4. Proof of convexity. We have shown that if every patch is nondegenerate, the surface $\mathcal{B}$ is locally convex, cusp free, tangent plane continuous, and locally non-self-intersecting.

LEMMA 7.6.1. *The pointwise limit of a convergent sequence of compact convex surfaces is a convex surface.*

Proof. See §4.6.4 of [10].

For the following discussion the reader may wish to refer to remarks by do Carmo [4, pp. 371–387, especially Remark 2, p. 387]. We now proceed to prove the convexity of the surface $\mathcal{B}$ with facet sphere radii as derived by the algorithm of §7.5. Because of the symmetric nature of the algorithm it is sufficient to show that the variation of the radius of *one* facet sphere leaves $\mathcal{B}$ convex, and we may begin by varying the known convex surface $\mathcal{B}^\infty$.

Let $\mathbf{x} : S \to \mathcal{B}$ be a bicontinuous and one-to-one map of the domain unit sphere (cf. §7.6) onto the surface. Fixing every facet sphere radius except $r = r_f$, $\mathbf{x}_r(s)$ is then a continuous function of r and s, where $s \in S$ and $|s| = 1$. Define $\bar{r} = \sup \{r \mid \mathbf{x}_r \text{ is self-intersecting}\}$. Trivially, if $r > \bar{r}$ then $\mathbf{x}_r$ does not self-intersect.

For our purposes, we define the *Gauss map* [cf. 4, p. 136] on the surface to be the map which carries each point on the surface to its unit normal vector (at the origin). The range of the Gauss map is the unit sphere.

LEMMA 7.6.2 [4, p. 374–375]. *Let $\pi : \tilde{B} \to B$ be a local homeomorphism, $\tilde{B}$ compact and B connected. Then π is a covering map.*

LEMMA 7.6.3 [4, p. 382]. *Let $\pi : \tilde{B} \to B$ be a covering map, $\tilde{B}$ arcwise connected, and B simply connected. Then π is a homeomorphism.*

Arcwise connected is also referred to as *pathconnected* in the literature. A set is arcwise connected if any two points in the set can be joined by a continuous path in the set.

LEMMA 7.6.4. *If $r > \max(\bar{r}, r_*)$, then $\mathbf{x}_r$ is convex.*

Proof. If the surface $\mathbf{x}_r (= \mathcal{B})$ is non-self-intersecting, equivalently if $\mathbf{x}_r$ is one to one, and tangent plane continuous (which requires that $r > r_*$) then the Gauss map [4, p. 136] of $\mathbf{x}_r (= \mathcal{B})$ onto the unit sphere is well defined. (Note that a well-defined tangent plane everywhere on a parameterized surface is not as strong as a well-defined tangent at every point on a surface in 3-space. If a parameterized surface self-intersects at a point, it may still have a well-defined tangent plane at this point for each domain point which maps to it. This is why we require the non-self-intersecting condition for this lemma.) Our goal is to show that the Gauss map is one to one. The surface $\mathbf{x}_r$ *may* have straight lines, or zero Gaussian curvature, on some flat facet patches, edge cones, or link edges which prevents the Gauss map from being one to one. We circumvent this by defining $\mathbf{x}_r^\delta$ by deforming these subsets of the surface by an arbitrarily small variation as follows. Let δ be some arbitrarily small but positive number. Study of the conditions of the algorithm of §7.5 reveals that there must exist a nonempty neighborhood on the surface of each initial tangent point $x_{fp}(\infty)$ such that we can choose a sufficiently small $\delta > 0$ and the radius of any flat facet sphere can be defined as $r_f = 1/\delta$ without violating the conditions. None of those restrictions forces any facet sphere radius to be infinite. This still leaves two types of subsets with zero Gaussian curvature, namely link edges and those facet spheres forced to remain planes (cf. remark of §7.3.2). For each such facet we can select one of its vertex spheres and change the radius slightly $\epsilon_p = \epsilon_p \pm \delta$ so that it is possible to choose a finite radius for the facet sphere $r_f = 1/\delta$. Each link edge is the surface of revolution of one of its straight lines about the line of centers of its vertex spheres. It is possible to replace this straight line by an arc of a circle of radius $1/\delta$ and generate a new link edge by rotating it about the same line of centers. We can choose δ small enough such that this new link edge is still convex and meets its vertex spheres in a convex, tangent plane continuous manner [10]. The new surface $\mathbf{x}_r^\delta$ can be made arbitrarily close to $\mathbf{x}_r$ by letting δ approach zero, and $\mathbf{x}_r^\delta$ is locally strictly convex everywhere (the Gaussian curvature is strictly positive where defined). Thus the Gauss map on $\mathbf{x}_r^\delta$ is locally one to one, or locally homeomorphic, for any sufficiently small $\delta > 0$.

If $\mathbf{x}_r^\delta$ is non-self-intersecting (which requires that $r > \bar{r}$) then, because $\mathbf{x}_r^\delta$ is compact and the unit sphere is connected, the Gauss map is a covering map by Lemma 2. (Note that the non-self-intersection of $\mathbf{x}_r^\delta$ is necessary to extend do Carmo's proof.) Furthermore, because the unit sphere is simply connected and $\mathbf{x}_r^\delta$ is connected, we have that the Gauss map is a homeomorphism, i.e., it is globally one to one from Lemma 7.6.3. Following the reasoning of a remark by do Carmo [4, p. 387, Remark 2] we see that since the Gauss map is a homeomorphism; each unit vector of $\mathcal{R}^3$ appears exactly once as a unit normal vector to $\mathbf{x}_r^\delta$. Taking a plane normal to this vector and away from the surface (i.e., which does not intersect the surface) and displacing it parallel to itself until it meets the surface, we observe that the plane must meet the point which is the inverse under the Gauss map of this vector before it meets any other point on the surface. Thus we conclude that $\mathbf{x}_r^\delta$ lies on one side of each of its tangent planes. Thus $\mathbf{x}_r^\delta$ is convex if r is greater than $\bar{r}$ and r_*. Finally, if we let δ approach zero then Lemma 7.6.1 implies that $\mathbf{x}_r$ is convex if $r > \max(\bar{r}, r_*)$, and the lemma is finished.

Because $\mathbf{x}_{\bar{r}} = \lim \mathbf{x}_r$ as r converges to $\bar{r}$ and $r < \bar{r}$ where each $\mathbf{x}_r$ is convex, Lemma 7.6.1 implies that $\mathbf{x}_{\bar{r}}$ is convex. In particular, $\mathbf{x}_{\bar{r}}$ does not self-intersect, i.e., it is one to one.

PROPOSITION 7.6.1. *If r_* is the new value of a facet sphere radius derived from one step of the algorithm of* §7.5 *then $\mathbf{x}_{r_*}$ is convex.*

Proof. By Lemma 7.6.2 and the above comment it suffices to show that $r_* \geq \bar{r}$. For a contradiction, assume that $r_* < \bar{r}$. Then by the definition of $\bar{r}$ and because $\mathbf{x}_{\bar{r}}$ does not self-intersect (i.e., because $\bar{r}$ is not in the set of which it is the sup), there exists a sequence (s_n, t_n, r_n) such that r_n converges to $\bar{r}$, $r_n < \bar{r}$, and $\mathbf{x}_{r_n}(s_n) = \mathbf{x}_{r_n}(t_n)$ where $s_n \neq t_n$ because $\mathbf{x}_{r_n}$ is self-intersecting. This is a bounded sequence in a compact space and therefore contains a convergent subsequence $(s_{n_k}, t_{n_k}, r_{n_k})$ which converges to $(\bar{s}, \bar{t}, \bar{r})$ where $\mathbf{x}_{\bar{r}}(\bar{s}) = \mathbf{x}_{\bar{r}}(\bar{t})$ because $\mathbf{x}_r$ is continuous with respect to r and s. But $\mathbf{x}_{\bar{r}}$ is one to one and thus $\bar{s} = \bar{t}$.

For the choice of r_* derived from the algorithm the surface $\mathcal{B} = \mathbf{x}_{r_*}$ is locally non-self-intersecting, i.e., there is a nonempty neighborhood N of the point $\bar{s}$ *on* the domain sphere such that the image of N is non-self-intersecting. From §7.6.3 we can take N to be the union of four domain patches sharing a common point (where $\bar{s}$ is in the interior of one domain patch, of the union of two domain patches, or of the union of the four domain patches). Then N does not change as we vary the facet sphere radii because the domain patches do not change. In particular, N does not become arbitrarily small. But $\bar{s} = \bar{t}$ implies that there exists s_{n_k} and t_{n_k} such that $s_{n_k} \neq t_{n_k}$, $s_{n_k}, t_{n_k} \in N$, and $r_{n_k} > r_*$ (since $r_{n_k} \to \bar{r} > r_*$), i.e., for which $\mathbf{x}_{r_{n_k}}$ self-intersects. This is a contradiction and we have that $r_* \geq \bar{r}$.

Hence the r_* found at each step of the algorithm guarantees that $\mathbf{x}_{r_*}$ is not self-intersecting and thus Lemma 7.6.4 implies that $\mathbf{x}_{r_*}$ is convex.

7.7. Example

The following set of shaded figures illustrates a particular example. In Fig. 7.15(a) we see the convex hull of ten points in $\mathcal{R}^3$. Then Fig. 7.15(b) shows the ten vertex spheres of maximum radii which satisfy the ϵ_p bounds of §§7.2 and 7.4. The vertices to be interpolated are on the surface of these spheres. The vertex, edge and facet patches which comprise the convex hull of these spheres, or the surface $\mathcal{B}^\infty$ with infinite radii facet spheres, are shown combined in Fig. 7.15(c). Deriving a set of finite facet sphere radii by the algorithm described herein yields the vertex, edge, and facet patches shown combined in Fig. 7.15(d).

7.7.1. Curvature. A curvature plot of such a surface is a plot of the curvature of the planar curve generated by intersecting the surface with a plane. Any plane must intersect our surface in a simple, closed, convex, planar curve. This curve consists only of circular arcs (of constant curvature) where the plane intersects vertex or facet patches, and of smooth arcs (of varying, but positive and monotonic curvature) where the plane intersects the cyclidal patches. Thus the most one can say about the curvature plot for an arbitrary plane is that it is a curve composed of constant segments and smooth monotonic segments.

7.7.2. Conclusion. We have evolved the description of an algorithm which creates an interpolating surface possessing the desired shape characteristics of tangent plane continuity and convexity. Over each patch the surface is infinitely differentiable. These qualities are obvious results of the same qualities

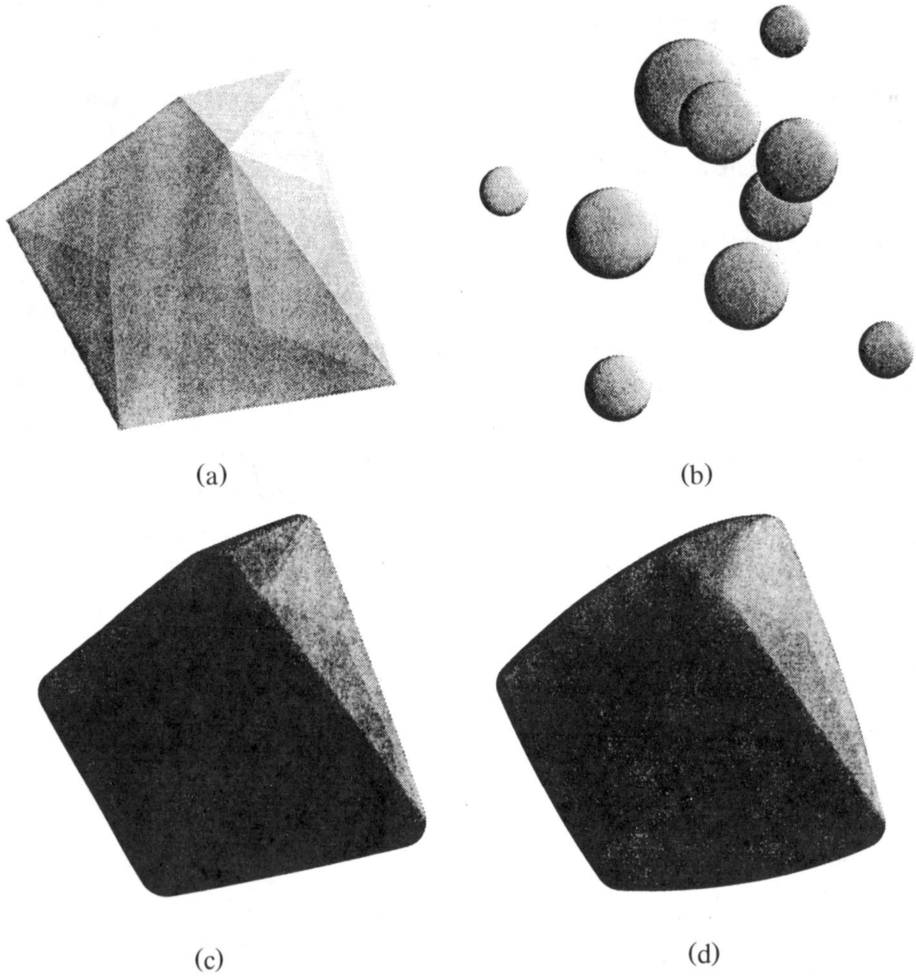

FIG. 7.15. (a) *The shaded image of the convex hull of ten points in* $\mathcal{R}^3$. (b) *The shaded image of the ten vertex spheres of maximum radii which satisfy the* ϵ_p *bounds of* §§7.2 *and* 7.4. (c) *The shaded image of the surface* $\mathcal{B}^\infty$ *which is the convex hull of the ten vertex spheres.* (d) *The shaded image of the surface with a derived set of finite facet sphere radii.*

in the surfaces from which the patches are taken and the method by which we assemble them. Convexity of the surface follows from the nondegeneracy of the patches used to construct the surface.

The representation is particularly simple since every patch has only circular lines of curvature. There is a natural link between spherical inversions and the use of cyclides for surface fitting because the inversions (1) always map lines of curvature to lines of curvature and (2) always map circles to circles, and the cyclides are exactly those surfaces with only circular lines of curvature (Nutbourne and Martin make extensive use of this fact and define their cyclides in this manner [12, p. 219]).

The radially monotonic behavior of each edge cyclide implies that the data points are at points of maximum curvature of the surface. Concentration of curvature at the vertex spheres is intentional because this provides designers with information on how the data points will be treated and even how to specify data points for the surface in mind. The distribution of curvature is fairly flexible within the bounds set by the tangent plane polyhedron and the algorithm of §7.5.4. The algorithm provides a symmetrical way to increase the curvature of each facet patch while maintaining convexity.

The symmetry of the bounds on the facet sphere radii in the algorithm of §7.5 insures that the algorithm will consistently generate the same surface given the same initial data. This is important because even if we could decide on criteria for an optimally minimal set of facet sphere radii, generally it would not be unique.

Acknowledgments
The first author thanks the Department of Defense and the Air Force Office of Scientific Research for support during this research.

References

[1] E. Andersson, R. Andersson, M. Boman, T. Elmroth, B. Dahlberg, and B. Johansson, *Automatic construction of surfaces with prescribed shape*, Comput. Aided Design, 20 (1988), pp. 317–324.

[2] M. S. R. Anjaneyulú, *Elements of Modern Pure Geometry*, Asia Publishing House, 1964.

[3] W. Boehm and D. Hansford, *Parametric representation of quadric surfaces*, Math. Model. Numer. Anal., 26 (1992), pp. 191–200.

[4] M. do Carmo, *Differential Geometry of Curves and Surfaces*, Prentice-Hall, Englewood Cliffs, NJ, 1976.

[5] S. Chern, *Curves and Surfaces in Euclidean Space* in Studies in Global Geometry and Analysis, The Mathematical Association of America, 4 (1967), pp. 16–56.

[6] J. L. Coolidge, *Treatise on the Circle and the Sphere*, Chelsea, New York, 1971 (Oxford, 1916).

[7] H. S. M. Coxeter, *Geometry Revisited*, Mathematical Association of America, Washington, DC, 1967.

[8] H. Eves, *Fundamentals of Geometry*, Allyn and Bacon, Boston, MA, 1969.

[9] G. Farin, *Editorial*, Comput. Aided Design, 20 (1988).

[10] T. Gallagher, Ph.D. dissertation, Department of Mathematical Sciences, Rensselaer Polytechnic Institute, Troy, NY, 1992.

[11] R. A. Johnsson, *Advanced Euclidean Geometry* (*Modern Geometry*), Dover, New York, 1960 (Houghton and Mifflin, 1929).

[12] A. Nutbourne and R. Martin, *Differential Geometry Applied to Curve and Surface Design Volume* 1: *Foundations*, Ellis Horwood Limited, 1988.

[13] F. Preparata and M. Shamos, *Computational Geometry*, Springer-Verlag, New York, 1985.

PART

3

Interactive Techniques for Aesthetic Surface Design

CHAPTER

8

The Highlight Band, a Simplified Reflection Model for Interactive Smoothness Evaluation

Klaus-Peter Beier and Yifan Chen

8.1. Introduction

The design of surfaces plays an important role in various industries, particularly in automotive design, ship design, and aerospace design [1]. Graphics-oriented computer-aided design tools are in use for the creation and manipulation of surfaces and a host of supporting mathematical methods has been developed in the past. Bézier and B-spline representations are most commonly used, and the NURBS (the Non-Uniform Rational B-Spline) has become almost an industry standard for curve and surface representation. The design of surfaces can be governed by various objectives ranging from satisfying geometric conditions or manufacturing aspects to aerodynamic, hydrodynamic, or aesthetical requirements. A common goal is a "smooth" surface, a design criterion which is difficult to define in quantitative terms. Various visual methods are in use, allowing the evaluation of smoothness and the detection of otherwise invisible shape irregularities. Such surface analysis methods include the display of curvature maps, geodesic paths, isophotes, reflection lines, and other indicators (e.g., [2]–[5]). The resulting patterns are sensitive indicators and become distorted in areas of surface irregularities. While some of these methods, e.g., curvature maps, display "artificial" indicators, reflection lines are more "natural" and easier to comprehend. By simulating the effect of a realistic lighting environment, the reflection of a linear light source on an assumed glossy surface is calculated and displayed for a given viewer's position.

In the following sections, a simplified reflection line model called highlight lines will be introduced and will be expanded into the concept of highlight bands. The model is viewer independent and suitable for fast, interactive smoothness evaluation. The properties of highlight bands are explained and application examples are given. A robust algorithm for the finding of these highlight bands is presented. The algorithm was designed for optimal performance on a given graphics supercomputer with vector and parallel processing capabilities [6]. The major objective in the algorithm's development was the

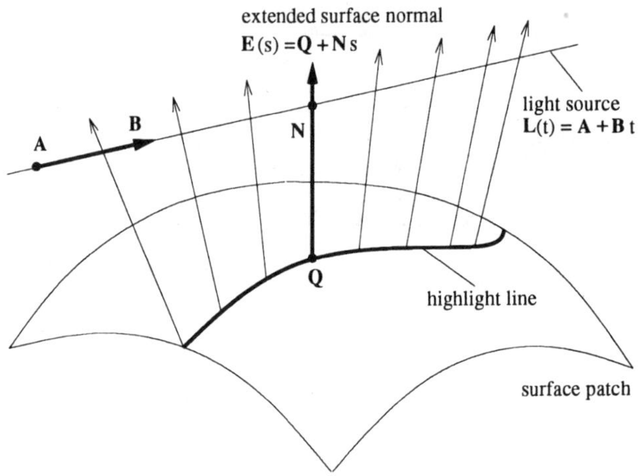

FIG. 8.1. *Definition of a highlight line.*

display of dynamically changing highlight bands in real time response to various user actions like light source manipulation or surface modification. This objective was achieved through stepwise improvements of the algorithm's performance and code optimization for the given supercomputer architecture.

8.2. Highlight Lines and Highlight Bands

A highlight line is created by an assumed linear light source idealized by a straight line with an infinite extension. The line is positioned somewhere above the surface under consideration. As illustrated in Fig. 8.1, the highlight line is defined as the collection of all surface points for which the extended surface normal passes through the linear light source.

A highlight line is a simplified reflection line model in the sense that it is viewer independent. The surface and light source may be rotated together in order to facilitate the inspection of the highlight line. In contrast to reflection lines, the path of a highlight line will not change if the user changes his or her viewpoint. The highlight line can be interpreted as an abstract "imprint" of the light source on the surface. The shape of this imprint is governed by the surface normal directions and is sensitive to the changes of these directions. Since the viewer is removed, the highlight line is controlled by fewer parameters. This simplifies the user interface and reduces the required calculation.

The criterion that a surface point belongs to a highlight line can be formulated as follows: assume that the linear light source (Fig. 8.1) is defined by

$$\mathbf{L}(t) = \mathbf{A} + \mathbf{B}t, \tag{8.1}$$

where $\mathbf{A}$ is a point on the light source and $\mathbf{B}$ is a vector defining the direction of the light source. For a given point $\mathbf{Q}$ on the surface, let $\mathbf{N}$ be the surface normal vector at point $\mathbf{Q}$. The extended surface normal $\mathbf{E}(s)$ is a line passing through $\mathbf{Q}$ with the direction given by $\mathbf{N}$ and defined by

$$\mathbf{E}(t) = \mathbf{Q} + \mathbf{N}s. \tag{8.2}$$

Point $\mathbf{Q}$ belongs to the highlight line if both lines $\mathbf{L}(t)$ and $\mathbf{E}(s)$ intersect or if the perpendicular distance d between both lines (see, e.g., [7])

$$d = \frac{|(\mathbf{B} \times \mathbf{N}) \cdot (\mathbf{A} - \mathbf{Q})|}{\|\mathbf{B} \times \mathbf{N}\|} \tag{8.3}$$

is zero.

The concept of highlight lines can be extended by replacing the idealized straight line with a cylinder of a given radius, thereby simulating a more realistic light source. This will expand the highlight line into a highlight band. The definition (8.1) of the light source is now restricted to the centerline of the cylinder. A point $\mathbf{Q}$ on the surface belongs to a highlight band if the extended surface normal at $\mathbf{Q}$ passes through the cylinder, i.e., if

$$d \leq r, \tag{8.4}$$

where d, as defined in (8.3), is the perpendicular distance of the extended normal from the center of the light source and r is the radius of the light source cylinder.

8.3. Highlight Band Properties and Examples

Compared with the highlight line, the highlight band provides additional information through its width. Figure 8.2 illustrates the manifestation of highlight bands in a plane normal to the linear light source $\mathbf{L}(t)$ for three different shapes. The convex shape (a) exposes a narrow band, the flat shape (b) exposes a wider band, and the concave shape (c) exposes the widest band. The band width is directly connected to the shape and to the local curvature of the surface in the intersecting plane normal to the light cylinder.

The graphical representation of varying band width can be further enhanced. Since each point of a highlight band carries a distance d within the

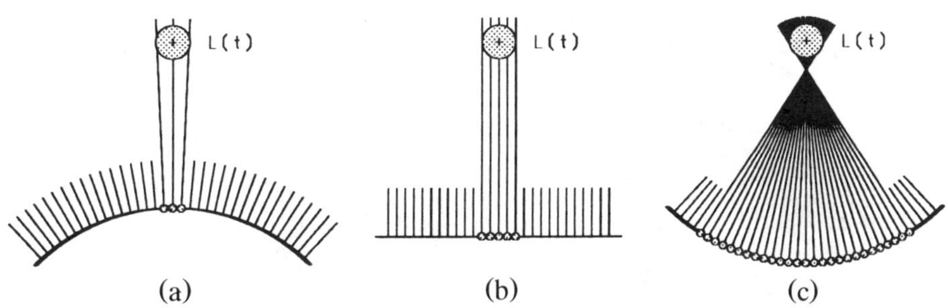

FIG. 8.2. *Manifestation of a highlight band in a planar surface intersection with a convex* (a), *flat* (b), *and concave* (c) *shape.*

range $0 \leq d \leq r$, this piece of information can be used to color-code or gray-shade the point. In all following highlight band graphs, a dark intensity is used for points where the normal extension passes through the center of the cylindrical light source ($d = 0$); a light intensity indicates that the normal extension is tangential to the light source cylinder ($d = r$). The color-coding or gray-shading scheme allows the viewer to recognize the path of the highlight line within the highlight band and provides additional information related to the normal changes within the band.

In Fig. 8.3, a cylindrical light source (represented by a straight line) is placed at the convex side of a cubic Bézier surface. The resulting highlight band indicates the transition from flat (at the patch boundaries) to increasingly convex in the center of the patch through a decreasing width of the band. Figure 8.4 shows the opposite effect. The light source is now placed at the concave side of the same patch and the resulting highlight band becomes wider at the center. Note that the band width at the patch boundaries in Fig. 8.3 is greater than shown in Fig. 8.4 due to the different location of the light source.

More complex highlight bands are shown in Figs. 8.5 and 8.6. These graphs illustrate the manifestation of intersecting bands and circular shapes created by a single light source. The resulting band patterns are typical for concave shapes where a highlight band may split into three bands as evident from

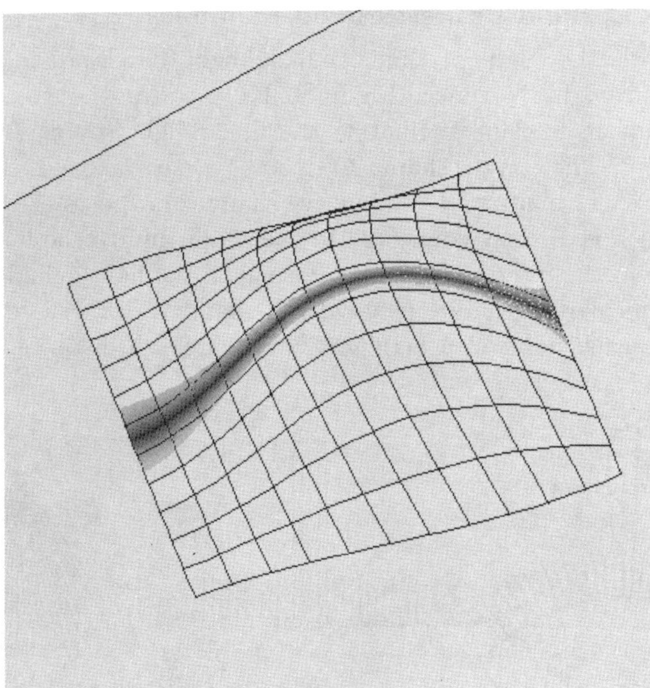

FIG. 8.3. *Highlight band created by a light source located off the convex side of a Bézier surface patch.*

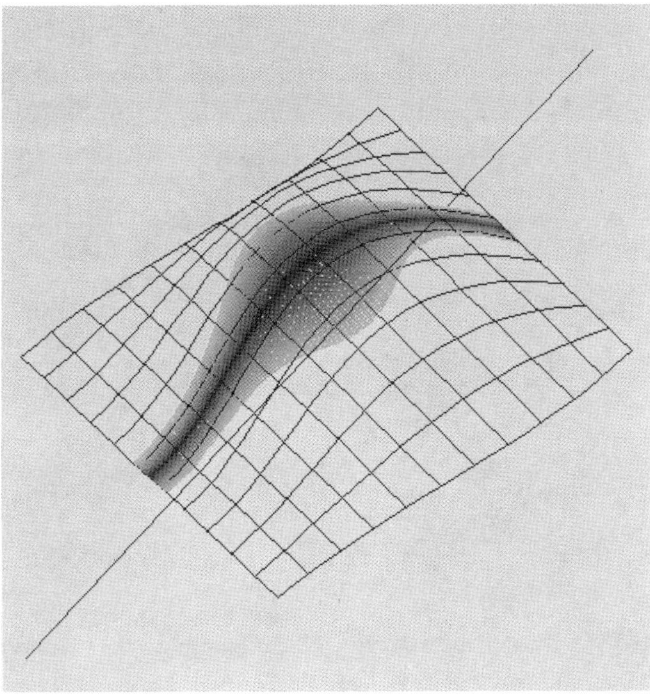

FIG. 8.4. *Highlight band created by a light source located off the concave side of a Bézier surface patch.*

Fig. 8.7. Circular highlight bands correspond to similar patterns observed in a real lighting environment. In automotive styling, for example, concave surface features are frequently found in transition areas (e.g., around a wheel well) and expose this type of reflection distortion.

Like the reflection lines and isophotes, the formulation of the highlight lines and highlight bands involves the evaluation of surface normals. The normal of a parametric surface $\mathbf{S}(u, v)$ is defined by the cross product of its two first derivatives with respect to u and v. For this reason, the highlight lines and highlight bands can be considered as the first-order smoothness indicators, as compared with the second-order indicators, such as a curvature plot. In other words, the highlight lines and highlight bands are capable of magnifying surface discontinuity by an order of one. This capability is exemplified in Figs. 8.8 and 8.9.

Figure 8.10 shows a real car surface patch with a minor dent in the middle. The wireframe rendering is not able to reveal the problem. The highlight band, however, clearly indicates the irregularity through a "glitch" lying across the dent, which corresponds to a dark strip on the curvature plot (see Fig. 8.11). The surface after refinement is shown in Fig. 8.12. Obviously, the highlight band becomes much better, as does the curvature distribution as shown in Fig. 8.13.

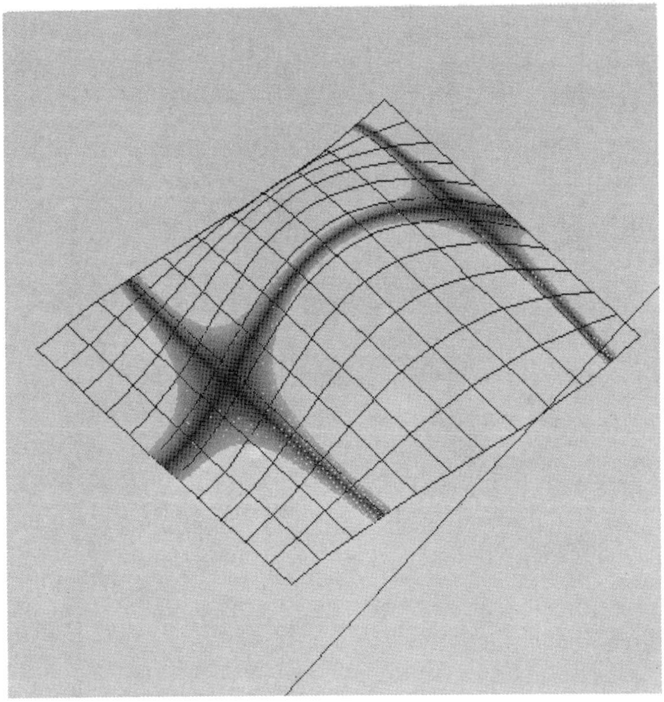

FIG. 8.5. *Self-intersecting highlight band created by a single light source.*

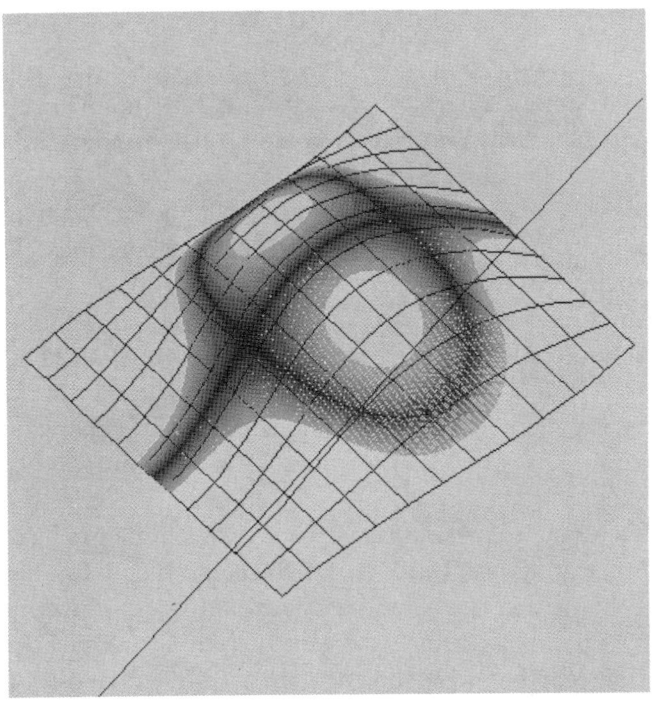

FIG. 8.6. *Highlight band with circular shape created by a single light source.*

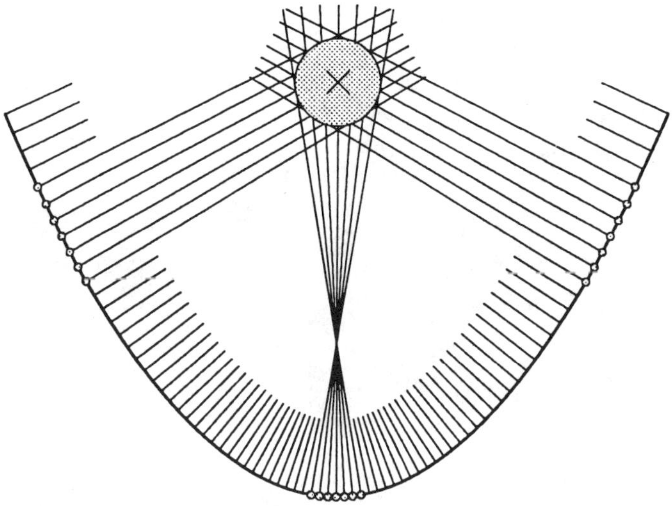

FIG. 8.7. *A highlight band may split into three bands when the light source is located off the concave side of a surface.*

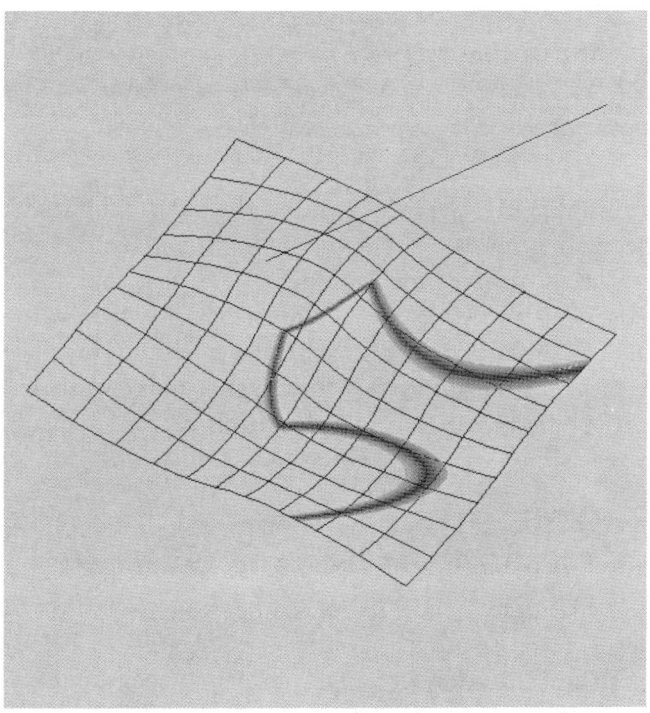

FIG. 8.8. *A quadratic* B-*spline surface defined by* 4×4 *control points. The surface has four internal patches connected with* C^1 *continuity. The highlight bands are matched along the patch boundaries with only* C^0 *continuity.*

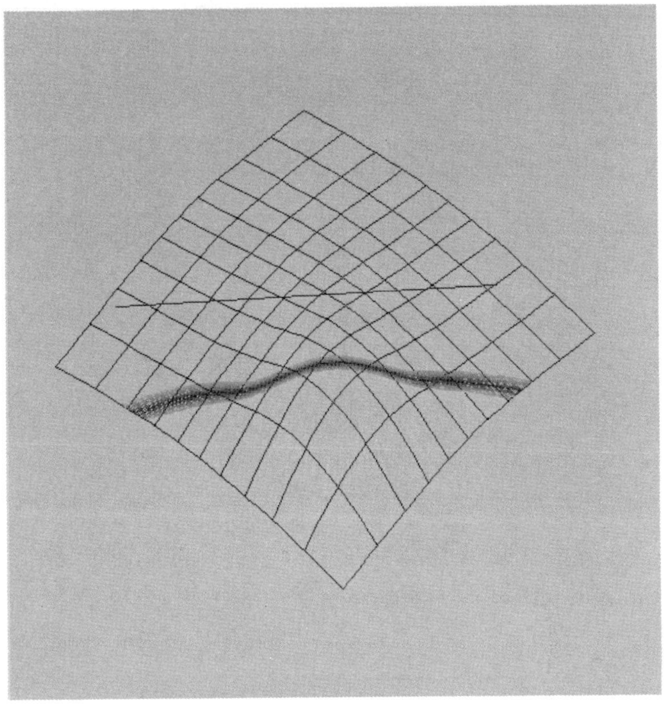

FIG. 8.9. *A cubic* B*-spline surface defined by* 5×5 *control points. The surface has four internal patches connected with* C^2 *continuity. The highlight bands are matched along the patch boundaries with only* C^1 *continuity.*

Another example (Fig. 8.14) illustrates the rather "dramatic" effect of a small surface concavity on the shape of the highlight band.

In an interactive graphics environment, the user can inspect the entire surface by translating and rotating the light source and thereby sweeping the highlight band over the surface. As an additional parameter, the radius r of the light cylinder can be changed in order to widen or narrow an existing highlight band. Finally, the change of the highlight bands can be displayed in response to surface modifications, i.e., the user can see the effect of any surface change immediately.

8.4. Principles of a Highlight Band Algorithm

The creation of highlight bands in real time response to any of the above-mentioned user actions requires significant computational power and a well optimized algorithm. In addition, high-performance graphics are needed to display the dynamically changing three-dimensional objects. To achieve the desired real time response, a graphics supercomputer with vector and parallel processing capabilities was chosen as the target system for the development of a highlight band algorithm. The system combines the power of a minisuper-computer with high-performance graphics and is mostly utilized in a single-

FIG. 8.10. *A bicubic B-spline surface with a small dent in the middle. The highlight band clearly indicates the problem.*

user setting. FORTRAN and the system's optimizing compiler were used. The graphics part was implemented in PHIGS+, an extension of the standardized graphics subprogram package PHIGS (Programmer's Hierarchical Interactive Graphics System).

Various approaches can be used to find the highlight bands. A strategic search method may be suggested which, after finding a first point on a highlight line, traces this line across the surface patch and, in a second step, expands the line into a highlight band. Such an algorithm has to deal with several numerical problems and is predominantly sequential, i.e., not suitable for vector or parallel processing. In general, the tracing of smoothness indicators over a surface requires the integration of coupled differential equations and the use of a numerical stepping procedure as, for example, described in [2] and [3].

The algorithm presented in this chapter uses an exhaustive search method. Its operations are rather simple (if compared with a strategic search method) but need to be executed in great numbers in order to achieve reliable results. The developed exhaustive search method covers the entire surface patch with a densely spaced grid of $n \times m$ points and determines for each point if the point belongs to a highlight band. According to the highlight band definition, the following basic steps are required:

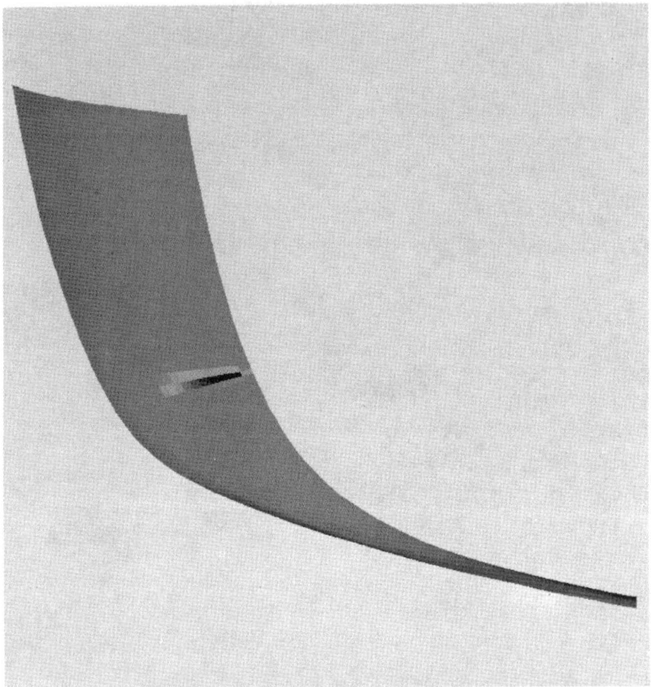

FIG. 8.11. *The average curvature plot of the surface shown in Fig.* 8.10.

FIG. 8.12. *The* B*-spline surface from Fig.* 8.10 *after refinement.*

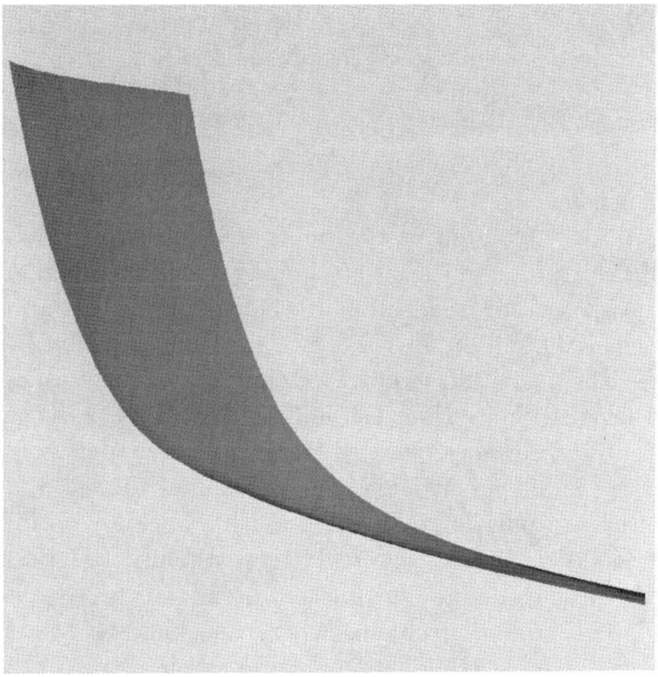

FIG. 8.13. *The average curvature plot of the refined surface in Fig.* 8.12.

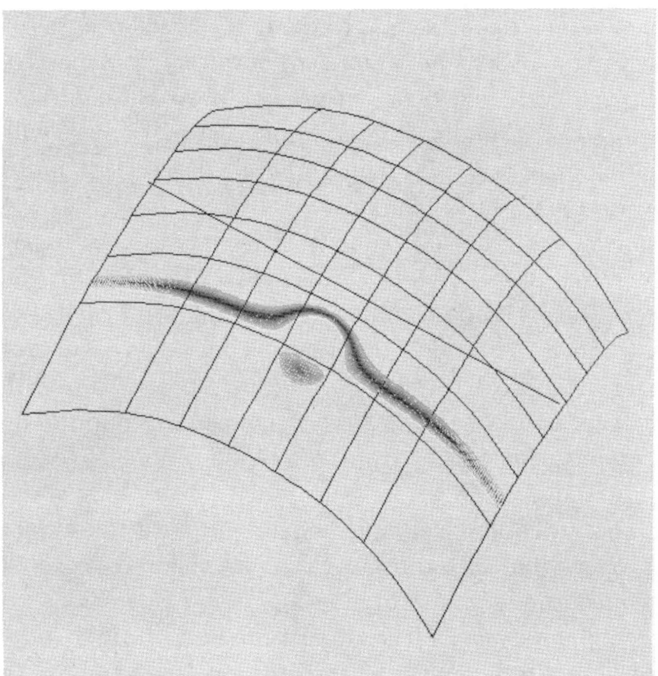

FIG. 8.14. *A small concavity on a surface dramatically affects the shape of the highlight band.*

Step (A): Find all points $\mathbf{Q}_{i,j}$ on a given point grid ($1 \le i \le n$; $1 \le j \le m$).
Step (B): Find all normals $\mathbf{N}_{i,j}$ for all points $\mathbf{Q}_{i,j}$.
Step (C): Find all distances $d_{i,j}$ between $\mathbf{N}_{i,j}$ and the light centerline.
Step (D): Select the points where $d_{i,j} \le r$.
Step (E): Display the selected points using a color-coding or gray-shading scheme.

The density or resolution of the point grid ($n \times m$) is instrumental for achieving reliable results and accurate highlight band shapes. A resolution is sufficient if the selected highlight points, when displayed, blend together into a continuously colored band. In Figs. 8.3–8.6, a resolution of 320×320 grid points has been used. The algorithm requires an enormous amount of operations. For the sample resolution, more than $100,000$ point vectors $\mathbf{Q}$, normal vectors $\mathbf{N}$, and distances d need to be calculated. However, these operations are identical within each step, independent from each other and, therefore, extremely well suited for vector and parallel processing. With additional optimization of the individual steps, the objective of real time response has been successfully achieved on the target system. The method described can serve as an example for the solution of similar problems where an exhaustive search with high resolution offers the most reliable and accurate results.

8.5. Development of an Optimal Performing Code

The exhaustive search method, characterized by large amounts of mostly identical and independent operations, promised to perform best on a supercomputer architecture. Optimal performance, however, does not come automatically. A multiphase code improvement process was required to gain the desired real time response. The underlying concept for this process will be described briefly in the following sections. A more detailed description of the entire optimization process is given in [8].

As the single criterion for code improvement, the elapsed wall clock time for executing steps (A)–(E) was used. This is in contrast to the performance measure Mflops (millions of floating point operation per second) commonly used to assess code optimization on a given supercomputer with a known theoretical peak performance (also measured in Mflops). Time-consuming floating point operations are the most critical parts in scientific and engineering computation and, therefore, code development for supercomputers concentrates around the Mflops measure.

The first task in code development is to estimate the number of required floating point operations in a first operational version of the code. The second task is to reduce the number of floating point operations to a minimum through a careful analysis of the algorithm and through appropriate code decomposition. The last task concentrates on executing as many floating point operations per time unit as possible by restructuring the code for optimal use of the vector and parallel processors. In this phase, the achieved

Mflops performance (together with other run-time statistics) is inspected as an indicator for unused system resources and the remaining potential for speed improvement.

Throughout the three stages, progress in code optimization was constantly measured in terms of wall clock time by using a bicubic Bézier patch (e.g., [9]) as a benchmark surface.

8.5.1. Analysis of floating point operations. The calculation of surface points (step (A)), surface normals (step (B)), and distances (step (C)) contain all the floating point operations of the algorithm. Step (D) (select points) consists only of a loop with an `IF` statement. The final step (E) deals predominately with the creation, transformation, and display of graphics data structures. Although time consuming, step (E) will not be discussed in this paper since the use of an existing PHIGS+ implementation restricts the optimization of the processes involved. However, the time used by step (E) must be included in the total response time assessment.

In a first operational code version, a point $\mathbf{Q}$ on the benchmark Bézier surface was found using

$$\mathbf{Q}(u,v) = \sum_{k=1}^{4}\sum_{l=1}^{4} B_k(u)B_l(v)\mathbf{P}_{k,l} \tag{8.5}$$

and the surface normal $\mathbf{N}$ from

$$\mathbf{N}(u,v) = \mathbf{Q}_u(u,v) \times \mathbf{Q}_v(u,v). \tag{8.6}$$

To avoid the computationally expensive square root operation in (8.3), the squared distance

$$d^2 = \frac{((\mathbf{B}\times\mathbf{N})\cdot(\mathbf{A}-\mathbf{Q}))^2}{(\mathbf{B}\times\mathbf{N})\cdot(\mathbf{B}\times\mathbf{N})} \tag{8.7}$$

was calculated and the selection criterion (8.4) was modified into

$$d^2 \le r^2. \tag{8.8}$$

Close to 500 floating point operations are required in steps (A)–(C) to find $\mathbf{Q}$, $\mathbf{N}$, and d^2 for a single surface point when using (8.5)–(8.7) directly. For the sample grid resolution of 320×320 points, approximately 50 million floating point operations need to be executed. This sizable number of operations must be executed several times per second in order to achieve real time response for the highlight band generation. A total time of 18.1 seconds for a single highlight band calculation on the benchmark surface was measured on a Stardent GS2000 graphics supercomputer. This is far from real time response. Yet code improvement with the goal to reduce the total time to below one second seems to be feasible: the vector processor of the target system has a theoretical peak performance of 80 Mflops which indicates a sufficient margin for speed improvement. Furthermore, the basic equations (8.5) and (8.6) can be processed in a more efficient way.

8.5.2. Code decomposition. A closer look at steps (A)–(D) reveals that not all operations need to be executed for each of the possible user actions. Moving the light source requires only steps (D) and (E) with the relative inexpensive calculations in (8.7). Changing the light source radius involves only a loop over the selection criterion (8.8). Steps (A) and (B) are critical and need to be executed during surface modification. The commonly used technique of precalculated blending functions [10], [11] was introduced to improve the performance of these steps. For each grid point, a set of specially derived mixed blending functions for use in point and normal evaluation is calculated at the beginning of the process and the initial surface points and normals are found. If the user modifies the surface by moving a control point, the following, which is a very efficient updating mechanism, can be applied:

$$\mathbf{Q}_{[\text{new}]}(u,v) = \mathbf{Q}_{[\text{old}]}(u,v) + (\mathbf{P}_{k,l[\text{new}]} - \mathbf{P}_{k,l[\text{old}]})BM_{k,l}(u,v), \tag{8.9}$$

where

$$BM_{k,l}(u,v) = B_k(u)B_l(v) \tag{8.10}$$

is one of the precalculated mixed blending functions. A similar updating mechanism can be derived for the tangent vectors $\mathbf{Q}_u$ and $\mathbf{Q}_v$ in (8.6).

Figure 8.15 illustrates the resulting modular structure of the algorithm. Besides the graphics part, up to five modules are executed in response to the four user actions listed at the bottom. The dashed line in Fig. 8.15 indicates a status where the surface points $\mathbf{Q}$ and the normals $\mathbf{N}$ are available. The modules above can be replaced if a different surface type is used. The modules below the dashed line are independent of the surface definition.

In the refined procedure, the maximum number of floating point operations required in response to interactive user actions (Modules 3, 4, and 5) has been significantly reduced to a total of 60 per surface point. The precalculation of the various mixed blending functions (Module 1) and the finding of the initial surface points and normals (Module 2) is still computationally intensive (340 floating point operations per surface point). Modules 1 and 2, however, are uncritical for the response time to the user actions considered. They are executed once during a startup phase at the beginning of the process and need to be repeated only if the user changes the grid resolution.

8.5.3. Code optimization for vector and parallel processing. In the final phase, the code was fine-tuned for optimal performance on the vector and parallel processing architecture of the target system. The algorithm consists almost entirely of FORTRAN `DO`-loops and is, therefore, well suited for vectorization. All loops were inspected separately and, if appropriate, restructured for improved vector performance. A host of well-known techniques for loop restructuring (e.g., [12] and [13]) has been applied in this study. After achieving best vector performance, further improvement was gained through parallel processing of vectorized loops in selected code segments.

The results of code optimization for a supercomputer architecture can be significant. For example, the speed of point evaluation in Module 2 was improved by more than twentyfold through loop unrolling and execution of loop subsets on the four available parallel processors. The simple loop in Module 5 (without any floating point operations) was executed 3.8 times faster after distributing the load equally among the four parallel processors.

8.5.4. Global results. Table 1 shows the finally achieved results for each module from Fig. 8.15 for three selected grid resolutions. The table shows clearly that after the initial calculations in Modules 1 and 2 are completed, the graphics part becomes the bottleneck of the algorithm. The system's response time for each of the four user actions can be found by accumulating the execution times according to the flow in Fig. 8.15. The results are presented in Table 2 and include the unavoidable time for the graphics part. The critical operation "modify surface" can be executed close to five times per second when using the low grid resolution. The user perceives an almost smooth and continuous system reaction in response to the move of a control point. With higher resolutions, the user encounters an increasingly "stubborn" behavior

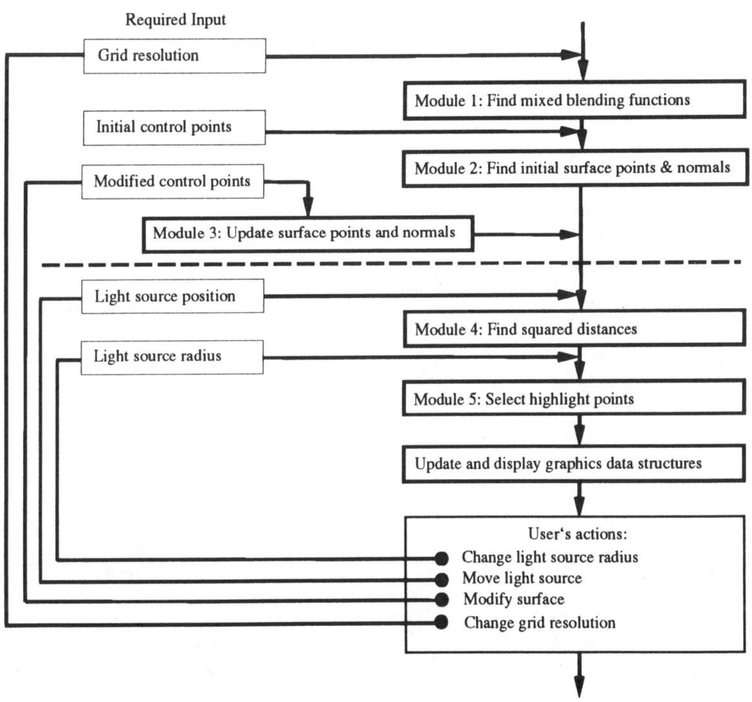

FIG. 8.15. *Final structure of refined procedure.*

TABLE 8.1

Final execution time (wall clock) in seconds of each module from Fig. 8.15 *for three different grid resolutions.*

Grid resolution $n \times m$: Total number of points:	200×200 40,000	300×300 90,000	400×400 160,000
Module 1: Find mixed blending functions	0.078	0.170	0.294
Module 2: Find initial points and normals	0.258	0.580	1.032
Module 3: Update points and normals	0.013	0.030	0.054
Module 4: Find squared distances	0.032	0.071	0.127
Module 5: Select highlight points	0.022	0.048	0.087
Update and display graphics structures	0.15	0.25	0.30

TABLE 8.2

Response time in seconds for the four user actions from Fig. 8.15.

Grid resolution $n \times m$: Total number of points:	200×200 40,000	300×300 90,000	400×400 160,000
Change grid resolution	0.54	1.12	1.84
Modify surface	0.22	0.40	0.57
Move light source	0.20	0.37	0.51
Change light source radius	0.17	0.30	0.39

of the system. Obviously, further attempts to improve the performance must concentrate on the graphics part, for example, by replacing the PHIGS+ calls through a direct use of the target system's low-level graphics primitives. In addition, Modules 1 and 2 (although uncritical for most of the user actions) could be further optimized by exploiting the coherence of the surface patch and the regularity of the point grid used.

8.6. Conclusion

The highlight band is a sensitive tool and, similar to other normal based smoothness indicators, magnifies surface discontinuities by an order of one. The changing band width, however, provides additional information and increases the sensitivity through supplementary pattern shapes. Furthermore, the highlight band is a viewer-independent, simplified reflection model and, therefore, easy to understand without special mathematical training. The tool is controlled by a minimum number of parameters and creates reflection patterns which correspond to patterns observed in a real lighting environment. This intuitive behavior combined with the ease of use in an interactive graphics environment has already gained appreciation and acceptance during a first introduction in automotive styling. Besides the detection of surface irregularities, the highlight bands have also been found useful in assessing and changing the overall character of an already smooth surface.

The presented exhaustive search algorithm is robust and can handle all cases of highlight band patterns. The algorithm's implementation for a graphics supercomputer provided real time response in the generation of the highlight bands. This goal, however, was not easily achieved. Obviously, the presented problem approaches the limits of the target system's capabilities. Future performance improvement in hardware and software will pave the road for further development of similar or even more complex algorithms.

Acknowledgments

The authors thank EDS Corporation and General Motors Corporation for sponsoring this study through generous research grants. Additional help from Ichiro Ueno and Dimitris Papageorgiou in preparing this paper is appreciated. Furthermore, the editor and the reviewers provided valuable and useful comments and suggestions.

References

[1] K.-P. Beier, *Free-Form Shape Design and Graphical Control of Shape Quality*, in Proceedings INCARF '89, Fourth International Conference on CAD, CAM, Robotics and Factories of the Future, Vol. 1, B. L. Juneja, Tate-McGraw-Hill, New York, pp. 7–16, New Delhi, India, December 1989.

[2] J. M. Beck, R. T. Farouki, and J. K. Hinds, *Surface Analysis Methods*, IEEE Computer Graphics and Applications, December (1986), pp. 18–36.

[3] T. Poeschl, *Detecting Surface Irregularities Using Isophotes*, Comput. Aided Geom. Design, 1 (1984), pp. 163–168.

[4] R. Klass, *Correction of Local Surface Irregularities Using Reflection Lines*, Comput. Aided Design, 12 (1980), pp. 73–76.

[5] H. Hagen, Th. Schreiber, and E. Gschwind, *Methods for Surface Interrogation*, Proceedings of the First IEEE Conference on Visualization, Visualization '90, A. Kaufman, ed., IEEE Computer Society Press, New York, 1990, pp. 187–193.

[6] R. Fichera, *The Graphics Supercomputers*, Computer Graphics World, 11 (1988), pp. 61–66.

[7] G. A. Korn and T. M. Korn, *Mathematical Handbook for Scientists and Engineers*, second ed., McGraw-Hill, New York, 1968, pp. 71–72.

[8] K.-P. Beier and Y. Chen, *Implementation of a Highlight Band Algorithm on a Graphics Supercomputer*, Technical Report, Department of Naval Architecture and Marine Engineering, University of Michigan, Ann Arbor, MI, August 1990.

[9] D. F. Rogers and J. A. Adams, *Mathematical Elements for Computer Graphics*, McGraw-Hill, New York, 1976.

[10] D. F. Rogers and S. G. Satterfield, *Dynamic B-spline Surfaces*, Proceedings of ICCAS '82, Fourth International Conference on Computer Applications in Automation of Shipyard Operation and Ship Design, North-Holland, Amsterdam, 1982, pp. 189–196.

[11] D. F. Rogers and L. A. Adlum, *Dynamic rational B-spline surfaces*, Comput. Aided Geom. Design, 22 (1990), pp. 609–616.

[12] J. M. Levesque and J. W. Williamson, *A Guidebook to Fortran on Supercomputers*, Academic Press, Inc., New York, 1989.

[13] J. J. Dongarra and S. C. Eisenstat, *Squeezing the Most out of an Algorithm in CRAY FORTRAN*, ACM Trans. Math. Software, 10 (1984), pp. 219–230.

CHAPTER

9

Interactive Design Using Partial Differential Equations

Malcolm I.G. Bloor and Michael J. Wilson

9.1. Introduction

Broadly speaking, the uses to which CAD (computer-aided design) systems are put lie somewhere in a spectrum, the extremes of which are shape design and shape representation. Although what often happens in practice is a process which lies somewhere in between, the aim of this paper is to describe a technique which is nearer to design than representation. Jensen, Petersen, and Watkins [1] characterise the current role of CAD in engineering design as the digitising of the clay models which are the first physical embodiments of a new design concept being developed by the stylist. However, they describe a future state in which stylists would be able to use CAD systems to directly create a model of their design concept, i.e., as though they were working with "electronic clay."

In CAD it is usual to generate curves or surfaces using mathematical representations based upon polynomial functions of a single parameter in the case of curves or two parameters in the case of surfaces. The type of polynomial functions used gives the various forms of curve and surface representations, e.g., Bézier curves and surfaces [2], [3], B-splines [2], [3], rational B-splines [4]; surface design using such curve/surface representations has been discussed by a number of authors. Tiller [5], for instance, describes a method of using rational B-spline curves to produce a variety of different surface shapes. He describes this method as "skinning." It creates a surface by defining a series of plane cross-sectional profiles through which the surface must pass; these can be arranged along a "path profile" which serves to orient the cross sections and order them in space. Woodward [6] also describes a variety of techniques for producing free-form surfaces which can be collectively labelled with the term "cross-sectional design." In a later work, Woodward [7] goes on to describe in some detail interactive skinning techniques using B-spline surface interpolation which, by defining an object in "outline" by a collection of projection and section curves, enables him to produce a wide variety of shapes. More recently, Jensen, Petersen, and Watkins [1] have described, using the example of automobile styling, a system for geometric modelling which aims at allowing a designer to create a shape by specifying a series of "feature lines" between which the system generates surfaces.

This paper discusses how the PDE (partial differential equation) method for surface design can be used for the interactive specification of the shape of new objects. The PDE method is a method for surface generation which produces functionally useful surfaces as the solutions of suitably chosen boundary-value problems, in particular as solutions to elliptic partial differential equations. The method was first discussed in the context of blend generation where the object was basically to find a function which was, in a certain sense, "smooth" or "fair," and which satisfied certain contintuity conditions along its boundaries [8]. The use of the method to generate sculptured surfaces has also been discussed and it has been demonstrated that surfaces serving a wide variety of functions could be produced, e.g., ship hulls and propeller blades [9]. This method is primarily a technique for surface generation rather than surface representation, and so the ultimate criteria by which its results are judged should be based upon how well the objects so described meet their functional objectives, whether they are aesthetic or more quantitative in nature. To this end, a certain amount of work has been carried out on the integration of the geometric design of PDE surfaces with their intended functionality [10].

The technique is a boundary-value approach whereby boundary conditions imposed around the edges of a surface patch control the internal shape of the surface, and this paper is concerned with the question of how the boundary conditions may be interactively chosen in order to produce desired surface shapes. The question of the choice of boundary condition has been partially addressed in earlier work [9], and this paper will further elaborate this theme by means of a qualitative discussion with examples drawn from a session at a workstation.

Previously, a variety of methods have been used to generate PDE surfaces. In some circumstances, for certain boundary conditions, it is possible to obtain surface solutions in closed form, and computationally this makes the calculation—and recalculation in response to user manipulation—of such surfaces very efficient. In the case of general boundary conditions, however, a numerical solution to a PDE must be sought, and a number of different techniques have been used, e.g., finite difference [9], finite element [11], and collocation [12]. Although some of these are very efficient as far as numerical methods go, given the present state of CPU performance in the workstations commonly available it does take longer than a fraction of a second to generate a solution and as far as interactive design is concerned this is not ideal. However, a method will be described elsewhere whereby solutions to the generating PDEs may be obtained in closed form for most cases of practical interest [13]. Thus, the interactive manipulation of closed form surfaces is a practical proposition, given a workstation of moderate computational power and provided the workstation's three-dimensional graphics performance is adequate. In the work described here it is important to bear in mind that the changes in a surface that result from user manipulation are to all intents and purposes instantaneous, thus allowing a designer to work in real-time.

9.2. The PDE Method

The PDE method produces a surface as a solution to a suitably chosen partial differential equation. Past work has concentrated upon solutions to the following equation:

$$\left(\frac{\partial^2}{\partial u^2} + \mathsf{a}^2 \frac{\partial^2}{\partial v^2}\right)^2 \underline{X} = 0. \tag{9.1}$$

This equation is solved over some finite region Ω of the (u, v) parameter plane, subject to boundary conditions on the solution which usually specify how $\underline{\mathrm{X}}$ and its normal derivative $\frac{\partial X}{\partial \mathbf{n}}$ vary along $\partial\Omega$. The three components of the function $\underline{\mathrm{X}}$ are the Euclidean coordinate functions of points on the surface given parameterically in terms of the two parameters u and v which define a coordinate system on the surface. Note that in the simplest cases (9.1) is solved independently for the x, y, and z coordinates.

The partial differential operator in (9.1) represents a smoothing process in which the value of the function at any point on the surface is, in a certain sense, an average of the surrounding values. In this way a surface is obtained as a smooth transition between the boundary conditions imposed on both the function and its first derivative. The parameter a controls the relative rates of smoothing between the u and v parameter directions, and for this reason has been called the smoothing parameter in earlier work [19], although this is really a smoothing bias parameter. To see that the partial differential operator in (9.1) constitutes a smoothing operation, consider (9.2):

$$\left(\frac{\partial^2}{\partial u^2} + \mathsf{a}^2 \frac{\partial^2}{\partial v^2}\right)^2 \phi = 0. \tag{9.2}$$

We can in principal write down the solution of this equation using a Green's function [18]. If we denote the dependent variable by ϕ, where ϕ may be either x, y, or z in the present context, then the value of ϕ at any point P in Ω is given by

$$\phi(P) = \frac{1}{2\pi} \int_{\partial\Omega} \phi_c \frac{\partial(\Delta G)}{\partial \mathbf{n}} - \Delta G \left(\frac{\partial \phi}{\partial \mathbf{n}}\right)_c ds, \tag{9.3}$$

where Δ is the "modified" Laplacian operator $(\frac{\partial^2}{\partial u^2} + \mathsf{a}^2 \frac{\partial^2}{\partial v^2})$ in u and v, the subscript c denotes values on the boundary $\partial\Omega$, and $\frac{\partial}{\partial \mathbf{n}}$ denotes the partial derivative in the direction of the outward normal to the contour $\partial\Omega$. The area Ω is enclosed by $\partial\Omega$. The Green's function G for the biharmonic equation at a fixed point P is a function of Q in Ω and must satisfy the following:

(1) $\Delta^2 G = 0$ in Ω.
(2) $G = 0$ and $\frac{\partial G}{\partial \mathbf{n}} = 0$ on $\partial\Omega$.
(3) $G(Q) \to r^2 \ln(r)/4$ as $r \to 0$, where $r = |PQ|$.
(4) $G(Q)$ has no other singularities in Ω or on $\partial\Omega$.

The integral on the right-hand side of (9.3) represents the smoothing of the function and its normal derivative specified on the boundary. That is, at any point P in Ω, ϕ is a weighted average of its value and the value of its normal derivative over the bounding contour.

When one is concerned with the interactive design of a surface it is essential that the designer sitting at a workstation gets an intuitive understanding of the way in which the shape of the surface may be changed by user control. In the case of B-spline surfaces, given an initial shape with its associated set of control points, the designer can think of the control points as being in some sense elastically linked with neighbouring regions of the surface, and the manipulation of the surface through changes in the position of the control points becomes intuitive and for simple systems, predictable. However as the surface becomes more complex the relationship between the surface and its control points is less easily discerned and for practical applications design from "scratch" using this approach is extremely difficult. Local limited modifications to existing surfaces represented in this way, however, can still be handled in this fashion.

For the case of surfaces developed as solutions to a boundary-value problem, it is clear that the designer cannot be expected to be familiar with the solution of elliptic partial differential equations and the mathematics of how the boundary conditions and equation parameters influence the solution. A methodology must therefore be developed which allows the designer to operate in an instinctive way in order to produce the desired surface shape. In other words the details of the mathematical resolution of the boundary-value problem must be hidden from view.

It has been demonstrated that complex practical surfaces can be generated by this method, frequently using only a single patch [9]. The scope for effective interactive control of the surface is great since only a limited number of parameters govern the rich variety of shapes which can be obtained by virtue of the boundary-value nature of the problem. Furthermore, the surfaces generated are smooth (in fact infinitely differentiable [18]) and fair: variation in the interior of the surface cannot exceed that imposed by the boundary conditions. This is indicated by the following argument. Writing $\Delta^2\phi = 0$ for the PDE in (9.2), note that a solution to this equation also satisfies the equation $\Delta\phi = f$, where the function f satisfies $\Delta f = 0$. Now by considering a small contour around a point in the (u, v) plane, Green's theorem indicates that f $(= \Delta\phi)$ is a measure of the average directional derivative of ϕ normal to the contour, i.e., the extent to which ϕ is locally "peaked"; and since $\Delta f = 0$ we know that the extremal values of f occur on the boundaries. Hence the tendency for extremal values of the coordinate functions of PDE surfaces to occur at the

FIG. 9.1. *The variation in mean curvature over the surface of a* PDE *blend between two intersecting cylinders.*

patch boundaries. Fig. 9.1 shows the distribution of mean curvature over the surface of a blend between two cylinders that intersect at right angles [8]. Note that the largest variations in curvature take place nearest the boundaries, in particular the bottom boundary.

In considering the design of surfaces it is sensible to distinguish between free-form surfaces and blend surfaces in the sense that for blend-like surfaces the freedom of action with the boundary conditions is rather more limited than in the free-form case. However, where a free-form surface comprises a number of solution patches joined together, apart from the first surface formed, this problem has an element of the restrictions of the blend problem. In view of these considerations, it is prudent to deal with the two cases separately.

First of all we consider a free-form surface generated as a single patch. Since the surface is to be a solution of (9.1) subject to appropriate boundary conditions, it is necessary to have a very fast solver so that as the smoothing parameters in the PDE and the boundary conditions are changed, the resulting surface can be viewed "immediately," by which we mean moving picture speed. It is not within the scope of this chapter to discuss solution methods; suffice it to say that we have a solver capable of producing the necessary data on a thousand topologically rectangular elements to produce a shaded image on a Silicon Graphics 4D/420VGX in about one hundredth of a second. In particular, we use solutions of a biharmonic-like (9.1), modified so that solutions satisfying the boundary conditions can be found in closed form. This means that we have an explicit expression for the the surface $\underline{X}$ at any value of u and v.

The main purpose of this paper, however, is to consider the imposition of the boundary conditions in graphical form such that the basic form of the resulting surface shape can be foreseen by the designer. For instance, the function value $\underline{X}$ can be dealt with straightforwardly since this is simply a question of drawing some boundary shape on the screen and imposing the parametrisation of these curves. For the purposes of illustration we shall consider a surface which is periodic in the parameter v. We will then go on to consider the interactive design of a blend surface between two simple primary objects, and then finally we will consider the free-form design of a yacht hull.

9.3. The Design and Manipulation of a Simple Surface

A simple example of the process is shown in Fig. 9.2, which shows the stage in the interactive generation of a surface at a workstation. We will assume that we are solving (9.1) over the (u, v) region $[0, 1] \times [0, 1]$, and that we are considering solutions that are periodic in the v direction.

In Fig. 9.2(a), two (closed) space curves are defined by input from the mouse; in this particular case the curves are defined in terms of periodic cubic B-splines. These curves are the "images" of the isoparametric lines $u = 0$ and $u = 1$, and are the curves that will bound the patch of surface we are about to create.

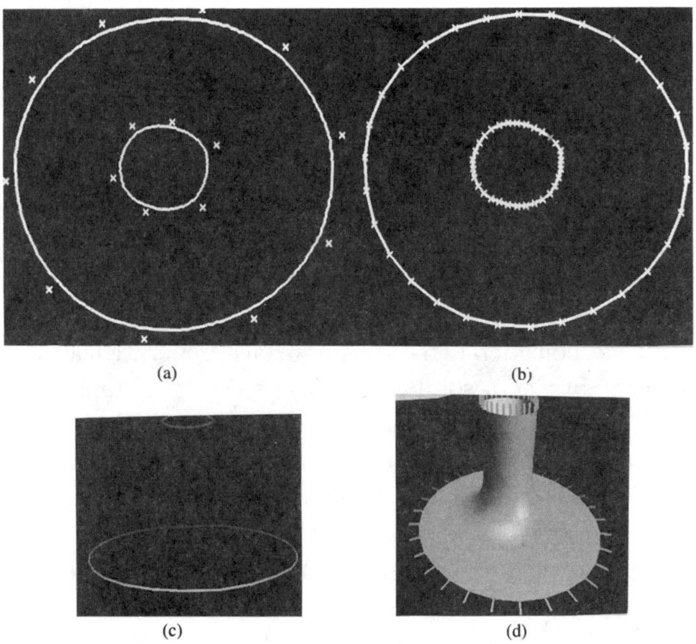

FIG. 9.2. (a)–(d) *The stages in the user specification of the function boundary conditions for a simple* PDE *surface.*

Figure 9.2(b) shows the parametrisation of these bounding curves in terms of v; the designer is free to choose this parametrisation (within the constraints of periodicity) and in this example, for simplicity, we have settled upon arc-length parametrisation, i.e., at all points on the boundary curves $|(dX(v))/dv| = 1$. Note that given the shape produced by an initial parametrisation, other parametrisations may be appropriate where the necessary changes to the original parametrisation can often be judged from the initial shape. For clarity the bounding curves $u = 0, 1$ have been defined as plane curves, and Fig. 9.2(c) shows the separation of these in 3-space (which the designer can change interactively).

So far in the design process we have effectively determined the boundary conditions on the function X, and we now need to define the boundary conditions on $\frac{\partial X}{\partial \mathbf{n}}$—which we will refer to as the derivative vector. Here, this entails defining the vector field $\frac{\partial X}{\partial u}$ along each of the curves $u = 0, 1$. The direction of the vector $\frac{\partial X}{\partial u}$ determines the direction in which the iso-v lines leave the patch boundaries, while its magnitude gives the "speed" of the iso-v lines. In the interest of clarity we will (initially) consider the case where the derivative vector field on $u = 0$ is independent of v in both magnitude and direction, and likewise the derivative vector field on $u = 1$. Note that in this example the derivative vectors on $u = 0$ are all "vertical" while the derivative vectors on $u = 1$ are all "horizontal."

The surface that results by solving (9.1) with these boundary conditions is shown in Fig. 9.2(d); note that a=1/2. The surface is rendered by calculating the function $\underline{\mathrm{X}}(u, v)$ over a rectangular (u, v) mesh, and thus generating a faceted approximation to the surface. Since the surface is given parameterically in terms of u and v, it is a simple task to calculate the surface normal at each point on the mesh. The graphics hardware of the Silicon Graphics workstation (4D/420VGX) on which these calculations were performed then produces the shaded image. A visual representation of the derivative vectors is shown by the lines drawn at certain points along the boundary curves.

Figure 9.3(a)–(d) illustrates the effect of the boundary derivative vectors on the surface by a sequence where their magnitude is varied on each boundary successively. The derivative magnitude is varied on a particular boundary by "clicking" on one of the individual vectors and then moving the mouse, the change in the derivative magnitudes being proportional to the displacement of the mouse. The relative magnitude of the derivative vectors can be judged from the length of the lines. Note that in Fig. 9.3(d) the direction of the derivative vectors on $u = 0$ has been reversed.

Figure 9.4(a)–(d) shows the effect of the smoothing parameter a upon the surface; the magnitude of a is low in Fig. 9.4(a) and high in Fig. 9.4(d). As discussed in earlier work this parameter controls the rate at which the boundary conditions are propagated into the interior of the surface patch. Where a is large the boundary conditions decay rapidly away from the boundary and a

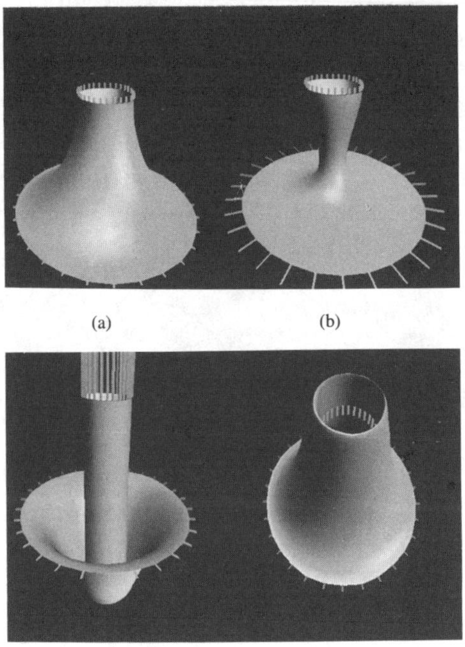

(c) (d)

FIG. 9.3. *The effect of the derivative vectors.*

(a) (b)

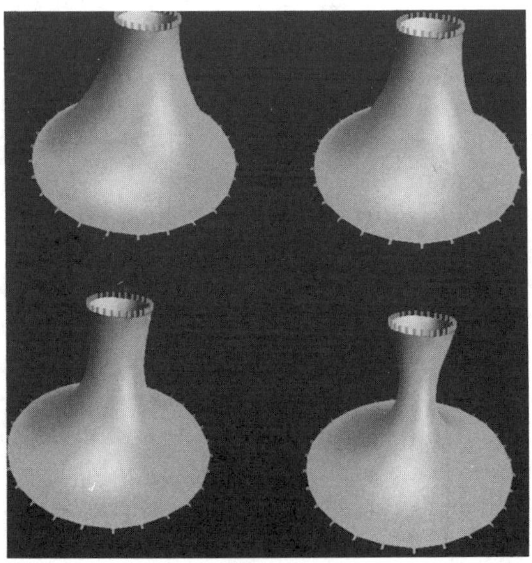

(c) (d)

FIG. 9.4. (a)–(d) *The effect of the smoothing parameter.*

"waisting" appearance in the surface is produced; whereas when a is small the influence of the boundary conditions is preserved for some distance away from the boundary. The particular workstation on which these calculations were performed comes equipped with a "dials-and-buttons" box, and the value of a was varied by movements of one of the dials.

In Fig. 9.5(a)–(d) a variation in the magnitude of the derivative vectors has been introduced on one the boundaries. In Fig. 9.5(a) because the value of the smoothing parameter is too large the effect of this variation is not apparent in the shape of the surface: the variations in derivative are being smoothed out in the vicinity of the boundary. Subsequently, in Figs. 9.5(b)–(d), the value of a is reduced and "ridges" appear in the surface adjacent to those regions of the boundary where the derivative magnitudes are relatively small.

Having introduced the sort of shape changes that can easily be effected by a user sitting at a workstation, we consider in the next section how such techniques may be used in the interactive design of a blend surface.

9.4. Interactive Blend Design

In blending we are often given the primary surfaces which define the bulk of an object's shape, and we seek secondary surfaces to form a smooth transition between the primary surfaces. This may be for aesthetic considerations, or

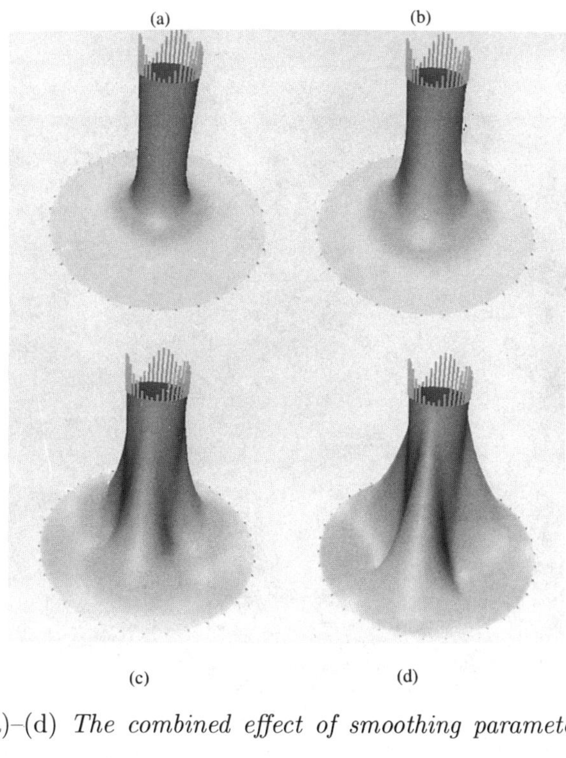

FIG. 9.5. (a)–(d) *The combined effect of smoothing parameter and derivative vectors.*

perhaps for more functional reasons, such as the relief of potential stress concentrations [14], or so that it can be manufactured.

Cheng et al. [15] have indicated how the PDE method may be used to generate PDE blends between simple quadric solids and also between parametric surfaces. In this work it was assumed that the primary surfaces were intersecting and that the boundary conditions for the PDE blend surfaces were automatically generated by offsetting from the intersection curve in the parameter planes of the intersecting surfaces. In some cirumstances, however, one can envisage that the designer may want to exercise more control over the creation of a blending surface, or perhaps the blend is of the "extended" type (see the discussion below) and so there are no intersection curves from which to start. In these cases the blend can be specified interactively, and one method by which this can be achieved is now described. Figure 9.6 shows a sphere and a cylinder. The sphere is positioned near the cylinder but does not intersect it. The purpose of what follows is to generate a connecting surface between these two objects which blends smoothly between them with continuity of tangent plane. Usually blends are between intersecting surfaces, but we can look at this surface as an example of an "extended" blend. This is a surface whose specific shape is not particularly important so long as it satisfies certain constraints (e.g., mass, strength, and clearance) and joins certain functional surfaces whose shape is determined by their interaction with adjacent parts with which they are in contact. It has been pointed out to the authors (by D.C. Gossard, Dept. of Mechanical Engineering, MIT) that many mechanical parts are of this type.

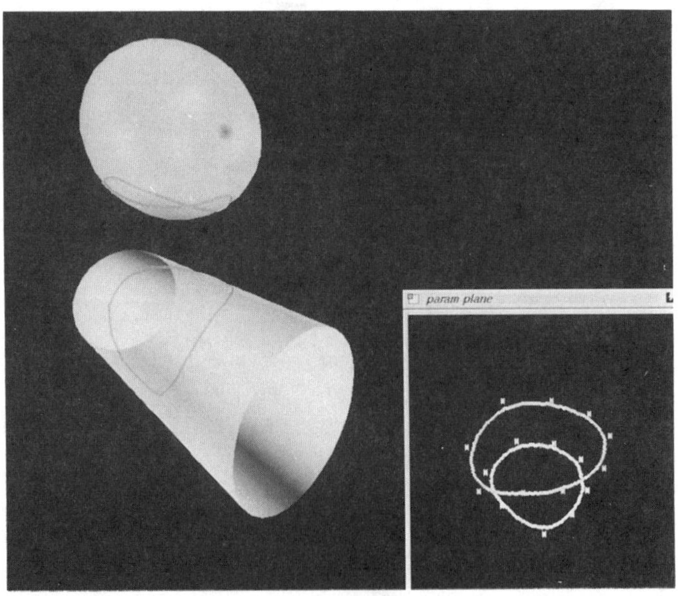

FIG. 9.6. *Primary surfaces to be blended between.*

Figure 9.6 also shows a window (inset) representing the individual parameter planes of the sphere and cylinder, which in this case are based upon the usual spherical polar and cylindrical polar coordinate systems. Note that for brevity both parameter planes are represented in the same window. To generate a blend between the sphere and cylinder we carry out the following steps, which are illustrated in Figs. 9.6–9.9. Just as in the previous example, we can input two B-spline curves using the mouse, but in this case one curve is defined in the parameter plane of the sphere while the other is defined in the parameter plane of the cylinder. As the figures show, the workstation used in this example allows us to display the curves in the two-dimensional space of the parameter planes, and also the curves they map to in E^3. Notice that since the curves are defined in the parameter planes of the primary surfaces, the curves bounding the blend are guaranteed to lie on the primary surfaces in E^3, thus satisfying the requirement of function continuity. As before we will assume that the curves bounding the blend are the iso-lines $u = 0, 1$. These boundary curves are defined in the parameter planes of both primitives by expressions of the form

(9.4) $$\theta = \theta(t), \phi = \phi(t),$$

where the surface parameters θ and ϕ are in this case expressed as periodic B-spline functions of the parameter t (note they are closed curves), and the primitives are given parametrically in terms of θ and ϕ thus

(9.5) $$\underline{X}_p = \underline{X}_p(\theta, \phi).$$

It is then convenient to generate the function boundary conditions $\underline{X}(0,v)$ and $\underline{X}(1, v)$ by assuming that around the relevant boundary curve $dv = dt$, e.g., for the boundary condition on $\underline{X}(1,v)$ we put

(9.6) $$\underline{X}(1, v + v_o) = \underline{X}_p(\theta(t), \phi(t)),$$

where v_o is a constant.

Figure 9.6 shows the two boundary curves (the blend trimlines as they are sometimes referred to) as initially defined by the mouse in the parameter planes of the two primitives. These curves then appear in E^3 on the surface of the primitives, and can be given the desired shape in E^3 by manipulating them in the parameter planes of the primitives. So that the user can see what is going on, in this part of the process the primitives can be translated and rotated in space (in this case by means of the dials-and-buttons), and have also been rendered transparent. Even though the boundary curves are manipulated by translating their control points about the parameter planes of the primitives, by getting real-time feedback from the screen the user can very quickly gain a feel of how the points should be moved in order to obtain a suitable shape for the curves in E^3. Figure 9.7 indicates how the shape of the boundary curves in E^3 are affected by changes to their shape in the parameter planes. This achievement is now described.

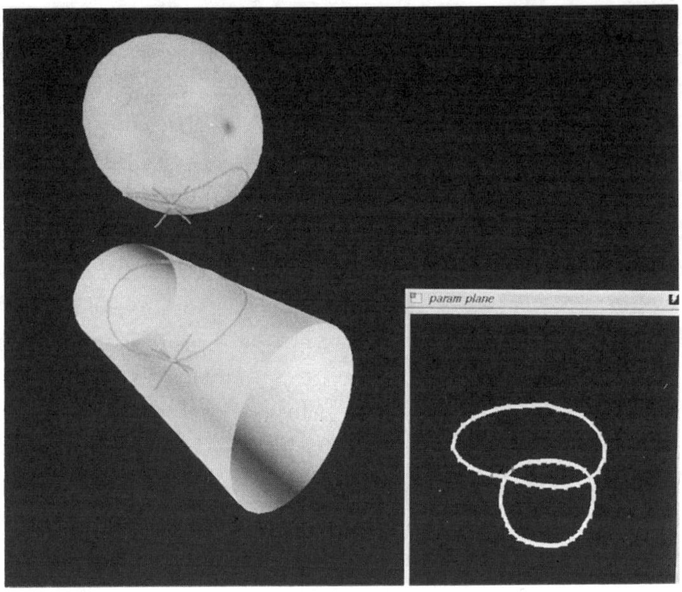

FIG. 9.7. *Modification of trimline shape.*

The next step in the process is to decide upon the parametrisation of the boundary curves. To some extent this has already been determined by the assumption that on each boundary curve $dv = dt$, and by the rotational sense in which we orginally defined the two curves in the parameter planes—in the same sense, preferably. We do need to decide, however, which points on the two curves have the same value of v. This can be achieved in the following way: on each boundary curve a number of "reference" points, equally spaced in v, are plotted; then a point on each of the two boundary curves is selected using the mouse, and it is specified that these two points have the same value of v. This process is illustrated in Fig. 9.8.

Note that the boundary curves have also been discretized. This is because a faceted approximation is used to render the blend surface, and the discrete points on the curves shown in Fig. 9.8 correspond to the corner vertices of the facets at edge of the blend.

Having decided how v varies around the boundary curves and hence the function boundary conditions, the next step is to generate the derivative boundary conditions. There are a variety of ways of doing this within the constraints set by the requirement of tangent plane continuity. One method is to take the boundary curve on each of the primitives

$$\underline{X}(t) = \underline{X}(\theta(t), \phi(t)) \tag{9.7}$$

(where for clarity we temporarily dropped the subscript p used in equation (9.5)), and then calculate its velocity vector

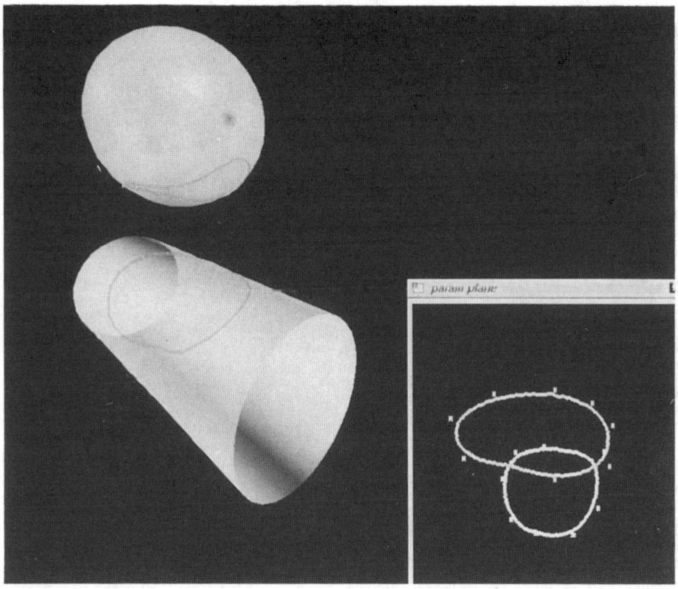

FIG. 9.8. *Choosing the parametrisation.*

(9.8)
$$\frac{d\underline{X}(t)}{dt} = \underline{X}_\theta \frac{d\theta}{dt} + \underline{X}_\phi \frac{d\phi}{dt}$$

where $\underline{X}_\theta$ and $\underline{X}_\phi$ are the coordinate vectors for the surface [16]. A suitable choice for the derivative boundary conditions $\underline{X}_u(0,v)$ and $\underline{X}_u(1,v)$ is any vector field that is nowhere parallel to the vector field defined in equation (9.8) and which also lies in the tangent plane of the primary surface. The choice that has been made here is to use on both boundaries a vector field of the following form:

(9.9)
$$X_u(v) = mX_\theta + nX_\phi,$$

where the relation between v and the curve parameter t is as defined above. The scalar quantities m and n are functions of position and are selected on the basis that

(9.10)
$$\frac{n}{m} = -\left(\frac{E + F\frac{d\phi}{d\theta}}{F + G\frac{d\phi}{d\theta}}\right),$$

where $E\,(=\underline{X}_\theta.\underline{X}_\theta)$, $F\,(=\underline{X}_\theta.\underline{X}_\phi)$, $G(=\underline{X}_\phi.\underline{X}_\phi)$ are the components of the first fundamental form for the surface [16]. This ensures that the derivative vectors X_u are perpendicular to the boundary curves; there is also the further condition on m and n that

(9.11)
$$\left(m\underline{X}_\theta + n\underline{X}_\phi\right) . \left(m\underline{X}_\theta + n\underline{X}_\phi\right) = 1.$$

Note that being a linear sum of the coordinate vectors ensures that the vector field defined in equation (9.9) lies in the tangent plane of the primary surface.

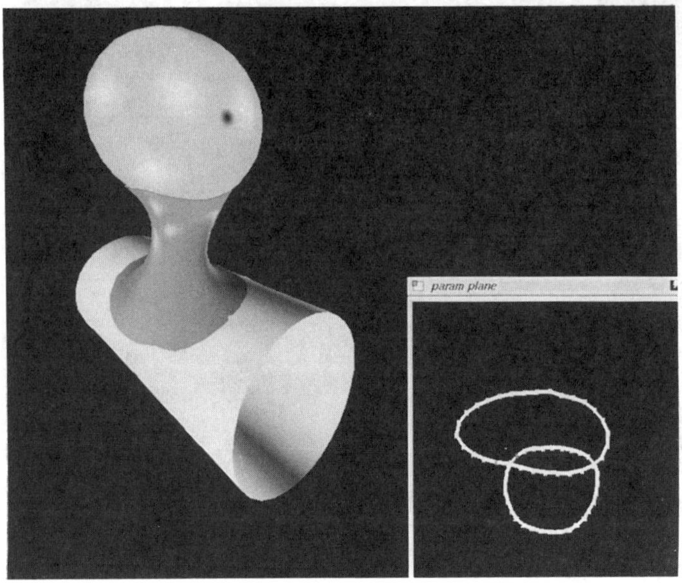

FIG. 9.9. *The final blend surface.*

Actually, to give the user more control over the shape of the blend, the derivative vector field used in this paper was based on a modified version of (9.9). Thus

$$\underline{X}_u(v) = S(m\underline{X}_\theta + n\underline{X}_\phi), \tag{9.12}$$

where S is a variable that controls the magnitude of the derivative vectors, and hence the shape of the blend in a fashion which is illustrated below. In this work S was taken to be a constant around each of the blend trimlines, i.e., independent of v, although it could be varied on each trimline independently. There is no reason, however, why this should be the case; allowing S to vary with v gives the user extra control over the blend. In fact, in certain circumstances, it would be sensible to adjust the value of S at a particular point automatically, based upon the physical separation of the point with the corresponding point on the other trimline with the same value of v.

Figure 9.9 shows the blend that results from the steps shown in Figs. 9.6–9.8 for the case where the derivative magnitude on each boundary curve is unity.

In Fig. 9.10 we can see the effect on the shape of the blend of varying the smoothing parameter, while in Fig. 9.11 we see the effect of varying the magnitude of the derivative vectors on each of the boundaries. In Fig. 9.11(a)–(b) the effect of increasing the derivative magnitude on the sphere from a low to a high value is shown, while in Figure 9.11(c)–(d) the effect of increasing the derivative magnitude on the cylinder from a low to a high value is shown.

(a)

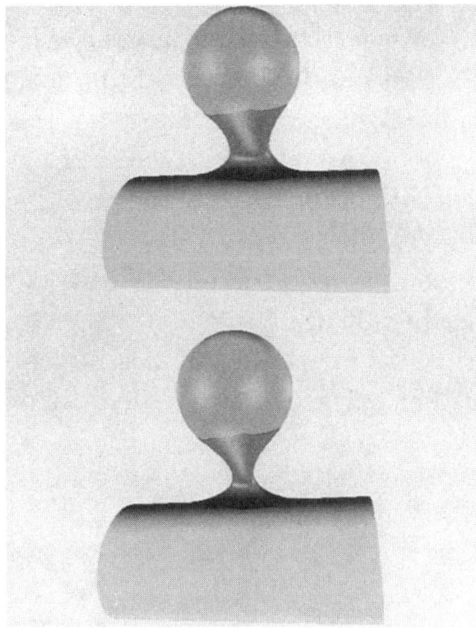

(b)

FIG. 9.10. (a)–(b) *The effect of the smoothing parameter on the blend surface.*

(a) (b)

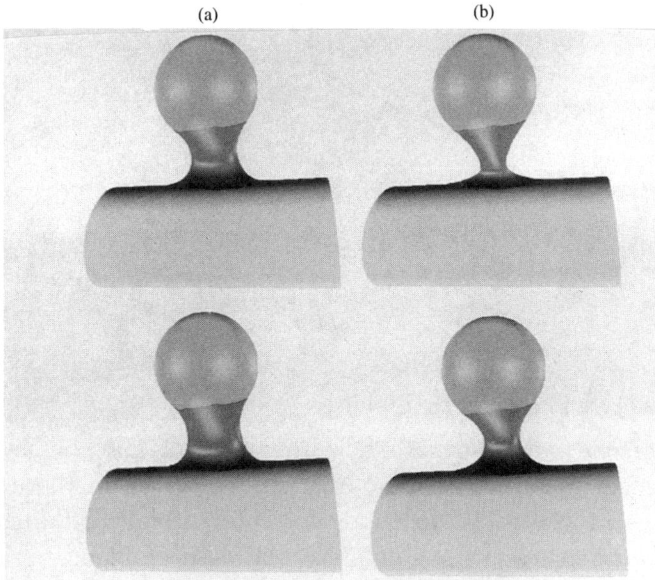

(c) (d)

FIG. 9.11. (a)–(d) *The effect of the derivative magnitude on the blend surface.*

Note that it is possible for the blend surface to intersect with the primary objects, so it may be necessary to vary the derivative magnitudes and also the smoothing parameter to prevent this from happening. Since we are blending between primary objects that can be described implicitly it is straightforward to test for the intersection of the blend surface with the primary objects.

Figure 9.12 shows two more blends, each having a different choice of bounding curves, in order to indicate the range of blend shapes that may be generated.

Note that although we have considered the case of blending between two surfaces that can be described by implicit equations, the fact that we have made use of parametric descriptions for these surfaces indicates that we could have blended equally well between surfaces that are described in terms of parametric patches.

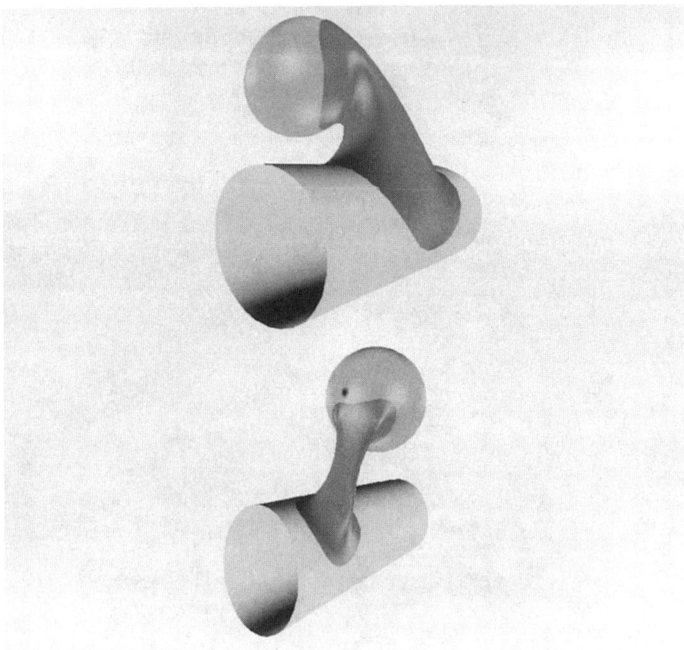

FIG. 9.12. *Alternative blend surfaces.*

9.5. Interactive Free-Form Design

In this section we will illustrate how the method may be used to design the entire surface of an object. Our discussion will be in the context of the design of a yacht hull. The methods for interactively defining the surface and manipulating it are very similar to the case of blend design discussed above; however, where there is a significant difference it is in the method for defining the derivative boundary conditions. As noted above, in the case of blend generation, although there is scope for choice within the constraints set by the requirements of continuity, we are not completely free in the choice of the derivative boundary conditions. When generating a free-form surface on the other hand, the only condition on our choice of derivative conditions is that we get a reasonable surface, which, in a sense, makes the problem of choosing these conditions rather harder.

At the start of the design process, it is an easy task to associate an initial derivative vector field around each of the boundary curves, but this may be in some way removed from the distribution of derivative vectors necessary to obtain a desirable surface shape. Even so, in our experience and in a surprising number of cases, one can get very close to the final surface shape with initial distributions that have a constant direction for the derivative vectors around each of the boundary curves, where to obtain the final shape is simply a matter of adjusting the magnitude of the derivative vectors on each of the trimlines. In other cases, however, the user is faced with the task of having to define a variable vector field in space around a curve. This section outlines a solution to this problem. We shall consider the design of a yacht hull from a periodic patch of PDE surface. We have considered this problem elsewhere from the point of view of an analytic specification of the boundary conditions, but here we describe for the first time the interactive specification of the hull shape. As in the case of blend generation, the function boundary conditions are obtained by specifying the shape of the curves that bound the surface using B-spline curves. These curves are shown in Fig. 9.13(a). The crosses mark the B-spline control points and the curves have been inputted and edited to obtain the appropriate shape. The boundary curves are taken to be plane curves, one of

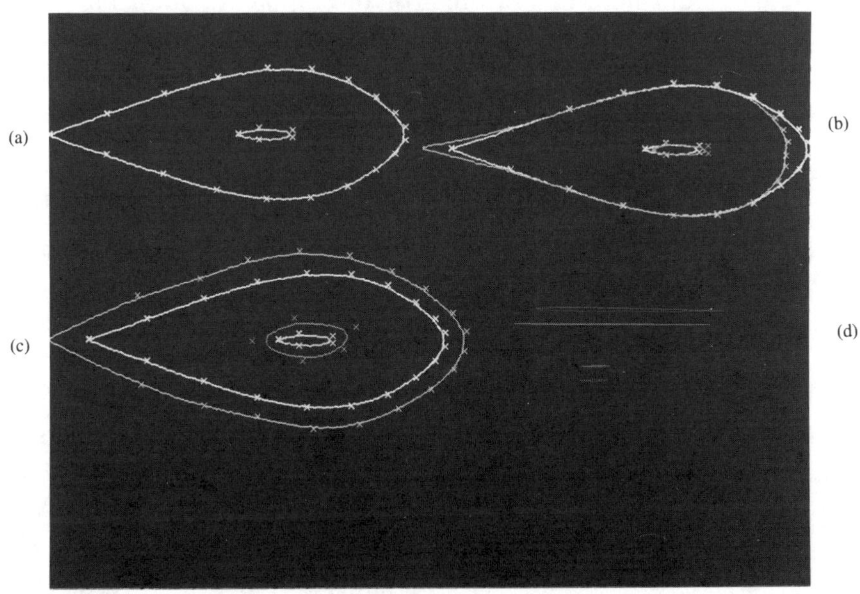

FIG. 9.13. (a)–(d) *The stages in the interactive design of a yacht hull.*

which defines the shape of the vessel's deck outline, while the other defines an elliptical curve lying at the base of the yacht's keel. The parametrisation of these curves is chosen in a similar fashion to the parametrisation of the blend trimlines in the previous section. Unlike the case of blending, however, we assume that these curves are defined in E^3 and for ease of definition we have taken them to be plane curves.

Now to define the derivative boundary conditions we proceed as shown in Fig. 9.13(b). We define two more B-spline curves which we shall refer to as derivative curves (for reasons that will become obvious), each of which is associated with one of the boundaries. These derivative curves are parametrized so that there is a one-to-one correspondence between each point on a particular derivative curve and each point on the associated boundary curve. The derivative boundary conditions at each point on the boundary curves are then obtained from the condition that at a point on a boundary curve the derivative vector is parallel to the displacement vector between it and the associated point on the derivative curve, and that the magnitude of the derivative vector is proportional to the magnitude of the displacment vector.

Figure 9.13(c) shows a suitable choice for the shape of the derivative curves. The derivative curve associated with the deck of the vessel has been chosen so that at the level of the deck the hull is raked backwards at the bow, amidships the hull is vertical, while at the stern the hull slopes backwards (a retroussé stern). The derivative curve associated with the bottom of the keel has been chosen so that the keel is raked forwards at all points around it. That these shapes of derivative curves do indeed result in an appropriate distribution of derivative vectors may not be immediately apparent until one considers Fig. 9.13(d), where it is shown that the boundary curves and derivative curves are now moved relative to each other in the direction parallel to the plane of the deck.

The surface shape that initially results is shown in Fig. 9.14(a) and by reducing the magnitude of the derivative vectors around the boundaries by a constant factor but still keeping their direction fixed, one obtains a surface shape shown in Fig. 9.14(b) which is a very passable first attempt at the design of a yacht hull. This may then be used as the initial shape in some optimisation procedure which seeks to improve its hydrodynamic characteristics [10], [17].

9.6. Conclusions

In past work we have tended to approach the generation of PDE surfaces from a rather analytical direction in that we have considered cases where the boundary conditions could be described by simple functions of the surface parameters. In this paper, by illustrating what appears upon the screen of a workstation we have tried to show how the boundary conditions can be specified in an obvious fashion using the mouse and other physical devices such as a dials-

(a)

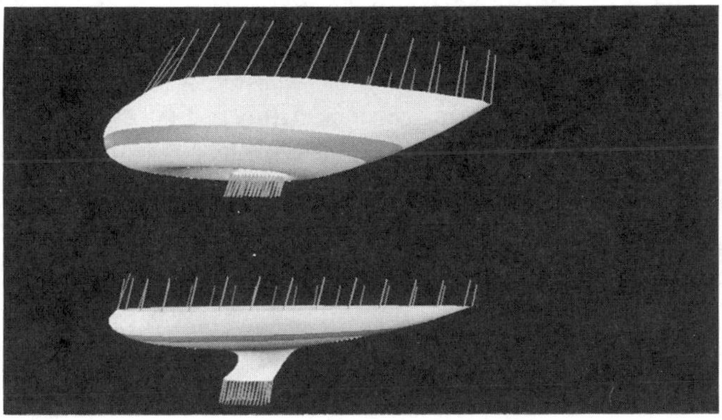

(b)

FIG. 9.14. (a)–(b) *Adjusting the derivative magnitudes to obtain the desired shape.*

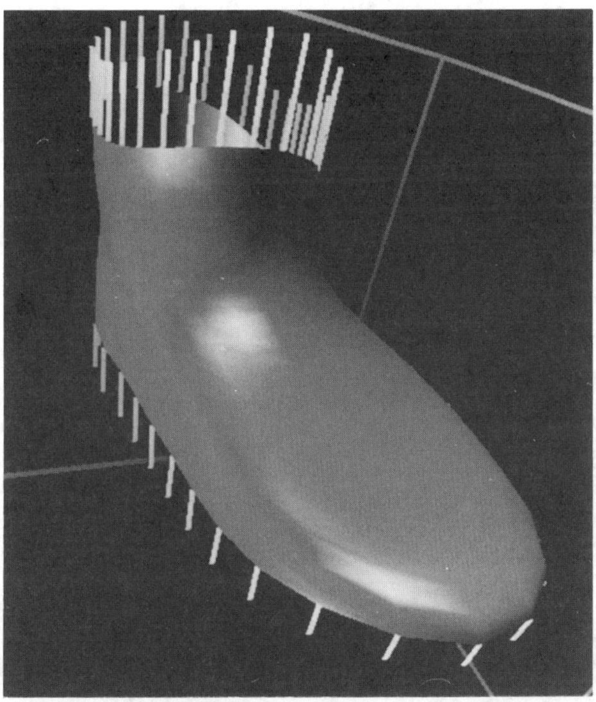

FIG. 9.15. *Boot shape designed interactively at workstation.*

and-buttons box, and that the reaction of the surface to the user manipulation is intuitively reasonable: in short, we have tried to show how a designer could use the system.

The problems and examples given so far have, for convenience, bounding curves which are planar. This is not at all necessary, and an example is shown in Fig. 9.15 where a boot shape has been designed interactively. However, there still may be difficulties in representing realistically on a screen the three-dimensionality of anything but very simple free curves drawn in space. This problem requires further work. Another example of free-form interactive aesthetic design is illustrated in Fig. 9.16. The shape of the wine glass is obtained from three boundary-value problems and parameters governing the shape are controlled through the dials-and-buttons box so that a whole variety of shapes may be produced until one suitable for selection is obtained. An added feature is that geometric constraints are easily imposed in this system so that, for example, the volume of a vessel may be held constant while the shape is changed. This is facilitated by the analytic nature of the solution to the problem, so that properties are readily calculated and constraints are used for given functional dependencies between parameters.

FIG. 9.16. *A wine glass designed interactively at a workstation; note that the wine glass has been rendered transparent.*

Acknowledgments

The authors thank the Science and Engineering Research Council for financial support, and Chris Parradine of the IBM Winchester Scientific Centre for interesting discussions.

References

[1] T. W. Jensen, C. S. Petersen, and M. A. Watkins, *Practical curves and surfaces for a geometric modeler*, Comp. Aided. Geom. Design, 8 (1991), pp. 357–369.

[2] I. D. Faux and M. J. Pratt, *Computational Geometry for Design and Manufacture*, Ellis Horwood, Chichester, UK, 1979.

[3] M. E. Mortenson, *Geometric Modeling*, Wiley-Interscience, New York, 1985.

[4] L. Piegl and W. Tiller, *Curve and surface constructions using rational B-splines*, Comput. Aided Design, 19 (1987), pp. 485–498.

[5] W. Tiller, *Rational B-splines for Curve and Surface Representation*, IEEE Comput. Graphics Applications, (1983), pp. 61–69.

[6] C. D.Woodward, *Cross-sectional design of B-spline surfaces*, Comput. and Graphics, 11, (1987), pp. 193–201.

[7] ——, *Skinning techniques for interactive B-spline surface interpolation*, Comput. Aided Design, 20 (1988), pp. 441–451.

[8] M. I. G. Bloor and M. J. Wilson, *Generating blend surfaces using partial differential equations*, Comput. Aided Design, 21 (1989), pp. 165–171.

[9] ——, *Using partial differential equations to generate free-form surfaces*, Comput. Aided Design, 22 (1990), pp. 202–212.

[10] T. W. Lowe, M. I. G. Bloor, and M. J. Wilson, *Functionality in blend design*, Comput. Aided Design, 22 (1990), pp. 655–665.

[11] J. M. Brown, M. I. G. Bloor, M. S. Bloor, and M. J. Wilson, *Generation and modification of non-uniform B-spline surface approximations to PDE surfaces using the finite-element method*, in Advances in Design Automation, Computer Aided and Computational Design, Vol. 1, B. Ravani, ed., American Society of Mechanical Engineering, (1990), pp. 265–272.

[12] M. I. G. Bloor and M. J. Wilson, *Representing PDE surfaces in terms of B-splines*, Comput. Aided Design, 22 (1990), pp. 324–331.

[13] ——, *A fast solver for the PDE method*, in preparation.

[14] J. R. Woodwark, *Blends in geometric modelling*, in The Mathematics of Surfaces II, R. R. Martin, ed., Oxford University Press, Oxford, UK, 1987, pp. 255–297.

[15] S. Y. Cheng, M. I. G. Bloor, A. Saia, and M. J. Wilson, *Blending between quadric solids using partial differential equations*, in Advances in Design Automation, Vol. 1 Computer Aided and Computational Design, B.Ravani, ed., ASME, (1990), pp. 257–263.

[16] J. J. Stoker, *Differential Geometry*, Wiley-Interscience, New York, 1969.

[17] T. W. Lowe, M. I. G. Bloor, and M. J. Wilson, *Functionality in surface design*, in Advances in Design Automation, Computer Aided and Computational Design, Vol. 1, B. Ravani, ed., ASME, (1990), pp. 43–50.

[18] E. Zauderer, *Partial Differential Equations of Applied Mathematics* Wiley-Interscience, New York, 1983.

[19] M. I. G. Bloor and M. J. Wilson, *Blend design as a boundary-value problem*, W. Straber and H.-P. Seidel, eds., in Theory and Practise of Geometric Modelling, Springer-Verlag, Berlin, 1989, pp. 221–234.

CHAPTER

10

Polynomial Splines of Nonuniform Degree: Controlling Convexity and Fairness

Alexandros I. Ginnis, Panagiotis D. Kaklis, and Nickolas S. Sapidis

10.1. Formulation of the Problem

The work presented in this paper is motivated by the following problem.

PROBLEM $(\mathcal{P})$. Let $\mathcal{D} = \{\mathbf{I_m} = (x_m, y_m)^T, m = 1(1)N\}$ be a set of points in the plane with $\mathbf{I_m} \neq \mathbf{I_{m+1}}$, $m = 1(1)N-1$, and $\mathcal{U} = \{u_1, u_2, \ldots, u_N\}$ a set of ordered real numbers with $u_1 < u_2 < \cdots < u_N$. Construct a C^2-continuous spline $\mathbf{Q}(u), u \in [u_1, u_N]$, which satisfies the following conditions: (i) $\mathbf{Q}$ interpolates the data set $\mathcal{D}$ with parametrization $\mathcal{U}$, i.e., $\mathbf{Q}(u_m) = \mathbf{I_m}$, $m = 1(1)N$; (ii) $\mathbf{Q}$ satisfies appropriate conditions $\mathcal{B}$ at the boundary points $\mathbf{I_1}$ and $\mathbf{I_N}$ of $\mathcal{D}$; (iii) $\mathbf{Q}$ preserves the "shape" implied by the data $\mathcal{D}$; (iv) $\mathbf{Q}$ is "fair" (or "visually pleasing").

The above problem is the combination of two important problems in computer-aided geometric design, namely the **shape-preserving interpolation problem** $(\mathcal{P}_{\text{shape}}) = \{\text{(i)–(iii)}\}$ and the **fairing problem** $(\mathcal{P}_{\text{fair}}) = \{(\text{i}_{\text{appr}}), \text{(iv)}\}$, where (i_{appr}) is a weak form of the interpolation constraint (i), that is, $(\text{i}_{\text{appr}}) = \{\mathbf{Q}$ *approximates the data set* $\mathcal{D}$ *with parametrization* $\mathcal{U}\}$. In the context of parametric interpolation, $(\mathcal{P}_{\text{shape}})$ can be phrased as the problem of ensuring that the interpolatory spline $\mathbf{Q}$ preserves the local convexity properties implied by $\mathcal{D}$ and $\mathcal{B}$ (see, e.g., Roulier [1], McAllister and Roulier [2], Goodman and Unsworth [3] and references therein). A solution to this problem, henceforth referred to as the convexity-preserving interpolation problem and denoted by $(\mathcal{P}_{\text{conv}}(\mathcal{U}, \mathcal{D}, \mathcal{B}))$, has been obtained by Kaklis and Sapidis [4], employing a new family of C^2-continuous parametric polynomial splines of nonuniform degree, i.e., polynomial splines where degree is permitted to vary from segment to segment (for the nonparametric case see Kaklis and Pandelis [5]). A simple algorithm is developed in [4] for automatically finding appropriate segment degrees for which the resulting $\mathbf{Q_{conv}}$ preserves the convexity properties implied by $(\mathcal{D}, \mathcal{B})$. Although a curve $\mathbf{Q_{conv}}$ produced by the above algorithm is usually visually pleasing (see the examples in [4] and [5]), one cannot guarantee that $\mathbf{Q_{conv}}$ satisfies any particular *fairness criterion*.

The present work aims at complementing the algorithm given in [4] with an interactive technique allowing the designer to remove local imperfections from the curvature $\kappa_{\text{conv}}(u)$ of $\mathbf{Q}_{\mathbf{conv}}(u)$, so that the final curve $\mathbf{Q}(u)$ remains convexity-preserving and satisfies the following criterion.

FAIRNESS CRITERION. *The curvature $\kappa(u)$ of $\mathbf{Q}(u)$, viewed as a function of the parameter u, must consist of as few monotone pieces as possible.* See, e.g., Farin and Sapidis [6], Sapidis and Farin [7], Roulier, Rando, and Piper [8] and the references therein. The technique is based on the assumption that the imperfection is located at an interior *offending node* $u_l, l \in \{2, \ldots, N-1\}$, specified either by the designer or by an appropriate criterion, e.g., defining the offending node to.be the one at which the first-order discontinuity of $\dot{\kappa}(u) = d\kappa(u)/du$ is maximised, which is the criterion of Sapidis and Farin [6]. The basic idea of the technique developed herein is to modify the curvature distribution $\kappa_{\text{conv}}(u)$ of $\mathbf{Q}_{\mathbf{conv}}(u)$ at the offending node u_l by suitably perturbing only the corresponding interpolation point $\mathbf{I_l}$ within the range of a small tolerance tol, i.e., the point set $\mathcal{D} \backslash \mathbf{I_l}$, the parametrization, and the segment degrees of the spline are not allowed to change. In this connection, we formulate the following problem.

PROBLEM ($\mathcal{P}_{\text{conv}}^{\text{fair}}$). Let $\mathbf{Q}_{\mathbf{conv}}$ be the solution of the convexity-preserving interpolation problem ($\mathcal{P}_{\text{conv}}(\mathcal{U}, \mathcal{D}, \mathcal{B})$) in the class of C^2-continuous parametric splines of nonuniform degree. Let $\mathbf{Q}_{\mathbf{conv}}$ possess a fairness imperfection at an interior offending node $u = u_l, l \in \{2, \ldots, N-1\}$ and let tol be a given small tolerance for the points in $\mathcal{D}$. Find the nearest point $\tilde{\mathbf{I}}_\mathbf{l}$ to $\mathbf{I_l}$ so that the following constraints hold true:

(i) $\|\mathbf{I_l} - \tilde{\mathbf{I}}_\mathbf{l}\| \leq$ tol, where $\|\cdot\|$ denotes the Euclidean distance in the plane.

(ii) If $\tilde{\mathcal{D}} = \{\mathbf{I_1}, \mathbf{I_2}, \ldots, \tilde{\mathbf{I}}_\mathbf{l}, \ldots, \mathbf{I_N}\}$ then $(\tilde{\mathcal{D}}, \mathcal{B})$ and $(\mathcal{D}, \mathcal{B})$ are equivalent from the convexity-information point of view.

(iii) Let $\tilde{\mathbf{Q}}$ be the C^2-continuous parametric polynomial spline of a nonuniform degree, which interpolates $\tilde{\mathcal{D}}$ with parametrization $\mathcal{U}$, satisfies the boundary conditions $\mathcal{B}$ and possesses the same segment-degree distribution with $\mathbf{Q}_{\mathbf{conv}}$. Then $\tilde{\mathbf{Q}}$ preserves the convexity properties of $(\tilde{\mathcal{D}}, \mathcal{B})$.

(iv) At the offending node u_l, $\tilde{\mathbf{Q}}$ is fairer than $\mathbf{Q}_{\mathbf{conv}}$ and, away from u_l, $\tilde{\mathbf{Q}}$ is at least as fair as $\mathbf{Q}_{\mathbf{conv}}$.

The contents of the paper are organised as follows. In §10.2 we develop the algorithm ($\mathcal{A}_{\text{conv}}$) for solving the convexity-preservation problem ($\mathcal{P}_{\text{conv}}$) in the class Γ of parametric polynomial splines of nonuniform degree (see (10.1) and (10.2)). The algorithm ($\mathcal{A}_{\text{conv}}$) is automatic, with computational cost of $O(N)$ operations per iteration, and produces a locally convex C^2-continuous curve (see Definition 10.2.1 and Theorem 10.2.1), which exhibits at most one inflection point in each parameter segment $[u_m, u_{m+1}]$, $m = 1(1)N-1$ (see Corollary 10.2.1). In §10.3 we develop the algorithm ($\mathcal{A}_{\text{conv}}^{\text{fair}}$) for solving the problem ($\mathcal{P}_{\text{conv}}^{\text{fair}}$) under the hypothesis that the offending node u_l, as well as the amount $\delta\kappa_l$ by which $\kappa_{\text{conv}}(u_l)$ should be modified, are given (see Hypothesis 10.3.1 and Problem 10.3.1). The resulting nonlinear curvature-control problem is solved to the leading order of approximation with respect to the small parameter tol. It is then shown that if tol and $|\delta\kappa_l|$ are sufficiently small, then the algorithm $(\mathcal{A}^{\text{fair}}_{\text{conv}})$ provides a solution to the problem $(\mathcal{P}^{\text{fair}}_{\text{conv}})$ (see Theorem 10.3.1 and Questions 10.3.1–10.3.3, as well as their corresponding answers). Under the hypothesis that u_l and $\delta\kappa_l$ are specified, $(\mathcal{A}^{\text{fair}}_{\text{conv}})$ is automatic with a computational cost of $O(N)$ operations. In §1.4 we discuss the numerical performance of $(\mathcal{A}^{\text{fair}}_{\text{conv}})$ in the case of four data sets. The work concludes with §10.5, which contains some brief remarks on our current work, aiming at improving the robustness of the algorithm $(\mathcal{A}^{\text{fair}}_{\text{conv}})$.

10.2. Constructing Convexity-Preserving Interpolants

In this section we develop a method for solving the convexity-preservation problem $(\mathcal{P}_{\text{conv}}(\mathcal{D},\mathcal{U},\mathcal{B}))$. To start with, let $\mathcal{K} = \{k_1, k_2, \ldots, k_{N-1}\}$ be a set of positive integers with $k_m \geq 3$, $m = 1(1)N-1$. We denote by $\Gamma(\mathcal{K},\mathcal{U})$ a new class of parametric polynomial splines of nonuniform degree $\mathbf{Q}(u) = (Q_1(u), Q_2(u))^T, u \in [u_1, u_N]$, defined by

$$\mathbf{Q}(u) = \mathbf{I_m}(1-t) + \mathbf{I_{m+1}}t + h_m^2\ddot{\mathbf{Q}}_\mathbf{m}F_m(1-t) + h_m^2\ddot{\mathbf{Q}}_\mathbf{m+1}F_m(t), \quad u \in [u_m, u_{m+1}], \quad t \in [0,1], \tag{10.1}$$

where

$$F_m(t) = \frac{t^{k_m} - t}{k_m(k_m-1)}, \qquad t = \frac{u-u_m}{h_m}, \qquad h_m = u_{m+1} - u_m, \tag{10.2a}$$

$$\ddot{\mathbf{Q}}_\mathbf{m} = (\ddot{Q}_{1m}, \ddot{Q}_{2m})^T, \qquad \ddot{Q}_{im} = \ddot{Q}_i(u_m), i = 1, 2, \tag{10.2b}$$

and the dot accent denotes differentiation with respect to the global parameter u. Note that if $k_m = 3$, $m = 1(1)N-1$, then $\mathbf{Q}$ reduces to the classical cubic spline.

The problem of finding a C^2-continuous spline in $\Gamma(\mathcal{K},\mathcal{U})$, which interpolates the data set $\mathcal{D}$ and satisfies conventional boundary conditions $\mathcal{B}$, namely type I (given $\dot{\mathbf{Q}}_\mathbf{1}, \dot{\mathbf{Q}}_\mathbf{N}$), type II (given $\ddot{\mathbf{Q}}_\mathbf{1}, \ddot{\mathbf{Q}}_\mathbf{N}$) or periodic boundary conditions in the case of closed curves ($\dot{\mathbf{Q}}_\mathbf{1} = \dot{\mathbf{Q}}_\mathbf{N}, \ddot{\mathbf{Q}}_\mathbf{1} = \ddot{\mathbf{Q}}_\mathbf{N}$), admits a unique solution (see Theorem 2.1 in [4]). More specifically, this problem can be equivalently formulated as a pair of two uncoupled linear systems with respect to the second-order parametric nodal derivatives $\ddot{Q}_{im}, m = 1(1)N, i = 1, 2$. Both systems share a common coefficient matrix $\mathbf{A}$, which is symmetric and strongly diagonally dominant. Moreover, $\mathbf{A}$ is tridiagonal in the case of type I and type II boundary conditions, whereas $\mathbf{A}$ is cyclic-tridiagonal in the case of periodic boundary conditions. In the remainder of this and the forthcoming sections, and without loss of generality, we restrict ourselves to type I boundary conditions only. If this is the case, the tangent vectors at $\mathbf{I_1}$ and $\mathbf{I_N}$ are specified by the designer, say,

$$\dot{\mathbf{Q}}_\mathbf{1} = \mathbf{s_0} = (s_{10}, s_{20})^T, \quad \dot{\mathbf{Q}}_\mathbf{N} = \mathbf{s_N} = (s_{1N}, s_{2N})^T, \tag{10.3}$$

and the abovementioned linear systems can be compactly written as follows:

$$e_{m-1}\ddot{\mathbf{Q}}_\mathbf{m-1} + (d_{m-1} + d_m)\ddot{\mathbf{Q}}_\mathbf{m} + e_m\ddot{\mathbf{Q}}_\mathbf{m+1} = \mathbf{b_m}, \quad m = 1(1)N, \tag{10.4}$$

where

$$e_0 = d_0 = 0, \quad e_N = d_N = 0, \tag{10.5a}$$

$$e_m = h_m/k_m(k_m - 1), \quad d_m = h_m/k_m, \quad m = 1(1)N-1, \tag{10.5b}$$

and

$$\mathbf{b_m} = (b_{1m}, b_{2m})^T, \quad b_{im} = s_{im} - s_{i,m-1}, \quad i = 1, 2, \tag{10.6a}$$

with

$$s_{1m} = (x_{m+1} - x_m)/h_m, \quad s_{2m} = (y_{m+1} - y_m)/h_m, \quad m = 1(1)N-1. \tag{10.6b}$$

We now proceed to investigate the possibility of constructing interpolatory splines $\mathbf{Q} \in \mathit{\Gamma} = \Gamma(\mathcal{K},\mathcal{U}) \cap C^2[u_1, u_N]$, which preserve the convexity information provided by the data set $\mathcal{D}$ and the boundary conditions $\mathcal{B}$. To do so, we first define the so-called augmented polygonal arc $\mathcal{L}_{\mathcal{DB}}$, which connects the points $\mathbf{I_0}, \mathbf{I_1}, \ldots, \mathbf{I_N}, \mathbf{I_{N+1}}$, where

$$\mathbf{I_0} = (x_0, y_0)^T, \quad x_0 = x_1 - h_0 s_{10}, \quad y_0 = y_1 - h_0 s_{20}, \tag{10.7a}$$

and

$$\begin{aligned} &\mathbf{I_{N+1}} = (x_{N+1}, y_{N+1})^T, \\ &x_{N+1} = x_N + h_N s_{1N}, \quad y_{N+1} = y_N + h_N s_{2N}, \end{aligned} \tag{10.7b}$$

with h_0, h_N being two arbitrary positive numbers. Furthermore, given a pair of two-dimensional vectors $\mathbf{a} = (a_1, a_2)^T$ and $\mathbf{b} = (b_1, b_2)^T$, we define the scalar quantity

$$\mathbf{a} \times \mathbf{b} := a_1 b_2 - a_2 b_1 = |\mathbf{a}||\mathbf{b}| \sin\theta, \tag{10.8}$$

where θ is the angle from $\mathbf{a}$ to $\mathbf{b}$ in an anticlockwise direction. Note that $\mathbf{a} \times \mathbf{b}$ is but the z coordinate of the standard vector product between the three-dimensional vectors $\tilde{\mathbf{a}} = (a_1, a_2, 0)^T$ and $\tilde{\mathbf{b}} = (b_1, b_2, 0)^T$. Then the convexity information contained in the augmented polygonal arc $\mathcal{L}_{\mathcal{DB}}$ can be adequately characterised by the sign properties of the quantities

$$(10.9) \qquad P_m = (\mathbf{I_m} - \mathbf{I_{m-1}}) \times (\mathbf{I_{m+1}} - \mathbf{I_m}), \qquad m = 1(1)N.$$

In the present work we shall restrict ourselves to data sets $\mathcal{D}$ and type I boundary conditions $\mathcal{B}$ for which $P_m \neq 0, m = 1(1)N$; in other words, the augmented polygonal arc $\mathcal{L}_{\mathcal{DB}}$ should not contain triplets of colinear vertices. We are now ready to precisely define the notion of convexity-preserving interpolation in Γ.

DEFINITION 10.2.1. *A spline $\mathbf{Q} \in \Gamma$, which interpolates $\mathcal{D}$ and satisfies $\mathcal{B}$, will be called convexity-preserving* (c-p) *provided that the curvature $\kappa(u)$ of $\mathbf{Q}(u)$ satisfies the following:* (i) $\kappa(u_m)P_m > 0$, $m = 1(1)N$, (ii) *if $P_m P_{m+1} > 0$ for some $m \in \{1, \ldots, N-1\}$, then $\kappa(u)P_m > 0$ for $u \in [u_m, u_{m+1}]$.*

The above definition can be alternatively stated in terms of the anticlockwise angle $\theta_{m-1,m}$ from $\mathbf{I_m} - \mathbf{I_{m-1}}$ to $\mathbf{I_{m+1}} - \mathbf{I_m}$ (see, e.g., Roulier [9]). Furthermore, Definition 10.2.1 preserves local, not global, convexity, and its apparent weakness is that it allows for more than one inflection point in the interior of a parameter segment $[u_m, u_{m+1}]$ for which $P_m P_{m+1} < 0$.[1] However, as it will be shown in the sequel, a c-p spline cannot possess more than one inflection point in $[u_m, u_{m+1}], m = 1(1)N-1$ (see Corollary 10.2.1). Finally, Definition 10.2.1 can be readily enriched to handle data sets containing triplets of colinear points (see [4, Def. 2.1]).

We shall now derive a set of sufficient conditions ensuring c-p interpolation in Γ in the sense of the above definition. For this purpose, we introduce the index sets $\mathcal{I} = \{1, 2, \ldots, N\}$,

$$(10.10) \qquad \mathcal{I}_1 = \{m \in \mathcal{I} \ : \ P_{m+n}P_{m+n+1} > 0 \text{ for either } n = -1 \text{ or } 0\},$$

and set the curvature $\kappa(u)$ of $\mathbf{Q}$ in the form

$$(10.11) \qquad \kappa(u) = \frac{\lambda(u)}{(\dot{Q}_1^2(u) + \dot{Q}_2^2(u))^{3/2}}, \qquad \lambda(u) = \dot{\mathbf{Q}}(u) \times \ddot{\mathbf{Q}}(u).$$

The denominator in the rational expression (10.11) for $\kappa(u)$ is nonnegative. Accordingly, any set of conditions, inducing the validity of Definition 10.2.1, should be expressed in terms of the numerator $\lambda(u)$. In this connection, we give the following alternative representation for

(10.12)
$$\lambda(u) = \lambda_m F_m''(1-t) + \lambda_{m+1} F_m''(t) + h_m \mu_{m,m+1} \varphi_m(t), \quad u \in [u_m, u_{m+1}],$$

[1] We thank Professor J. Roulier for raising this issue.

where

(10.13a)
$$\lambda_m = \lambda(u_m), \quad \mu_{m,m+1} = \ddot{\mathbf{Q}}_{\mathbf{m+1}} \times \ddot{\mathbf{Q}}_{\mathbf{m}},$$

(10.13b)
$$\varphi_m(t) = \frac{t^{k_m-2}(1-t)^{k_m-2}}{k_m - 1}, \qquad t \in [0,1].$$

The prime in (10.12) denotes differentiation with respect to the local parameter t. Combining Definition 10.2.1 with formula (10.12) permits obtaining a set of sufficient conditions for c-p interpolation in Γ in terms of nodal quantities only, namely the quantities λ_m and $\mu_{m,m+1}$. More specifically, one can prove (see [4, Lemma 3.2]) the following theorem.

THEOREM 10.2.1. *Let* $\mathbf{Q} \in \Gamma$ *interpolate the data set* $\mathcal{D}$ *and satisfy the boundary conditions* $\mathcal{B}$ *and the following conditions:* (i) $\lambda_m P_m > 0$ *for* $m \in \mathcal{I}$, (ii) *if* $m \in \mathcal{I}_1$ *and* $\mu_{m+n,m+n+1} P_m < 0$ *for* $n = -1$ *and/or* 0, *then*

(10.14)
$$|\mu_{m+n,m+n+1}| < h_{m-n}^{-1} 2^{k_{m+n}-1}(k_{m+n} - 1) \cdot \min\left\{|\lambda_{m+n}|, |\lambda_{m+n+1}|\right\}.$$

Then $\mathbf{Q}$ *is c-p.*

Furthermore, the following result holds true.

COROLLARY 10.2.1 *If* $\mathbf{Q} \in \Gamma$ *satisfies the conditions of Theorem* 10.2.1, *then it can possess at most one inflection point in* $[u_m, u_{m+1}], m = 1(1)N-1$.

Proof. Given that in the present work, $P_m \neq 0, m = 1(1)N$, and $\mathbf{Q}$ satisfies the conditions of Theorem 10.2.1, it is easily seen that inflection points may occur only in the interior of parameter segments $[u_m, u_{m+1}]$ for which $P_m P_{m+1} < 0$, equivalently $\lambda_m \lambda_{m+1} < 0$. Let, e.g., $\lambda_m > 0, \lambda_{m+1} < 0$ and $\mu_{m,m+1} \leq 0$. Then, if $u = u^*$ is an inflection point of $\mathbf{Q}(u)$ in (u_m, u_{m+1}), the numerator $\lambda(u)$ of the curvature $\kappa(u)$ must vanish at this point, i.e., $\lambda(u^*) = 0$. Using now (10.12) and (10.13), the previous equality can be written as

(10.15)
$$\lambda_m = g(t^*) \quad \text{with} \quad g(t) = -\lambda_{m+1}\left(\frac{t}{1-t}\right)^{k_m-2} - h_m \mu_{m,m+1} \frac{t^{k_m-2}}{k_m - 1},$$
$$t^* = (u^* - u_m)/h_m, \quad t, t^* \in [0,1].$$

Since $g'(t) > 0$, we readily conclude that the equation $\lambda_m = g(t^*)$ possesses only one solution t^* in $[0,1]$. This completes the proof of Corollary 10.2.1. □

The splines of the family Γ possess a very important asymptotic property, which implies a similarity with the various families of splines in tension. This property may be summarised as follows: as the segment degrees increase globally, i.e., $k_m \to \infty$, $m = 1(1)N-1$, $\mathbf{Q} \in \Gamma$ tends to the corresponding linear interpolant with the best approximation rate for a Lipschitz continuous function (see [4, Thm. 2.2]). Using this property, one can easily construct an automatic, iterative algorithm which, after a finite number of iterations, yields a set $\mathcal{K}_{\text{conv}}$ of segment degrees for which the corresponding spline $\mathbf{Q}_{\mathbf{conv}}$ satisfies the conditions of Theorem 10.2.1 and, accordingly, is c-p in the sense of Definition 10.2.1. This algorithm, henceforth referred to as the algorithm $(\mathcal{A}_{\text{conv}})$, follows.

The Algorithm $(\mathcal{A}_{\text{conv}})$.

STEP 0. Determine the set $\mathcal{I}_1$ and set $j = 0$, $k_m^{(j)} = 3$, $m = 1(1)N-1$.
STEP 1. Construct the spline $\mathbf{Q}^{(\mathbf{j})} \in \mathit{\Gamma} = \Gamma(\mathcal{K}^{(j)}, \mathcal{U}) \cap C^2[u_1, u_N]$, $\mathcal{K}^{(j)} = \{k_1^{(j)}, k_2^{(j)}, \ldots, k_{N-1}^{(j)}\}$, which interpolates the data set $\mathcal{D}$ and satisfies the boundary conditions $\mathcal{B}$.
STEP 2. Compute the quantities $\lambda_m^{(j)}$, $m \in \mathcal{I}_1$ and $\mu_{m,m+1}^{(j)}, m, m+1 \in \mathcal{I}_1$. If these quantities satisfy the conditions of Theorem 10.2.1, then $\mathbf{Q}^{(\mathbf{j})}$ is convexity-preserving in the sense of Definition 10.2.1 and **stop**.
STEP 3. If, for some $m \in \mathcal{I}$, $\lambda_m^{(j)}$ does not fulfill condition (i) of Theorem 10.2.1, then increase $k_{m-1}^{(j)}$ and $k_m^{(j)}$ by one. Otherwise, go to step 4.
STEP 4. If, for some pair $m, m+1 \in \mathcal{I}_1$, $\mu_{m,m+1}^{(j)}$ does not fulfill condition (ii) of Theorem 10.2.1, then increase $k_{m-1}^{(j)}, k_m^{(j)}$ and $k_{m+1}^{(j)}$ by one. Set $j = j+1$ and go to step 1.

From the computational cost point of view, the most expensive step of $(\mathcal{A}_{\text{conv}})$ is step 1, which involves the solution of the linear system (10.4), whose associated matrix is tridiagonal and positive definite. Accordingly, step 1 can be accomplished at the expense of $O(N)$ operations.

To illustrate the working mode of the algorithm $(\mathcal{A}_{\text{conv}})$, we consider a data set $\mathcal{D}_9$ consisting of nine $(N = 9)$ points with periodic boundary conditions and chord-length parametrization. Analytically, $\mathcal{D}_9 = \{(5.0, 1.0), (2.0, 1.5), (-2.0, 1.5), (-5.0, 1.0), (-5.0, -1.0), (-2.0, -1.5), (2.0, -1.5), (5.0, -1.0), (5.0, 1.0)\}$; see points marked by • in Fig. 10.1(a). The algorithm $(\mathcal{A}_{\text{conv}})$ converges after two iterations, giving $\mathcal{K}_{\text{conv}} = \{5, 5, 5, 3, 5, 5, 5, 3\}$. Fig. 10.1(a) depicts the final curve $\mathbf{Q}_{\mathbf{conv}}$ (solid line) along with the first iteration of the algorithm $(\mathcal{A}_{\text{conv}})$ (dashed line), which is but the C^2-continuous cubic spline $\mathbf{Q}_{\mathbf{cubic}}$ interpolating $\mathcal{D}_9$. The corresponding curvature distributions $\kappa_{\text{conv}}(u)$ and $\kappa_{\text{cubic}}(u)$, viewed as functions of the parameter u, are depicted in Fig. 10.1(b) as a solid and a dashed line, respectively.

10.3. Controlling the Curvature of a Convexity-Preserving Interpolant

In this section we develop a method for handling the problem $(\mathcal{P}_{\text{conv}}^{\text{fair}})$ (see §10.1), which occurs when the curvature distribution $\kappa_{\text{conv}}(u)$ of a c-p interpolant $\mathbf{Q}_{\mathbf{conv}} \in \mathit{\Gamma}_{\text{conv}} = \mathit{\Gamma}(\mathcal{K}_{\text{conv}}, \mathcal{U}) \cap C^2[u_1, u_N]$ exhibits a local imperfection

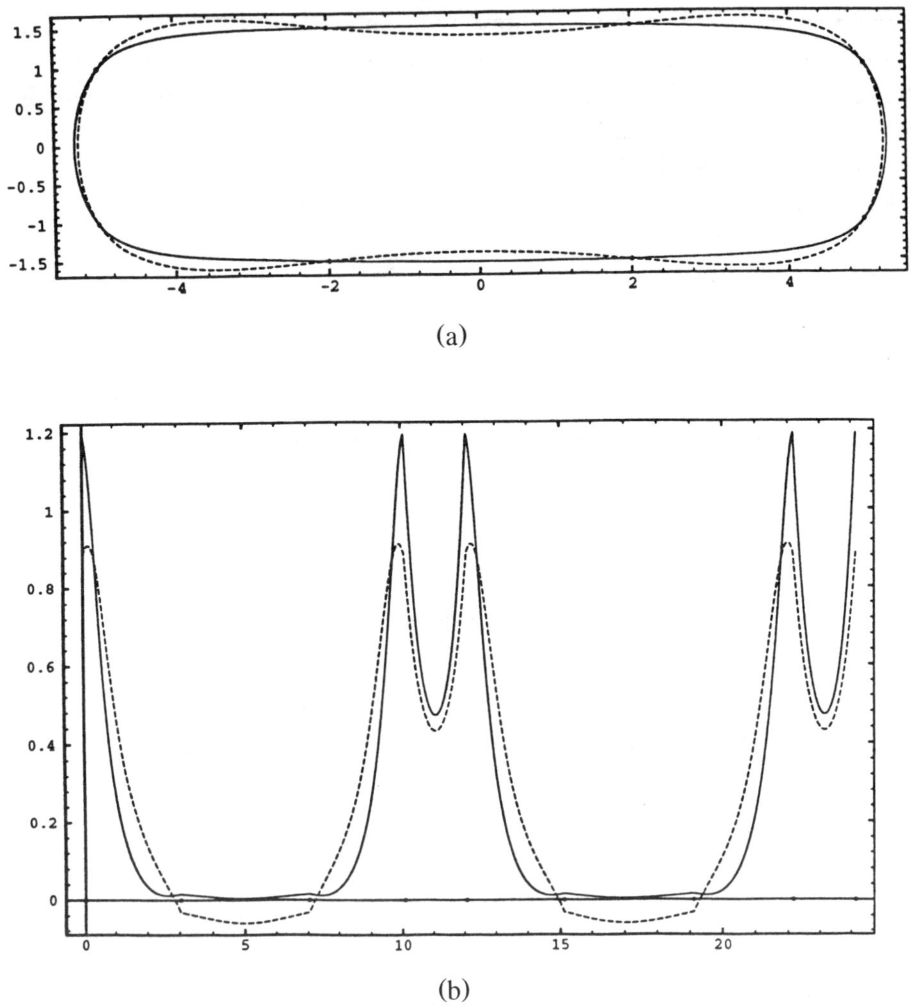

FIG. 10.1. (a) *Application of the algorithm* ($\mathcal{A}_{\text{conv}}$) *on the data set* $\mathcal{D}_9$. (b) *The curvature distributions of the curves shown in Fig.* 10.1(a).

at an interior offending node $u_l, l \in \{2, \ldots, N-1\}$. This method relies on the following hypothesis.

HYPOTHESIS 10.3.1. The offending node u_l and the amount $\delta\kappa_l$, by which $\kappa_{\text{conv}}(u)$ should be modified at $u = u_l$, are given.

Under the above hypothesis, we formulate the next problem.

PROBLEM 10.3.1. Find the nearest point $\mathbf{I}_l^*$ to $\mathbf{I}_l$, for which $\|\mathbf{I}_l - \mathbf{I}_l^*\| \leq \text{tol}$ and

$$\kappa^*(u_l) - \kappa_{\text{conv}}(u_l) = \delta\kappa_l, \tag{10.16}$$

where $\kappa^*(u)$ denotes the curvature distribution of the spline $\mathbf{Q}^* \in \Gamma_{\text{conv}}$, which interpolates the perturbed data set $\mathcal{D}^* = \{\mathbf{I_1}, \mathbf{I_2}, \ldots, \mathbf{I}_l^*, \ldots, \mathbf{I_N}\}$ and satisfies the boundary conditions $\mathcal{B}$.

In the sequel we shall construct an approximate, regarding condition (10.16), solution to the above nonlinear problem, and then use it for handling $(\mathcal{P}_{\text{conv}}^{\text{fair}})$. To start with, we note that since $\mathbf{Q}_{\mathbf{conv}}$ and $\mathbf{Q}^*$ belong to the same family Γ_{conv}, they share the same parametrization $\mathcal{U}$ and the same segment-degree distribution $\mathcal{K}_{\text{conv}}$. As a consequence, the linear systems (10.4) specifying the vectors $(\{\ddot{Q}_{\text{conv},im}\}_{m=1}^{N})^T$ and $(\{\ddot{Q}_{im}^*\}_{m=1}^{N})^T, i = 1, 2$, have the same left-hand side matrix $\mathbf{A}$. Furthermore, given that the interpolated data $\mathcal{D}$ and $\mathcal{D}^*$ differ in the lth point only, we have that the right-hand side vectors of the aforementioned systems differ in the elements of the $(l-1)$th, lth, and $(l+1)$th rows. Taking into account the above remarks and setting $\delta\ddot{Q}_{im} = \ddot{Q}_{im}^* - \ddot{Q}_{\text{conv},im}$, and $(\delta x_{1l}, \delta x_{2l})^T = \mathbf{I}_l^* - \mathbf{I}_l$, we arrive at the following formula:

$$\delta\ddot{Q}_{im} = D_m \delta x_{il}, \quad l, m \in \{2, \ldots, N-1\}, i = 1, 2, \tag{10.17a}$$

where

$$D_m = A_{m,l-1}^{-1}\frac{1}{h_{l-1}} - A_{ml}^{-1}\left(\frac{1}{h_{l-1}} + \frac{1}{h_l}\right) + A_{m,l+1}^{-1}\frac{1}{h_l}, \tag{10.17b}$$

with A_{ml}^{-1} denoting the element appearing in the mth row and lth column of the inverse $\mathbf{A}^{-1}$ of matrix $\mathbf{A}$.

The formula analogous to (10.17a) for the first-order nodal parametric derivatives $\delta\dot{Q}_{im}$ can be obtained by appealing to representation formulae (10.1), which, after differentiating with respect to the global parameter u, yields, in combination with formulae (10.17), the following result:

$$\begin{aligned} \delta\dot{Q}_{im} &= \delta s_{im} - (d_m D_m + e_m D_{m+1})\,\delta x_{il}, \\ &\qquad l, m \in \{2, \ldots, N-1\}, i = 1, 2, \end{aligned} \tag{10.18a}$$

where

$$\delta s_{im} = \left\{ \begin{array}{ll} 0 & \text{for } m \neq l-1, l, \\ \dfrac{\delta x_{il}}{h_{l-1}} & \text{for } m = l-1, \\ \dfrac{-\delta x_{il}}{h_l} & \text{for } m = l, \end{array} \right\} \tag{10.18b}$$

and the quantities d_l, e_l are defined by (10.5).

Assuming that $\mathbf{Q}^*$ and $\mathbf{Q}_{\mathbf{conv}}$ represent regular curves and using the fact that, due to the constraint, $\|\mathbf{I}_l^* - \mathbf{I}_l\| \leq \text{tol}$, the quantities $\delta x_{il}, i = 1, 2$, are as small as the tolerance tol, we arrive after some straightforward asymptotic algebra at the following asymptotic approximation for the difference

$$\kappa^*(u_l) - \kappa_{\text{conv}}(u_l) = \rho_1 \delta x_{1l} + \rho_2 \delta x_{2l} + O(\delta x_{il}\delta x_{jl}, i, j = 1, 2), \tag{10.19a}$$

where

$$\rho_i = -\left(\frac{1}{h_l} + D_l d_l + D_{l+1} e_l\right) \nu_{il} + D_l \xi_{il}, \quad i = 1, 2, \tag{10.19b}$$

with

$$\begin{aligned} \nu_{1l} &= \alpha_l^{-3/2} \ddot{Q}_{2l} - 3\alpha_l^{-1} \kappa_l \dot{Q}_{1l}, \\ \nu_{2l} &= -\alpha_l^{-3/2} \ddot{Q}_{1l} - 3\alpha_l^{-1} \kappa_l \dot{Q}_{2l}, \quad \kappa_l = \kappa(u_l), \end{aligned} \tag{10.20a}$$

and

$$\xi_{1l} = -\alpha_l^{-3/2} \dot{Q}_{2l}, \quad \xi_{2l} = \alpha_l^{-3/2} \dot{Q}_{1l}, \quad \alpha_l = \dot{Q}_{1l}^2 + \dot{Q}_{2l}^2. \tag{10.20b}$$

Using now (10.19a), the basic constraint (10.16) of Problem 10.3.1 can be approximated by the linear one,

$$\rho_1 \delta x_{1l}^{(0)} + \rho_2 \delta x_{2l}^{(0)} - \delta \kappa_l = 0, \tag{10.21}$$

which is accurate up to terms of order $O(\delta x_{il}^{(0)}, i = 1, 2)$ or equivalently $O(\text{tol})$. In this approximate form, Problem 10.3.1 is geometrically equivalent to finding the intersection between the line (10.21) and its perpendicular from the point $(0, 0)$, and then checking if this intersection point lies within the disk $(\delta x_{1l}^{(0)})^2 + (\delta x_{2l}^{(0)})^2 \le \text{tol}^2$. It is then easy to prove the following theorem.

THEOREM 10.3.1. *If* $|\rho_1| + |\rho_2| \neq 0$ *and* $|\delta\kappa_l| \le (\rho_1^2 + \rho_2^2)^{1/2}\text{tol}$, *then*

$$\begin{aligned} &\mathbf{I}_l^* = \mathbf{I}_l^{(0)} + o(\text{tol}) \quad \textit{where} \quad \mathbf{I}_l^{(0)} = \mathbf{I}_l + (\delta x_{1l}^{(0)}, \delta x_{2l}^{(0)})^T \\ &\textit{with} \quad \delta x_{il}^{(0)} = \frac{\rho_i \delta\kappa_l}{\rho_1^2 + \rho_2^2}, \quad i = 1, 2. \end{aligned} \tag{10.22}$$

Let us now comment on the conditions $|\rho_1| + |\rho_2| \neq 0$ and $|\delta\kappa_l| \le (\rho_1^2 + \rho_2^2)^{1/2}\text{tol}$, which are necessary for the validity of Theorem 10.3.1. The first constraint is a "hard" one, for if $\rho_1 = \rho_2 = 0$, then (10.22) becomes meaningless. To handle such a case, one should enrich (10.19a) with higher-order terms. However, $\rho_1 = \rho_2 = 0$ means that a small perturbation of $\mathbf{I}_l$ produces only higher-order changes to the curvature of $\mathbf{Q}_{\mathbf{conv}}$. Since this strongly disagrees with the numerical experience (that the curvature of $\mathbf{Q}_{\mathbf{conv}}$ is strongly affected when a single point $\mathbf{I}_l$ is moved), we can safely assume that in general, $|\rho_1| + |\rho_2| \neq 0$. On the other hand, the constraint $|\delta\kappa_l| \le (\rho_1^2 + \rho_2^2)^{1/2}\text{tol}$ is "soft" in the sense that if the tolerance tol is given and the quantities $\rho_i, i = 1, 2$, are calculated, we can provide the designer with an upper bound for the permitted $|\delta\kappa_l|$.

The question if the approximate solution $\mathbf{I}_l^{(0)}$ of Problem 10.3.1 is of any use for handling the problem $(\mathcal{P}_{\text{conv}}^{\text{fair}})$ arises. This question can be broken down into the following three questions.

QUESTION 10.3.1. *Let* $\mathcal{D}^{(0)} = \{\mathbf{I_1}, \mathbf{I_2}, \ldots, \mathbf{I}_\mathbf{l}^{(\mathbf{0})}, \ldots, \mathbf{I_N}\}$. *Are* $(\mathcal{D}^{(0)}, \mathcal{B})$ *and* $(\mathcal{D}, \mathcal{B})$ *equivalent from the convexity-information point of view?*

QUESTION 10.3.2. *Let* $\mathbf{Q^{(0)}}$ *be the spline, which interpolates* $\mathcal{D}^{(0)}$*, satisfies* $\mathcal{B}$*, and belongs to* Γ_{conv}*. Is* $\mathbf{Q^{(0)}}$ *c-p?*

QUESTION 10.3.3. *Is* $\mathbf{Q^{(0)}}$ *fairer than* $\mathbf{Q_{conv}}$?

Regarding Question 10.3.1, we first recall that the initial configuration $(\mathcal{D}, \mathcal{B})$ satisfies the condition $P_m \neq 0$, $m = 1(1)N$ (see equation (10.8)). Then since $\mathcal{D}^{(0)}$ and $\mathcal{D}$ differ in the lth point only, it is easy to show that the inequalities

$$P_m P_m^{(0)} > 0, \quad m = l-1, l, l+1, \tag{10.23}$$

are necessary and sufficient conditions for $(\mathcal{D}^{(0)}, \mathcal{B})$ and $(\mathcal{D}, \mathcal{B})$ to be equivalent from the convexity-information point of view. Now, using an alternative definition for P_m, namely

$$P_m = X_{m-1}Y_m - X_m Y_{m-1}, \quad X_m = x_{m+1} - x_m, \quad Y_m = y_{m+1} - y_m, \tag{10.24}$$

inequalities (10.23) can be expressed in terms of the quantities $\delta x_{il}^{(0)}, i = 1, 2,$ as follows:

$$-P_{l-1}Y_{l-2}\,\delta x_{1l}^{(0)} + P_{l-1}X_{l-2}\,\delta x_{2l}^{(0)} + P_{l-1}^2 > 0, \tag{10.25a}$$

$$P_l(Y_l + Y_{l-1})\,\delta x_{1l}^{(0)} - P_l(X_l + X_{l-1})\,\delta x_{2l}^{(0)} + P_l^2 > 0, \tag{10.25b}$$

$$-P_{l+1}Y_{l+1}\,\delta x_{1l}^{(0)} + P_{l+1}X_{l+1}\,\delta x_{2l}^{(0)} + P_{l+1}^2 > 0. \tag{10.25c}$$

We can now give the following answer.

ANSWER TO QUESTION 10.3.1. $(\mathcal{D}^{(0)}, \mathcal{B})$ *and* $(\mathcal{D}, \mathcal{B})$ *are equivalent from the convexity-information point of view if and only if the linear inequalities* (10.25) *hold true, which is the case if the tolerance* tol *is sufficiently small.*

Assuming that the answer to Question 10.3.1 is in the affirmative, we proceed to investigate Question 10.3.2. Since $(\mathcal{D}^{(0)}, \mathcal{B})$ and $(\mathcal{D}, \mathcal{B})$ are equivalent from the convexity-information point of view, we have that $\mathcal{I}_1^{(0)} = \mathcal{I}_1$ (see (10.10)). Recall that Theorem 10.2.1 provides a set of inequalities on $\lambda_m^{(0)}, m \in \mathcal{I}_1$, and $\mu_{m,m+1}^{(0)}, m, m+1 \in \mathcal{I}_1$ (see (10.10) and (10.13)), ensuring that $\mathbf{Q^{(0)}}$ is c-p in the sense of Definition 10.2.1. Using (10.17a) and (10.18a) we arrive, after some elementary algebra, at the following expression for

$$\lambda_m^{(0)} = \lambda_{\text{conv},m} + \sigma_1 \delta x_{1l}^{(0)} + \sigma_2 \delta x_{2l}^{(0)} + \sigma_{12} \delta x_{1l}^{(0)} \delta x_{2l}^{(0)}, \quad m \in \mathcal{I}_1, \tag{10.26}$$

$$\mu_{m,m+1}^{(0)} = \mu_{\text{conv},m,m+1} + \tau_1 \delta x_{1l}^{(0)} + \tau_2 \delta x_{2l}^{(0)} + \tau_{12} \delta x_{1l}^{(0)} \delta x_{2l}^{(0)}, \quad m, m+1 \in \mathcal{I}_1, \tag{10.27}$$

where $\lambda_{\text{conv},m}, \mu_{\text{conv},m,m+1}, \sigma_i, \tau_i, i = 1,2, \sigma_{12}$ and τ_{12} are constants depending on $\mathbf{Q}_{\mathbf{conv}}$ only. By means of (10.26) and (10.27) the sufficient conditions of Theorem 10.2.1 can finally be expressed, after some lengthy but straightforward algebra, as a set of M bilinear inequalities of the form

$$(10.28) \qquad \xi_{0j} + \xi_{1j}\delta x_{1l}^{(0)} + \xi_{2j}\delta x_{2l}^{(0)} + \xi_{12j}\delta x_{1l}^{(0)}\delta x_{2l}^{(0)} > 0, \quad j = 1,\dots,M,$$

with $\xi_{0j}, \xi_{ij}, i = 1,2$ and ξ_{12j} being constants depending on $\mathbf{Q}_{\mathbf{conv}}$ only. Given now that $\mathbf{Q}_{\mathbf{conv}}$ is c-p, it is easy to prove that (10.28) holds true if $\delta x_{il}^{(0)}, i = 1,2,$ are sufficiently small. Thus we can give the following answer.

ANSWER TO QUESTION 10.3.2. *Let* $(\mathcal{D}^{(0)}, \mathcal{B})$ *and* $(\mathcal{D}, \mathcal{B})$ *be equivalent from the convexity-information point of view. If* (10.28) *holds true then* $\mathbf{Q}^{(0)}(u)$ *is c-p, which is the case if the tolerance* tol *is sufficiently small.*

Let us now consider Question 10.3.3. In order to compare the fairness of $\mathbf{Q}^{(0)}$ with that of $\mathbf{Q}_{\mathbf{conv}}$, we investigate the behaviour of the difference $\mathbf{Q}_{\mathbf{conv}}(u) - \mathbf{Q}^{(0)}(u)$ as u moves away from the offending node u_l. Since both $\mathbf{Q}^{(0)}$ and $\mathbf{Q}_{\mathbf{conv}}$ belong to the class Γ_{conv}, this problem reduces to the problem of studying the difference $|\ddot{Q}_{im}^{(0)} - \ddot{Q}_{\text{conv},im}|$ as $|l-m|$ increases. Using a result due to Kershaw [10], who derives sharp estimates for the elements of the inverse of a class of tridiagonal matrices, we get the following inequalities:

$$(10.29) \qquad |\ddot{Q}_{\text{conv},im} - \ddot{Q}_{im}^{(0)}| \le c_i^{(0)} \frac{k_{\max}}{h_{\min}^2} 2^{-|l-m|}, \quad i = 1,2,$$

where

$$(10.30) \qquad \begin{aligned} c_i^{(0)} = O(\delta x_{il}^{(0)}),\ i = 1,2, \quad k_{\max} &= \max\{k_{\text{conv},m}, m = l-1, l\}, \\ h_{\min} &= \min\{h_m, m = l-1, l\}. \end{aligned}$$

Thus we have the following answer.

ANSWER TO QUESTION 10.3.3. *If the tolerance* tol *is sufficiently small, then* (a) *at the offending node* u_l *the curvature of* $\mathbf{Q}^{(0)}$ *has the desired value,* (b) *in parameter segments* $[u_m, u_{m+1}]$, *located sufficiently far from the offending node* u_l, $\mathbf{Q}^{(0)}(u)$ *will be at least as fair as* $\mathbf{Q}_{\mathbf{conv}}(u)$ *is.*

Adopting Hypothesis 10.3.1 and taking into account Theorem 10.3.1, as well as the material collected in the form of questions and answers (see Questions 10.3.1–10.3.3 and their corresponding answers), we propose the following algorithm $(\mathcal{A}_{\text{conv}}^{\text{fair}})$ for solving the problem $(\mathcal{P}_{\text{conv}}^{\text{fair}})$.

The Algorithm $(\mathcal{A}_{\text{conv}}^{\text{fair}})$.

STEP 0. Let the data set $\mathcal{D}$, the parametrization $\mathcal{U}$, the boundary conditions $\mathcal{B}$, and the tolerance tol be given.

STEP 1. Solve the convexity-preserving interpolation problem $\mathcal{P}_{\text{conv}}(\mathcal{D}, \mathcal{U}, \mathcal{B})$ using the algorithm $(\mathcal{A}_{\text{conv}})$. Let $\mathbf{Q}_{\mathbf{conv}}(u) \in \Gamma_{\text{conv}} = \Gamma(\mathcal{K}_{\text{conv}}, \mathcal{U}) \cap C^2[u_1, u_N]$

be the convexity-preserving spline produced by $(\mathcal{A}_{\text{conv}})$ and $\kappa_{\text{conv}}(u)$ its curvature distribution.

STEP 2. Locate the offending node $u_l, l \in \{2, \ldots, N-1\}$ and specify the amount $\delta\kappa_l$ by which $\kappa_{\text{conv}}(u_l)$ has to be modified. Evaluate the quantities ρ_1 and ρ_2 and, if $|\delta\kappa_l| > (\rho_1^2 + \rho_2^2)^{1/2}\text{tol}$, modify $\delta\kappa_l$ so that $|\delta\kappa_l| = (\rho_1^2 + \rho_2^2)^{1/2}\text{tol}$.

STEP 3. Move $\mathbf{I}_l$ to its new position $\mathbf{I}_l^{(0)}$ according to Theorem 10.3.1. Calculate the spline $\mathbf{Q}^{(0)}(u) \in \Gamma_{\text{conv}}$, which interpolates $\mathcal{D}^{(0)}$ and satisfies the boundary conditions $\mathcal{B}$, and **stop**.

From the computational cost point of view, the most expensive step of $(\mathcal{A}_{\text{conv}}^{\text{fair}})$ is step 2, where the quantities ρ_1 and ρ_2 are to be evaluated. Recalling their defining formula (10.19b), we readily see that these quantities require the calculation of the elements of, at most, two rows of the inverse of the matrix associated with the systems (10.4). Accordingly, the most expensive step of $(\mathcal{A}_{\text{conv}}^{\text{fair}})$ has a computational cost of $O(N)$ operations.

10.4. Numerical Results

We shall now present and discuss the results of applying the algorithm $(\mathcal{A}_{\text{conv}}^{\text{fair}})$, developed in the preceding section, to three data sets. All calculations were done in double precision on an AT-486 PC-system by using the computer-algebra system MATHEMATICA and its graphics utilities.

Each example is documented by a list, a table, and two figures. The contents of the list are described analytically as follows:

(1) **N**: the number of points of the data set $\mathcal{D}$.

(2) tol: the tolerance for the points in $\mathcal{D}$.

(3) $\mathbf{x_m}$, $\mathbf{m = 1(1)N}$: the x-coordinates, with respect to an arbitrary coordinate system $O'xy$, of the interpolation points of the initial data set $\mathcal{D}$.

(4) $\mathbf{y_m}$, $\mathbf{m = 1(1)N}$: the y-coordinates, with respect to the coordinate system $O'xy$, of the interpolation points of $\mathcal{D}$.

(5) **btans**: the boundary tangent vectors $\mathbf{s_0}$ and $\mathbf{s_N}$ (recall that in the present work we deal with type I boundary conditions only).

(6) **partype**: the type of parametrization used, e.g., partype = chord length.

(7) **niter**: the number of iterations performed by the algorithm $(\mathcal{A}_{\text{conv}})$ until the convexity-preserving spline $\mathbf{Q_{conv}}$ is obtained.

(8) $\mathcal{K}_{\mathbf{conv}}$: the degrees distribution of $\mathbf{Q_{conv}}$.

(9) **l**: the order of the offending node u_l.

(10) $\delta\kappa_{\mathbf{l}}$: the amount by which the curvature $\kappa_{\text{conv}}(u)$ of $\mathbf{Q_{conv}}(u)$ should be modified at $u = u_l$.

(11) $\mathbf{x_m^{(0)}}$, $\mathbf{m = 1(1)N}$: the x-coordinates, with respect to the coordinate system $O'xy$, of the interpolation points of the final data set $\mathcal{D}^{(0)}$ produced by the algorithm $(\mathcal{A}_{\text{conv}}^{\text{fair}})$.

(12) $\mathbf{y_m^{(0)}}$, $\mathbf{m = 1(1)N}$: the y-coordinates, with respect to the coordinate system $O'xy$, of the interpolation points of $\mathcal{D}^{(0)}$.

As noted above, each example is further documented by a table. This table provides the values of the local-fairness indicators

$$\Delta\dot{\kappa}_{\text{conv},m} = |\dot{\kappa}_{\text{conv}}(u_m+) - \dot{\kappa}_{\text{conv}}(u_m-)|, \qquad (10.31)$$
$$\Delta\dot{\kappa}_m^{(0)} = |\dot{\kappa}^{(0)}(u_m+) - \dot{\kappa}^{(0)}(u_m-)|, \quad m = 2(1)N-1.$$

Finally, the documentation of each example is completed with a figure. Panel (a) depicts the curve $\mathbf{Q}_{\mathbf{conv}}$ (dashed line) produced by the algorithm $\mathcal{A}_{\text{conv}}$ and the curve $\mathbf{Q}^{(0)}$ (solid line) produced by the algorithm ($\mathcal{A}^{\text{fair}}_{\text{conv}}$), the points of $\mathcal{D}$ and $\mathcal{D}^{(0)}$ being depicted in both cases by •. Panel (b) depicts the curvature distributions $\kappa_{\text{conv}}(u)$ (dashed line) and $\kappa^{(0)}(u)$ (solid line) of $\mathbf{Q}_{\mathbf{conv}}$ and $\mathbf{Q}^{(0)}$, respectively, viewed as functions of the parameter $u \in \mathcal{U}$, where $\mathcal{U}$ is the chord-length parametrization obtained from the initial data set $\mathcal{D}$.

Example 10.1. (See Table 10.1 and Fig. 10.2.)

(1) $N = 3$,
(2) tol $= 0.1$,
(3) $x_m, m = 1,2,3 : -5.00,\ 0.00,\ 5.00$,
(4) $y_m, m = 1,2,3 : 0.00,\ 1.15,\ 0.00$,
(5) $\mathbf{s_0} = (1.0,\ 1.9)$, $\mathbf{s_3} = (1.0,\ -1.9)$,
(6) partype = chord length,
(7) niter = 9,
(8) $\mathcal{K}_{\text{conv}} = \{9,\ 9\}$,
(9) l=2,
(10) $\delta\kappa_2 = 0.025$,
(11) $x_m^{(0)}, m = 1,2,3 : -5.00000,\ 0.00000,\ 5.00000$,
(12) $y_m^{(0)}, m = 1,2,3 : 0.00000,\ 1.08963,\ 0.00000$.

Table 10.1

	u_2(off.node)
$\Delta\dot{\kappa}_{\text{conv},2}$	0.075583
$\Delta\dot{\kappa}_2^{(0)}$	0.007364

The offending node in this example is the interior node u_2, where $\kappa_{\text{conv}}(u)$ possesses a local minimum (see Fig. 10.2(b)). This minimum violates the fairness criterion of §10.1, namely that the curvature of a fair curve should consist of as few as possible monotone pieces. This imperfection, which can be hardly detected in Fig. 10.2(a), has been almost eliminated by the algorithm ($\mathcal{A}^{\text{fair}}_{\text{conv}}$), which succeeds in considerably decreasing ($\approx 90\%$) the discontinuity of $\dot{\kappa}_{\text{conv}}(u)$ at $u = u_2$ (see Table 10.1 and Fig. 10.2(b)). This is done at the expense of a rather negligible increase ($\approx 2.9\%$) of the absolute value of the

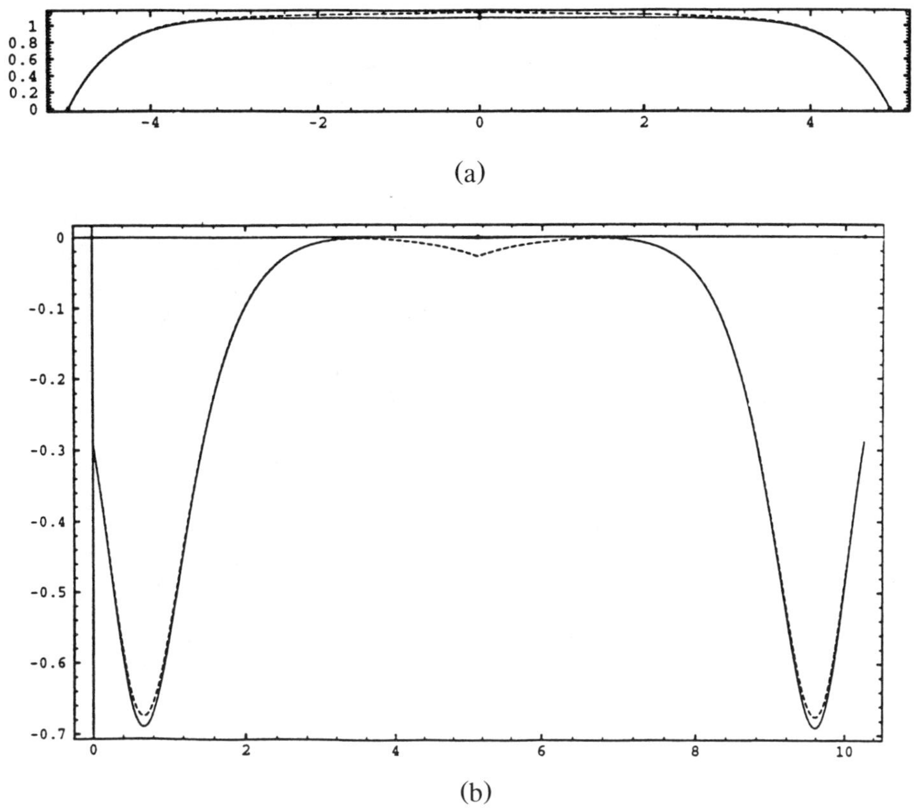

FIG. 10.2. (a) *Example* 10.1: *the curves* $\mathbf{Q_{conv}}$ *and* $\mathbf{Q}^{(0)}$. (b) *The curvature distributions of the curves shown in Fig.* 10.2(a).

two global maxima of $\kappa_{\text{conv}}(u)$, which are located symmetrically with respect to the offending node at $u \approx 0.7$ and $u \approx 9.6$. Finally, it is worth noticing that the algorithm $(\mathcal{A}_{\text{conv}}^{\text{fair}})$ preserves the symmetry of the data by giving $\delta x_{12}^{(0)} = 0$.

Example 10.2. (See Table 10.2 and Fig. 10.3.) The data are taken from McAllister, Passow, and Roulier [11].

(1) $N = 4$,

(2) $\text{tol} = 0.1$,

(3) $x_m, m = 1(1)4 : -2.00, \ -1.00, \ -0.30, \ -0.20,$

TABLE 10.2

	u_2(off.node)	u_3
$\Delta\dot{\kappa}_{\text{conv},m}$	0.290447	0.004098
$\Delta\dot{\kappa}_m^{(0)}$	0.087905	0.006704

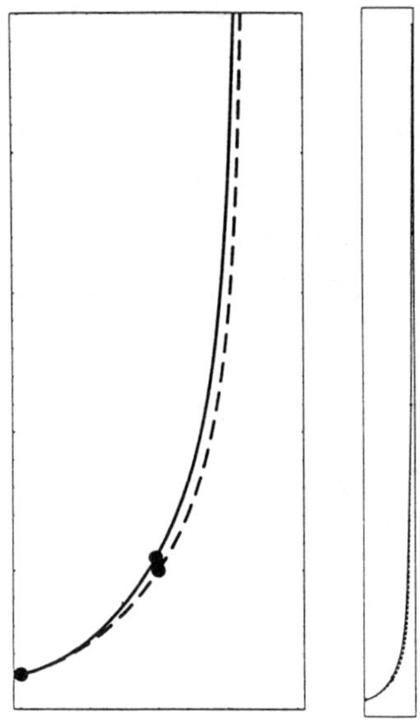

(a)

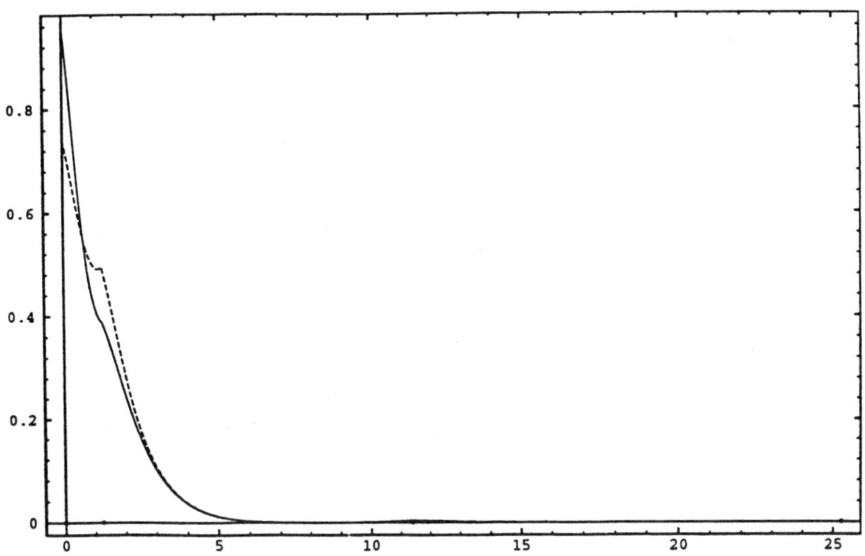

(b)

FIG. 10.3. (a) *Example* 10.2: *the curves* $\mathbf{Q_{conv}}$ *and* $\mathbf{Q^{(0)}}$. (b) *The curvature distributions of the curves shown in Fig.* 10.3(a).

(4) $y_m, m = 1(1)4 : 0.25,\ 1.00,\ 11.11,\ 25.00,$

(5) $\mathbf{s_0} = (1.0,\ 0.25),\ \mathbf{s_4} = (0.01,\ 2.50),$

(6) partype = chord length,

(7) niter = 7,

(8) $\mathcal{K}_{\text{conv}} = \{3,\ 10,\ 10\},$

(9) $l = 2$,
(10) $\delta\kappa_2 = -0.12$,
(11) $x_m^{(0)}, m = 1(1)4 : -2.00000,\ -1.01883,\ -0.30000,\ -0.20000$,
(12) $y_m^{(0)}, m = 1(1)4 : 0.25000,\ 1.09362,\ 11.11000,\ 25.00000$.

Table 10.2 and Fig. 10.3 readily imply that $(\mathcal{A}_{\text{conv}}^{\text{fair}})$ has removed the local maximum of $\kappa_{\text{conv}}(u)$ at the offending node $u = u_2$. On the other hand, $\Delta\dot{\kappa}_3^{(0)}$ is greater than $\Delta\dot{\kappa}_{\text{conv},3}$, which is however insignificant, for both these discontinuities are very small with respect to the corresponding ones at the offending node. In any case, one could further reduce $\Delta\dot{\kappa}_3^{(0)}$ by applying $(\mathcal{A}_{\text{conv}}^{\text{fair}})$ at $u = u_3$, although $u = u_2$ is still the offending node for $\mathbf{Q}^{(0)}(u)$ (see the next example where $(\mathcal{A}_{\text{conv}}^{\text{fair}})$ is successively applied at three neighbouring nodes). Finally, it should be noted that the global-fairness indicator

$$\Delta\dot{\kappa} = \sum_{m=2}^{N-1} \Delta\dot{\kappa}_m \tag{10.32}$$

of $\mathbf{Q}^{(0)}$ is obviously better than that of $\mathbf{Q}_{\text{conv}}$ ($\Delta\dot{\kappa}^{(0)} = 0.094609 \ll \Delta\dot{\kappa}_{\text{conv}} = 0.294545$.)

TABLE 10.3

	u_2 (2nd off.node)	u_3 (1st off.node)	u_4 (3rd off.node)
$\Delta\dot{\kappa}_{\text{conv},m}$	0.02965	0.01617	0.00386
$\Delta\dot{\kappa}_m^{(0)}$	0.00509	0.00102	0.00171

Example 10.3. (See Table 10.3 and Fig. 10.4.)
(1) $N = 5$,
(2) $\text{tol} = 0.1$,
(3) $x_m, m = 1(1)5 : 0.00000,\ 2.73000,\ 7.70181,\ 11.20070,\ 15.00000$,
(4) $y_m, m = 1(1)5 : 1.15000,\ 1.95000,\ 2.66064,\ 2.82994,\ 2.45000$,
(5) $\mathbf{s_0} = (0.91,\ 0.39)$, $\mathbf{s_5} = (1.00,\ -0.50)$,
(6) partype = chord length,
(7) niter = 2,
(8) $\mathcal{K}_{\text{conv}} = \{3,\ 3,\ 4,\ 4\}$,
(9) $l = 3,\ 2,\ 4$,
(10) $\delta\kappa_3 = 0.015$, $\delta\kappa_2 = -0.015$, $\delta\kappa_4 = 0.015$,
(11) $x_m^{(0)}, m = 1(1)5 : 0.00000,\ 2.72000,\ 7.70464,\ 11.20250,\ 15.00000$,
(12) $y_m^{(0)}, m = 1(1)5 : 1.15000,\ 1.98914,\ 2.60583,\ 2.79255,\ 2.45000$.

In this example, the algorithm $(\mathcal{A}_{\text{conv}}^{\text{fair}})$ has been successively applied at the interior nodes $u = u_3$, $u = u_2$ and $u = u_4$, which happen to be the

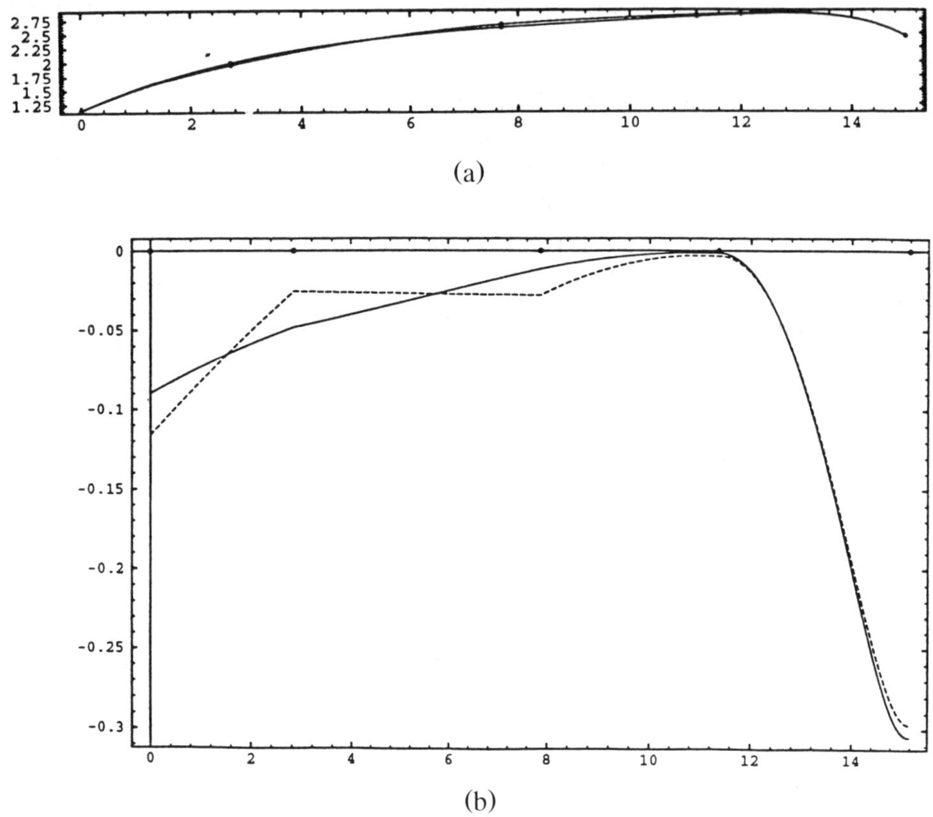

FIG. 10.4. (a) *Example* 1.3: *The curves* $\mathbf{Q}_{\mathbf{conv}}$ *and* $\mathbf{Q}^{(0)}$. (b) *The curvature distributions of the curves shown in Fig.* 10.4(a).

offending nodes at each of the three stages of the fairing process. Note that the parametrization $\mathcal{U}$ has been kept constant throughout the fairing process, being equal to the chord-length parametrization obtained from the initial data set $\mathcal{D}$. As a consequence, the matrix $\mathbf{A}^{-1}$, whose elements are required for the calculation of the quantities ρ_1 and ρ_2 (see step 2 of $(\mathcal{A}^{\text{fair}}_{\text{conv}})$), has been evaluated only once. Looking now at Fig. 10.4(b), it is readily seen that $(\mathcal{A}^{\text{fair}}_{\text{conv}})$ succeeds in giving a monotonically increasing curvature for $\mathbf{Q}^{(0)}(u)$ from $u = u_1$ until almost $u = u_4$. The value of the curvature is affected at $u = u_1$, but this cannot be avoided for $u = u_1$ is adjacent to the offending node during the second stage of the fairing process. Finally, $\Delta\dot{\kappa}^{(0)} = 0.00782 \ll \Delta\dot{\kappa}_{\text{conv}} = 0.04968$, which implies that $\mathbf{Q}^{(0)}$ is globally fairer than $\mathbf{Q}_{\mathbf{conv}}$.

We conclude this section by revisiting the example of §10.2, which provides a case where the algorithm $(\mathcal{A}^{\text{fair}}_{\text{conv}})$ is nonapplicable.[2] Due to the symmetry of $\mathcal{D}_9$, one can easily show that the ellipse $\mathbf{Q}_{\mathbf{ellipse}}(u) =$

[2] We thank an anonymous referee for this observation.

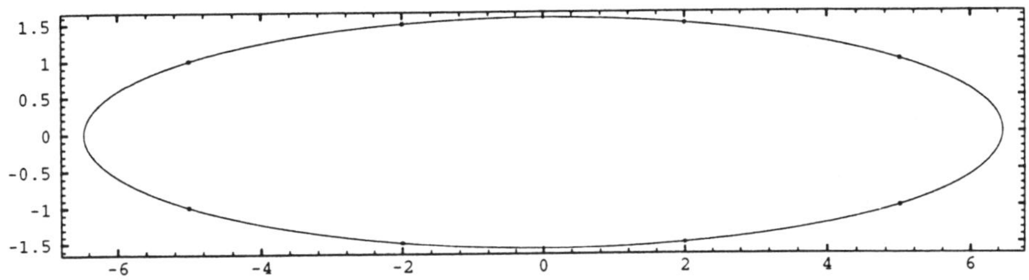

(a)

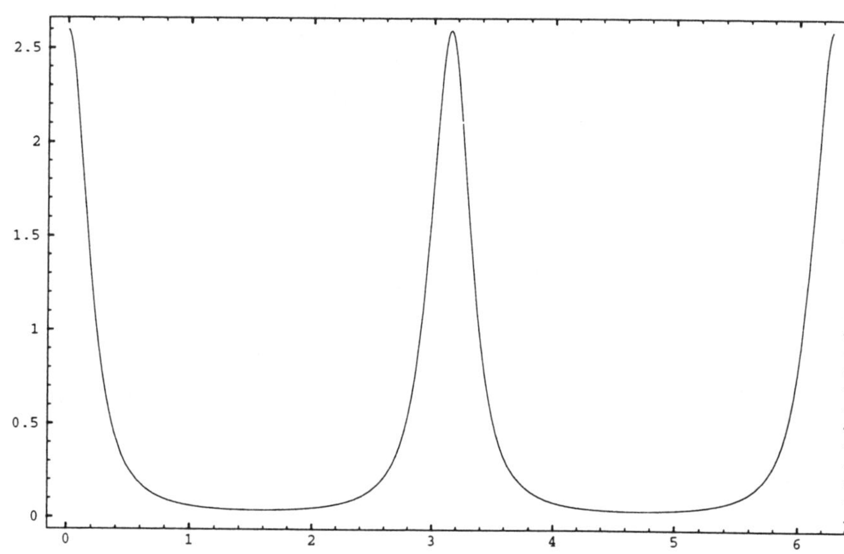

(b)

FIG. 10.5. (a) *The ellipse interpolating the data set* $\mathcal{D}_9$ *of* §10.2. (b) *The curvature distribution of the ellipse depicted in Fig.* 10.5a.

$(6.46529799\cos(u),\ 1.5773674\sin(u))^T$, $u \in [0,\ 2\pi]$, interpolates, $\mathcal{D}_9$ and has a curvature distribution $\kappa(u)$ with four monotone pieces (see Fig. 10.5). Recalling the herein adopted fairness criterion (see §10.1) and comparing $\kappa_{\text{ellipse}}(u)$ (see Fig. 10.5b) with $\kappa_{\text{conv}}(u)$ (see Fig. 10.1b), we readily see that u_4, u_5, and u_8, u_9 are indeed offending nodes for $\mathbf{Q_{conv}}(u)$. However, one can hardly characterize as local imperfections of $\kappa_{\text{conv}}(u)$ its four global maxima at $u = u_l, l = 4,5,8,9$. In fact, if one applies $(\mathcal{A}_{\text{conv}}^{\text{fair}})$ at each of the aforementioned offending nodes consecutively, the fairing process will not converge, due to nonlocal changes on the curvature distribution of the curve.

10.5. Concluding Remarks

Our current work focuses on improving the robustness of the algorithm $(\mathcal{A}_{\text{conv}}^{\text{fair}})$ by ensuring that the constructed curve $\mathbf{Q^{(0)}}(u)$ retains the convexity

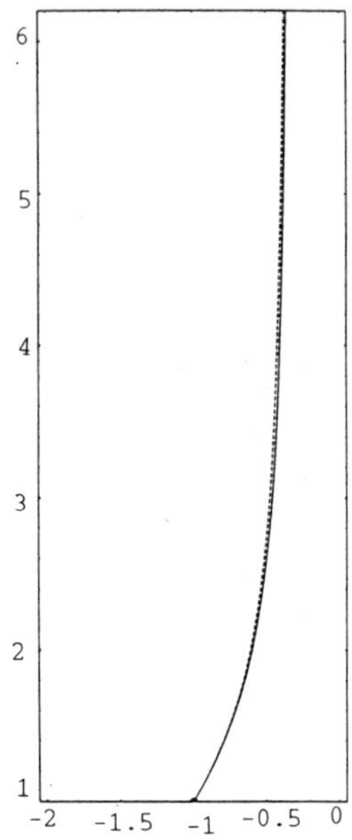

(a)

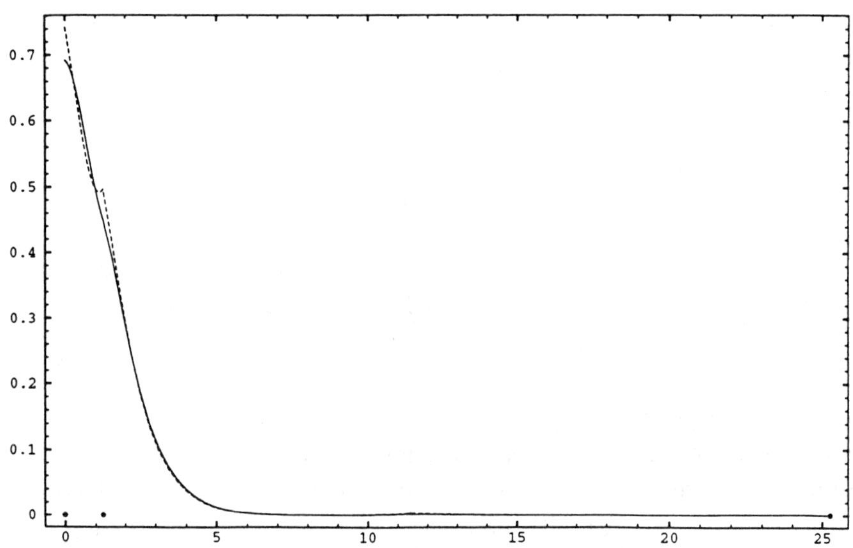

(b)

FIG. 10.6. (a) *Example* 10.2: *the curves* $\mathbf{Q_{conv}}$ *and* $\mathbf{Q_{conv}^{GC^2}}$. (b) *The curvature distributions of the curves depicted in Fig.* 10.6(a).

properties of $\mathbf{Q_{conv}}(u)$ even if tol (and thus $|\delta\kappa_l|$) is not so small. For this purpose, one has to incorporate the bilinear inequalities (10.25) and (10.28) into the formulation of Problem 10.3.1. This would transform it into a more complicated but still tractable quadratic programming problem, from both the theoretical and numerical point of view.

Furthermore, we intend to improve $(\mathcal{A}_{\text{conv}})$ and $(\mathcal{A}^{\text{fair}}_{\text{conv}})$ by replacing the requirement of C^2-continuity, adopted in §10.2, by the weaker and more geometric one of GC^2-continuity. Recall that the GC^2-continuity conditions have as follows: $\dot{\mathbf{Q}}(u_m+) = \alpha_m \dot{\mathbf{Q}}(u_m-)$, $\alpha_m > 0$, $\ddot{\mathbf{Q}}(u_m+) = \alpha_m^2 \ddot{\mathbf{Q}}(u_m-) + \beta_m \dot{\mathbf{Q}}(u_m-)$, $m = 2(1)N-1$. The idea is to use, in addition to the interpolation points $\mathbf{I_m}$, the shape parameters (α_m, β_m) for constructing GC^2-continuous convexity-preserving and fair interpolants. We expect that by introducing two additional degrees of freedom for each offending node we will be able to diminish the deviation of the interpolation points from their initial position. To illustrate this possibility, we go back to Example 10.2 and only change the shape parameter β_2 from zero to $\beta_2 = -0.2$ Although the resulting curve $\mathbf{Q}^{\mathbf{GC^2}}_{\mathbf{conv}}(u)$ (solid line in Fig. 10.6(a)) slightly deviates from $\mathbf{Q}_{\mathbf{conv}}(u)$ (dashed line in Fig. 10.6(a)), $\mathbf{Q}^{\mathbf{GC^2}}_{\mathbf{conv}}(u)$ is fairer than $\mathbf{Q}_{\mathbf{conv}}(u)$, which can be readily seen from the corresponding curvature plots $\kappa^{GC^2}_{\text{conv}}(u)$ (solid line in Fig. 10.6(b)) and $\kappa_{\text{conv}}(u)$ (dashed line in Fig. 10.6(b)). In fact, $\kappa^{GC^2}_{\text{conv}}(u)$ is monotonic along $[u_1, u_4]$, that is, $\mathbf{Q}^{\mathbf{GC^2}}_{\mathbf{conv}}(u)$ satisfies the fairness criterion adopted in §10.1 and thus u_2 is no longer an offending node for it. Finally, one could also argue that $\mathbf{Q}^{\mathbf{GC^2}}_{\mathbf{conv}}(u)$ is fairer than $\mathbf{Q}^{(\mathbf{0})}(u)$, produced by the algorithm $(\mathcal{A}^{\text{fair}}_{\text{conv}})$ (compare $\kappa^{GC^2}_{\text{conv}}(u)$ with $\kappa^{(0)}(u)$ in Fig. 10.3(b))!

Acknowledgments

The authors express their sincere thanks to Professor G. Farin for many useful discussions on curve-fairing and his critical remarks on an earlier version of this paper. Very special thanks are also due to Professor J. Roulier and an anonymous referee for the thorough review of the manuscript.

References

[1] J. A. Roulier, *Constrained interpolation*, SIAM J. Sci. Statist. Comput., 1 (1980), pp. 333–344.

[2] D. F. McAllister and J. A. Roulier, *An algorithm for computing shape-preserving osculatory quadratic spline*, ACM Trans. Math. Software, 7 (1981), pp. 331–347.

[3] T. N. T. Goodman and K. Unsworth, *Shape-preserving interpolation by curvature continuous parametric curves*, Comput. Aided Geom. Design, 5 (1988), pp. 323–340.

[4] P. D. Kaklis and N. S. Sapidis, *Convexity-preserving interpolatory parametric splines of non-uniform polynomial degree*, Comput. Aided Geom. Design, to appear.

[5] P. D. Kaklis and D. G. Pandelis, *Convexity-preserving polynomial splines of non-uniform degree*, IMA J. Numer. Anal., 10 (1990), pp. 223–234.

[6] G. Farin and N. S. Sapidis, *Curvature and the fairness of curves and surfaces*, IEEE Comput. Graphics Applications, 9 (1989), pp. 52–57.

[7] ———, *Automatic fairing algorithm for B-spline curves*, Comput. Aided Design, 22 (1990), pp. 121–129.

[8] J. Roulier, T. Rando, and B. Piper, *Fairness and monotone curvature*, in Approximation Theory and Functional Analysis, C. K. Chui, ed., Academic Press, Boston, 1990.

[9] J. A. Roulier, *Bézier curves of positive curvature*, Comput. Aided Geom. Design, 5 (1988), pp. 59–70.

[10] D. Kershaw, *Inequalities of the elements of the inverse of a certain tridiagonal matrix*, Math. Comp., 24 (1970), pp. 155–158.

[11] D. F. McAllister, E. Passow, and J. A. Roulier, *Algorithms for computing shape preserving spline interpolants to data*, Math. Comp., 31 (1977), pp. 717–725.

PART

4

Special Applications

CHAPTER

11

Constructing C^1 Surfaces of Arbitrary Topology Using Biquadratic and Bicubic Splines

Jörg Peters

11.1. Introduction

The algorithm of this chapter combines two approaches to building smooth free-form surfaces, generalized subdivision and composing surfaces from patches. The result is a low degree polynomial surface representation with B-spline-like control via the input mesh of points. The simplicity of the construction allows for detailed control of the surface shape.

Cutting off corners and edges to smooth a polytope is a geometrically intuitive design paradigm. Algorithmically, it is realized by generalized subdivision. Given an input mesh of points delineating a surface, the algorithms of [4], [6], [14], [8], etc., create at each stage a refined mesh of points by averaging neighboring points of the current mesh. Where the mesh is *regular*, i.e., where each mesh point is surrounded by exactly four quadrilateral cells, the mesh can be interpreted as a B-spline control mesh and the refinement of the mesh as knot insertion. Thus the limit surface has a standard parametrization of low degree. Unfortunately, regular meshes can only model very restricted geometric configurations. Modeling real world objects without singularity, and such that the mesh conforms with the features, generally requires an irregular mesh; for such meshes generalized subdivision methods do not provide an explicit parametrization of the limit surface. This makes it tricky to establish elementary properties like tangent plane continuity of the limit surface (see, e.g., [7] and [1]–[3]), and represents a major obstacle for integrating these techniques with other computer-aided design representations.

Assembling surfaces from patches is a second design paradigm. A number of surface constructions for irregular meshes have been derived over the last decade [10]. Compared to the standard tensor-product B-spline representation, these constructions either sacrifice the low degree of the surfaces (e.g., [20] and [12]) or depart from the standard polynomial representation (e.g., [5] and [16]). However, the main drawback of the patching approach seems to arise from the very tool that allows it to model complex smooth surfaces, namely reparametrization when crossing from one patch to the next.

This shifts the focus from the geometry of the modeling problem to clever uses of the chain rule. Thus the algorithms of [9], [13], and [17] fix the reparametrizations a priori dependent only on the number of the patches joining at a mesh point; in order to match geometric data, a large sparse linear system is solved. This makes it tricky to reason a priori about the shape of the surface.

The algorithm proposed here reconciles the subdivision paradigm with the parametric approach. It generalizes the standard biquadratic tensor-product spline representation to splines over irregular meshes. In particular, there are no regularity restrictions on the input mesh as in [21]. The underlying idea is to use a small number of subdivisions to give the surface its rough shape and separate irregular mesh regions. Then a biquadratic C^1 spline can be fit over the regular mesh regions and the remaining mesh holes can be covered by a composite of bicubic patches. The result is a polynomial surface of low degree. Since no system of equations has to be solved to determine the parametrization, but rather the input mesh points are averaged such that de Casteljau's algorithm generates a C^1 surface in the limit, the method is a generalized subdivision algorithm. Additionally, one can interpolate the vertices of the input mesh and normals at the vertices without solving systems of equations.

The structure of the chapter is as follows. Section 2 states the algorithm. Section 3 establishes the consistency and continuity of the resulting surfaces and proves simple shape properties. Section 4 gives examples and §5 summarizes the findings.

11.2. The Algorithm

The three steps of the algorithm are as follows:

- Refining the input mesh to isolate irregular mesh points by regular B-spline control points.
- (Optional) Converting the quadratic B-spline surface into Bernstein–Bézier form to make the representation uniform.
- Covering the irregular mesh cells with bicubic patches.

The *input* is any mesh of points such that at most two cells abut along any edge. The mesh cells need not be planar, and there is no constraint on the number of edges to a cell or the number of cells meeting at a vertex. The mesh may model a bivariate surface of arbitrary topological structure and with or without boundary. Each cell f of the input mesh has a shape parameter α_f, called *blend ratio*. The blend ratio is a number between zero and one. A smaller ratio results in a surface that follows the input mesh more closely and changes the normal direction more rapidly close to the mesh edges. The *output* of the algorithm are the Bernstein–Bézier coefficients of biquadratic and bicubic patches that parametrize a tangent-plane continuous surface. The surface interpolates the centroid, the average of the vertices, of each cell of the input mesh.

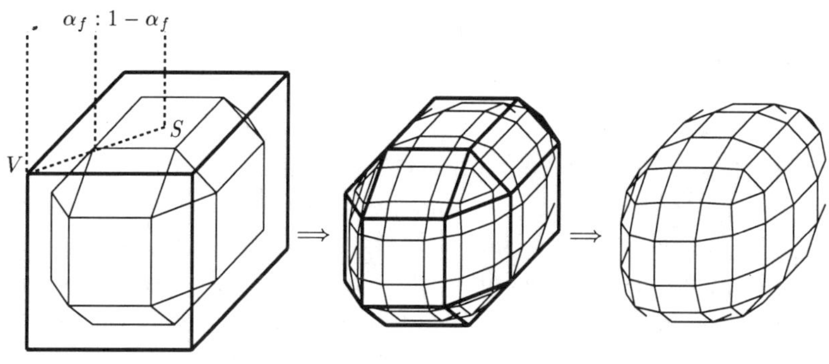

FIG. 11.1 *Two steps of the Doo–Sabin algorithm.*

The end of this section gives two simple extensions of the algorithm that guarantee interpolation of the mesh vertices and allow for a rational surface representation.

A1. Refining the input mesh to separate irregular mesh points by layers of regular B-spline control points. The refinement gives the object its rough shape and isolates nonquadrilateral mesh cells. It consists of two steps of Doo and Sabin's averaging procedure [7]. At each step, s new points are created for each s-sided cell. Each new point connects to four new points, two generated at the two adjacent vertices of the same cell and two corresponding to adjacent cells of the same old mesh point (cf. Fig. 11.1). A new point corresponding to a vertex V of the cell f with centroid S and n edges has the coordinates $(1-\alpha_f)V + \alpha_f S$ in the first step and

$$(1-\alpha)V + \alpha S, \alpha := \alpha_f(1 - \cos(2\pi/n))$$

in the second step.

After refining the mesh, each mesh point is surrounded by four cells. If the mesh point is *regular*, that is, if all four cells have exactly four edges, then the nine mesh points defining the cells can be interpreted as the control mesh of a quadratic spline. To guarantee a C^1 construction also at the irregular mesh points, the refined mesh has to be perturbed in some cases. Let $B_{i,j}$ be the mesh points edge-adjacent to the vertex C_i of an n-sided cell of the refined mesh (cf. Fig. 11.3) and define

$$E := \frac{-1}{2n} \sum_{i=1}^{s} \sum_{j=1}^{2} (-1)^{i+j} B_{i,j}.$$

If n is even and greater than four, then the $B_{i,j}$ are replaced by $B_{i,j} + (-1)^{i+j}E$.

If $\alpha_f = \frac{1}{2}$ uniformly, the cells of the refined quadrilateral mesh correspond to the same quadratic and can be stored more efficiently [6, p. 163]. If the input mesh already satisfies some or all conditions (R1)–(R3) of Proposition 11.3.1, then only one or no refinement step is necessary.

A2. (Optional) Converting the biquadratic B-spline surface into Bernstein–Bézier form. Any submesh of regular points of the refined mesh can serve as the control mesh of a biquadratic tensor-product spline surface. The spline can be expressed in Bernstein–Bézier form in order to unify the representation with the representation at the irregular mesh sites. By symmetry, it suffices to give the conversion formulas for the following four Bernstein–Bézier coefficients Q_{ij} in terms of the spline control points C_k (cf. Fig. 11.2):

$$\begin{aligned} Q_{00} &= (C_{00} + C_{10} + C_{01} + C_{11})/4, \\ Q_{10} &= (C_{11} + C_{10})/2, \\ Q_{01} &= (C_{11} + C_{01})/2, \\ Q_{11} &= C_{11}. \end{aligned}$$

For example, the mesh cell $A_i B_{i,1} C_i B_{i,2}$ in Fig. 11.3 yields $L_i = Q_{00} = (A_i + B_{i,1} + C_i + B_{i,2})/4$, $Q_{01} = (C_i + B_{i,2})/2$ and $Q_{11} = B_{i,2}$ for the coefficients Q_{ab} of $q_{i,2}$.

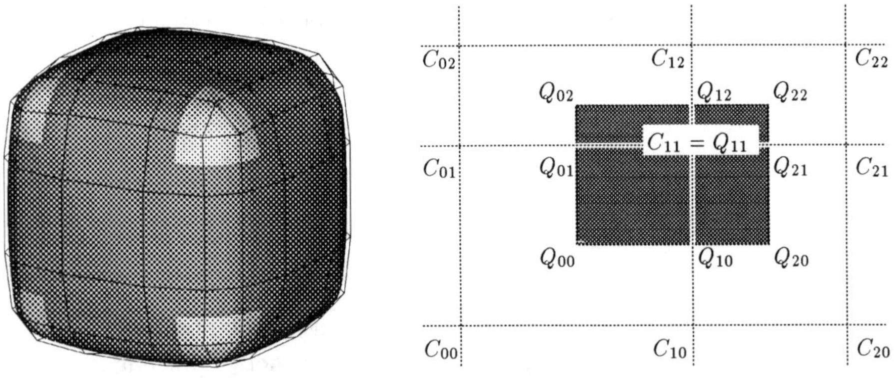

FIG. 11.2 *Refined mesh and C^1 surface: darker regions are parametrized by biquadratics.*

A3. Covering the irregular mesh cells with bicubic patches. The refinement step and the interpretation of the regular mesh points of the refined mesh at B-spline control points create a parametrically C^1 biquadratic tensor-product surface with nonquadrilateral holes. Each hole is split into quadrilaterals as indicated in Fig. 11.3. Each quadrilateral is covered by a bicubic patch. The ith patch, p_i, has coefficients $P_{kl,i}$, $k, l = 0 \cdots 3$ such that $P_{00,i}$ corresponds to the corner point L_i of the biquadratic surface and $P_{33,i}$ corresponds to the central point S where the bicubic patches join.

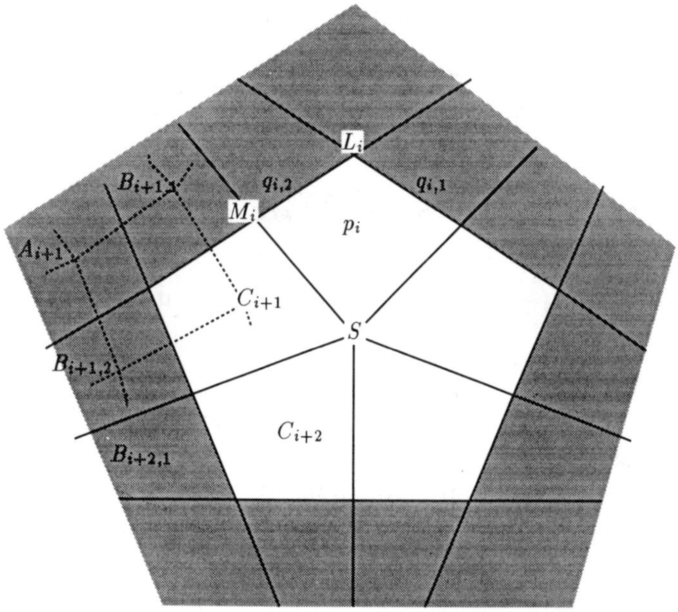

FIG. 11.3 *Control points A_i, B_{ij}, C_i, quadratic patches q_{ij}, and cubic patches p_i.*

To join the bicubic patch smoothly to the biquadratic surface, the boundary curves are the degree-raised quadratics

$$P_{00,i} = \frac{1}{4}(B_{i,2} + B_{i,1} + C_i + A_i), \quad P_{20,i} = \frac{1}{12}(5B_{i,2} + B_{i+1,1} + 5C_i + C_{i+1}),$$

$$P_{10,i} = \frac{1}{12}(5B_{i,2} + B_{i,1} + 5C_i + A_i), \quad P_{30,i} = \frac{1}{4}(B_{i,2} + B_{i+1,1} + C_i + C_{i+1}),$$

and the coefficients defining the transversal derivatives are chosen as

$$P_{11,i} = \frac{1}{36}(5B_{i,2} + 5B_{i,1} + A_i + 25C_i) + \frac{a}{36}(B_{i,2} - B_{i,1} + 5C_i - 5A_i),$$

$$P_{21,i} = \frac{1}{36}(5B_{i,2} + 5C_{i+1} + B_{i+1,1} + 25C_i)$$
$$+ \frac{a}{36}(-11B_{i,2} + 3B_{i+1,1} + 3C_{i+1} + 5C_i),$$

$$P_{31,i} = \frac{1}{12}(B_{i,2} + B_{i+1,1} + 5C_{i+1} + 5C_i) + \frac{a}{6}(C_{i+1} + C_i - B_{i,2} - B_{i+1,1}),$$

where $c := \cos(\frac{2\pi}{n})$ and $a := \frac{c}{1-c}$. The coefficients $P_{0k,i}$ and $P_{1k,i}$, $k = 0 \cdots 3$ (and hence the scaling of the α_f in the second refinement) are obtained by symmetry. For example, $P_{12,i} = \frac{1}{36}(5B_{i,1} + 5C_{i-1} + B_{i-1,2} + 25C_i) + \frac{a}{36}(-11B_{i,1} +$

$3B_{i-1,2} + 3C_{i-1} + 5C_i)$. The definition of the surface is completed by setting

$$P_{33,i} = S = \frac{1}{n}\sum_{i=1}^{n} C_i,$$

$$P_{32,i} = P_{23,i+1} = S + \frac{4\alpha}{3n}\sum_{l=1}^{n}\cos\left(\frac{2\pi}{n}l\right)P_{31,i+l},$$

$$P_{22,i} = \begin{cases} -\sum_{j=1}^{n}(-1)^j E_{i+j} & \text{if } n \text{ is odd,} \\ \frac{-2}{n}\sum_{j=1}^{n}(n-j)(-1)^j E_{i+j} & \text{if } n \text{ is even.} \end{cases}$$

The scalar $\alpha \in [0,1]$ is a shape parameter proportional to the diameter of the tangent plane of first differences (cf. [15, Fig. 3]) and therefore similar to the blend ratio of the Doo–Sabin algorithm. Its default setting is $\alpha = 1$. The vector $E_i := (1-c)P_{32,i} + cP_{31,i}$ is an average of the interior coefficients of the boundary curve.

11.2.1. Extensions of the algorithm.

A1a. Interpolation at the vertices of the input mesh. After the first of two refinements in step A1, move the control points C_i, $i = 1 \cdots n$ whose construction involves S by $S - \frac{1}{n}\sum_{i=1}^{n} C_i$ for each vertex S of the input mesh. Then S is the centroid of the resulting cell and will be interpolated. Similarly, one can interpolate normals at the vertices.

A2a. Conic blends and rational patches. To obtain conic blends and, more generally, rational surfaces, treat the control points as vectors in $\mathbb{R}^4$ and the fourth coordinate as an additional shape parameter. If $P, Q : [0,1]^2 \mapsto \mathbb{R}^3$ and $p, q : [0,1]^2 \mapsto \mathbb{R}$ and $P = Q$ and $p = q$ along a boundary shared by the functions P/p and Q/q, then the continuity conditions (see §11.3.2)

$$D_1\left(\frac{Q}{q} \circ \phi\right) = D_1\frac{P}{p} \qquad \iff$$

$$q(D_1P - D_1QD_1\phi^{[1]} - D_2QD_1\phi^{[2]}) = Q(D_1p - D_1qD_1\phi^{[1]} - D_2qD_1\phi^{[2]})$$

hold along that boundary if the masks of the algorithm are applied to the coefficients of $(P,p) \in \mathbb{R}^4$. That is, the fourth coordinate corresponds to the rational weight function. For example, choosing the weight component of two neighboring control points in the regular mesh to be 3 rather than 1 (and keeping the other weight components at 1) yields two circular arcs with weights (112) and one conic with weights (232) in between:

weights of the control points	1		1		3		1		1
weights of three quadratic boundary curves		1	1	2	3	2	1	1	
weights of the control points	1		1		3		1		1

In general, the algorithm works for constructing bivariate surfaces in $\mathbb{R}^n$.

A3a. An alternative choice of the tangent plane at S. By default, the tangent coefficients $P_{32,i}$ at S should be constructed symmetrically from the existing data. Since the tangent coefficients P_{31} at M are constructed by applying a mask to the control points, B_{ij} and C_i, one can simplify the formulae by making $P_{32,i}$ depend only on the C_i and ignoring the contribution of the B_{ij}.

$$P_{32,i} = S + \frac{4\alpha}{3n} \sum_{l=1}^{n} \cos\left(\frac{2\pi}{n} l\right) \frac{C_{i+l} + C_{i+1+l}}{2}.$$

11.3. Structure, Smoothness, and Shape of the Surface

This section analyzes the construction. The first part examines the refinement process. The second part proves consistency and smoothness of the surface. The third part discusses shape properties of the resulting surface. The fourth part considers the special case of triangular mesh cells.

11.3.1. Mesh structure. Denote as c-c every pair of non-four-sided cells arising from two adjacent non-four-sided cells or from two adjacent non-four-valent vertices, and denote as v-c every pair of non-four-sided cells arising from a non-four-sided cell and its non-four-valent vertex.

PROPOSITION 11.3.1. *The two refinements in step* A1 *generate a mesh with the following properties:* (R1) *every interior control point has four neighbors,* (R2) *every* c-c *pair is separated by three layers of quadrilateral cells, and* (R3) *every* v-c *pair is separated by one layer of quadrilateral cells.*

Proof. Since every new control point is connected to two new points on the same original cell and across to the edges of that cell, (R1) already holds after one step. Each step separates any two cells by one layer of quadrilateral cells. This implies (R2) and (R3) follows since every v-c pair still has a common vertex after the first refinement.

COROLLARY 11.3.2. *The control points $B_{i,j}$ of a* c-c *pair of cells are distinct. The perturbation of any $B_{i,j}$ does not alter the control points of a cell that contains the centroid of a cell of the input mesh.*

Concerning the shape, we observe that since the endpoints of the new and old edges together with the centroid form similar triangles, the refinement leaves all facet-cell edges parallel to the original facet cells.

11.3.2. Smoothness. This section shows that splines generated by the surface form a vector space. Unless $\alpha_f \notin (0,1)$ the surface pieces join with oriented tangent plane continuity, which is characterized as the agreement of the derivatives of two maps p and q from $\mathbb{R}^2$ to $\mathbb{R}^n$ after reparametrization by a map φ from $\mathbb{R}^2$ to $\mathbb{R}^2$ that connects the domains Ω_p and Ω_q of p and q:

$$p = q \circ \varphi \quad \text{and} \quad D_1 p = D_1(q \circ \varphi) \quad \text{along } E_p,$$

where $\varphi(E_p) = E_q$, E_p and E_q are edges of Ω_p and Ω_q, respectively. D_1 denotes differentiation in the direction perpendicular to E_p and the connecting map φ maps interior points of Ω_q to exterior points of Ω_p, thus avoiding cusps. The components of φ are $\varphi^{[1]}$ and $\varphi^{[2]}$.

THEOREM 11.3.3. *The surfaces generated by the algorithm are tangent plane continuous. Surfaces generated from input meshes with the same connectivity and the same blend ratio for corresponding cells form a vector space.*

The proof of the first part of the theorem is divided into four lemmas. The first lemma proves smoothness between the biquadratic patches, the second and third lemmas between biquadratics and bicubics at L_i and M_i, and the fourth lemma between bicubics at S (cf. Fig. 11.3). It suffices to show that the cubic polynomials $D_1(p_i - q_{i,j} \circ \phi_{ij})$ and $D_1(p_{i-1} - p_i \circ \psi_i)$ and their D_2 derivatives vanish at L_i, M_i and S since this implies that they vanish identically. The vector space property follows from the linearity of the algorithm for fixed blend ratios and connecting maps and also from the observation that two surfaces generated from input meshes with the same connectivity have a natural correspondence of abutting patches and hence are joined by the same connecting map.

LEMMA 11.3.4. *The conversion of the regular mesh points in step* A2 *generates the Bernstein–Bézier coefficients of a parametrically C^1 surface equivalent to the biquadratic tensor-product spline.*

Proof. Since the centroid of any four points is also the intersection of lines through opposite midpoints, the transversal derivatives across any boundary agree pairwise. That is, we have $D_1 q_1 = D_1 q_2$ and $\varphi = \text{id}$ for adjacent patches q_1 and q_2 as claimed.

LEMMA 11.3.5. *The extensions of the biquadratic patches $q_{i,1}$ and $q_{i,2}$ define the mixed derivative of p_i at L_i consistently.*

Proof. Choose the coordinate system such that $p_i(0,0) = L_i$ and $p_i(1,0) = M_i$. Then for each patch $q_{i,j}$, $j = 1,2$, the patch p_i is chosen such that $p_i = q_{i,j} \circ \phi_{i,j}$ and $D_j p_i = D_j(q_{i,j} \circ \phi_{i,j})$ along the common boundary, where $a := \frac{c}{1-c}$, $c := \cos(\frac{2\pi}{n})$, and

$$\phi_{i,1} := \text{id} + t_1 t_2 \begin{bmatrix} a \\ a(1-t_2) \end{bmatrix}, \quad \phi_{i,2} := \text{id} + t_1 t_2 \begin{bmatrix} a(1-t_1) \\ a \end{bmatrix}.$$

For simplicity, we drop the subscript i and do not explicitly mention that we evaluate at the origin.

We need to show that

$$\text{(L)} \qquad D_1 D_2 (q_1 \circ \phi_1) = D_1 D_2 p = D_1 D_2 (q_2 \circ \phi_2).$$

Expanding the left-hand side of (L) according to the chain rule and noting that $D_2 \phi_1 = (0,1)$ and $D_1 \phi_2 = (1,0)$, we have

$$D_1 D_2 q_1 + D_1 q_1 D_1 D_2 \phi_1^{[1]} + D_2 q_1 D_1 D_2 \phi_1^{[2]} = D_1 D_2 q_1 + a_1 D_1 q_1 + a_2 D_2 q_1.$$

The expression for the right-hand side is obtained by symmetry. By construction of the complex, $D_1D_2q_1 = A+C-B_1-B_2 = D_1D_2q_2$ and $D_jq_2 = D_jq_1$. Thus the left-hand side of (L) equals the right-hand side and uniqueness of the mixed derivative D_1D_2p follows.

LEMMA 11.3.6. *The patches join smoothly at M_i.*

Proof. We choose the coordinate system such that $p_i(0,0) = M_i$ and $p_i(1,0) = L_{i+1}$. Then

$$D_2^m p_i = D_2^m(q_{i,1} \circ \phi'_{i,1}), \quad m = 0,1,$$

$$D_2^m p_{i-1} = D_2^m(q_{i-1,2} \circ \phi'_{i-1,2}), \quad m = 0,1,$$

along the common boundary, where $a := \frac{c}{1-c}$, $c := \cos(\frac{2\pi}{n})$, and

$$\phi'_{i-1,2}(-t_1,t_2) = \phi'_{i,1}(t_1,t_2) = \mathrm{id} + at_2(1-t_1)\begin{bmatrix} t_1 \\ 1 \end{bmatrix}.$$

We show that this choice of reparametrization is consistent with the smooth join of the bicubic patches

$$\text{(M)} \qquad D_1^m p_{i-1} = D_1^m(p_i \circ \psi'_i), \quad m = 0,1, \quad \psi'_i := \mathrm{id} + t_1t_2\begin{bmatrix} 0 \\ -2c \end{bmatrix}.$$

For simplicity, we do not mention explicitly in the following that we evaluate at the origin. For $m = 0$, we check $p_{i-1} = p_i \circ \psi'_i$, $D_2p_{i-1} = (1+a)D_2q_{i-1,2} = (1+a)D_2q_{i,1} = D_2(p_i \circ \psi'_i)$, and $D_1p_{i-1} = D_1q_{i-1,2} = D_1q_{i,1} = D_1p_i = D_1(p_i \circ \psi'_i)$. For $m = 1$, we observe that $D_1D_2q_{i-1,2} = D_1D_2q_{i,1}$, $D_1q_{i-1,2} = D_1q_{i,1}$, and hence

$$\begin{aligned} D_2D_1p_{i-1} &= D_2D_1(q_{i-1,2} \circ \phi'_{i-1,2}) \\ &= (1+a)D_1D_2q_{i-1,2} + a(D_1q_{i-1,2} + D_2q_{i-1,2}) \\ &= D_2D_1(q_{i,1} \circ \phi'_{i,1}) + 2aD_2q_{i,1} \\ &= D_2D_1p_i + \frac{2a}{1+a}D_2p_i. \end{aligned}$$

The claim follows since $D_2D_1(p_i \circ \psi'_i) = D_2D_1p_i - 2cD_1p_i$ and $\frac{2a}{1+a} = -2c$.

LEMMA 11.3.7. *The bicubic patches meet smoothly at S.*

Proof. We choose the coordinate system such that $p_i(0,0) = p_{i-1}(0,0) = S$ and $p_i(1,0) = M_i$ and show that

$$\text{(I)} \qquad D_1p_{i-1} = D_1(p_i \circ \psi_i), \quad \text{where} \quad \psi_i := \mathrm{id} + t_1(1-t_2)\begin{bmatrix} 0 \\ 2c \end{bmatrix}.$$

In Bernstein–Bézier form this yields four constraints. Since two of these hold at M by the previous lemma, we need only show that

$$(\mathrm{I}_3) \qquad 2(1-c)S = P_{32,i-1} - 2cP_{32,i} + P_{32,i+1},$$

$$(\mathrm{I}_4) \qquad 2((1-c)P_{32,i} + cP_{31,i}) = P_{22,i-1} + P_{22,i} = 2E_i,$$

to establish tangent plane continuity between adjacent bicubics p_i and p_{i-1}. Constraint (I_3) holds since

$$\sum_{i=1}^{n} \cos\left(\frac{2\pi}{n}i\right)(P_{31,i-1} - 2cP_{31,i} + P_{31,i+1})$$

$$= \sum_{i=1}^{n} P_{31,i}\left(\cos\left(\frac{2\pi}{n}(i-1)\right) + \cos\left(\frac{2\pi}{n}(i+1)\right) - 2\cos\left(\frac{2\pi}{n}\right)\cos\left(\frac{2\pi}{n}i\right)\right) = 0.$$

If n is odd, then

$$P_{22,i-1} + P_{22,i} = -\sum_{j=1}^{n}(-1)^j(E_{i+j-1} + E_{i+j}) = 2E_i$$

as required by (I_4). If n is even, then

$$\sum_{j=1}^{n}(-1)^j P_{32,j} = \frac{\alpha}{n}\sum_{j=1}^{n}(-1)^j \sum_{l=1}^{n}\cos\left(\frac{2\pi}{n}l\right)P_{31,j+l} = 0$$

since $\sum_{j=1}^{n}(-1)^j\cos(\frac{2\pi}{n}j) = 0$ and $\sum_{j=1}^{n}(-1)^j P_{31,j} = 0$ since the C_i cancel and the perturbation forces $E := \sum_{i=1}^{n}\sum_{j=1}^{2}(-1)^{i+j}B_{i,j} = 0$. ($E = 0$ is also a necessary constraint for solvability.) Therefore $\sum_{j=1}^{n}(-1)^j E_j = 0$ and

$$P_{22,i-1} + P_{22,i} = \frac{-2}{n}\sum_{j=1}^{n}(n-j)(-1)^j(E_{i+j-1} + E_{i+j})$$

$$= \frac{-2}{n}\left(\sum_{j=1}^{n}(-1)^j E_j - nE_i\right) = 2E_i$$

as required.

Remark. The join between the bicubic patches with the above connecting map leaves three coefficients at S free to choose. Since $E = 0$, it is always possible to solve the least squares problem

$$\min \sum_{i=1}^{s} \|P^*_{32,i} - P_{32,i}\|^2 + \sum_{i=1}^{s}\|P^*_{22,i} - P_{22,i}\|^2$$

subject to the constraints (I_3) and (I_4), where $P^*_{32,i}$, $P^*_{22,i}$ are desirable locations obtained, say, from degree-raising. The particular solution used in the algorithm has the advantage of being explicit and symmetric. We also note that if $n = 4$, and hence $a = 0$, then the patches generated by step A3 are biquadratic rather than bicubic.

11.3.3. Shape considerations. This section shows that the surface is locally flat exactly when the input mesh is locally flat. It also characterizes monotonicity for symmetric data and proves that the surface tautly interpolates the input mesh when the blend ratios are zero.

PROPOSITION 11.3.8. *A biquadratic patch with center coefficient Q_{11} generated in step A2 has zero curvature if and only if the four cells surrounding the meshpoint $C_{11} = Q_{11}$ are coplanar. Two adjacent coplanar cells give rise to a linear boundary curve.*

Proof. The 9 by 9 system of equations relating the control points to the Bernstein–Bézier coefficients is of full rank. While the linearity of the boundary curve follows from coplanarity, the reverse does not hold, because the system is 3 by 6.

PROPOSITION 11.3.9. *The curvature at S is zero if and only if $P_{31,i}, i = 1 \cdots s$ and S lie in the same plane.*

Proof. According to Lemma 11.3.7, all $P_{32,i}$ and S lie in the same tangent plane. Let $P(n)$ be the component of P normal to that plane. If some $P_{31,i}$ do not lie in that plane, then the curvature of the ith boundary curve is nonzero and the $P_{22,i}$ do not all lie in the plane either. The latter follows by contradiction from I_4: $2cP_{31,i} = 2((1-c)P_{32,i}(n)+cP_{31,i}(n)) = P_{22,i-1}(n)+P_{22,i}(n) = 0$. Conversely, if all $P_{31,i}$ lie in the tangent plane, then the normal components of all $P_{32,i}$ and $P_{22,i}$ are zero by construction.

COROLLARY 11.3.10. *If all C_i and $(B_{i,1} + B_{i,2})/2$ are in the same plane, then the normal curvatures at S are zero. If additionally all A_i and $(B_{i,1} - B_{i,2})/2$ are in the common plane, then the bicubic patch is flat.*

To analyze monotonicity of the surface, define the averages $S_P := \frac{1}{s}\sum_{i=1}^{s} P_{31,i}$ and $S_C := \frac{1}{s}\sum_{i=1}^{s} C_i$. Assume *local symmetry* of the control mesh at S, i.e., A_{i+1}, $B_{i+1,j}$, and C_{i+1} can be obtained from A_i, B_{ij}, and C_i by a rotation by $\frac{2\pi}{n}$ and *local convexity* of the control mesh, i.e., the extensions of the cells spanned by A_i, B_{ij}, and C_i $i = 1 \cdots s$ form a convex cone. Let $\beta^* \geq 1$ be the smallest value such that $S_C + \beta^*(S_P - S_C)$ intersects the cone. Then we can more generally choose $P_{33,i} = S = S_C + \beta(S_P - S_C)$. Setting $\beta = 0$ guarantees interpolation at the centroid.

LEMMA 11.3.11. *If the control mesh is locally convex and symmetric, then for $s > 4$ the boundary curves are convex if $1 \leq \beta \leq \beta^*$; for $s = 3$, $\beta = 1$ is sufficient. If $s > 4$, then the components $P_{ij}(n)$ of the coefficients normal to the tangent plane are monotonically increasing with $i + j$ if and only if $A(n) > \frac{4+8a}{5a-1}B(n)$.*

Proof. Due to symmetry the C_i are coplanar and we may denote the normal distance of a point P from that plane by $P(n)$. Dropping the subscripts of the control points we have $A(n) < B(n) < C(n) = 0 < \beta = P_{32,i}(n) = S(n)$. By symmetry, the curve with coefficients $[P_{30,i}, 3P_{31,i}, 3P_{32,i}, P_{33,i}]$ constructed in step A3 falls into a perpendicular plane. Since $1 < a = \frac{1}{1-c}$ for $s > 4$ and $B(n) < 0$,

$$P_{31,i}(n) = \frac{1-2a}{6}B(n) > 0.$$

That is, the plane through the C_i separates the M_i from the plane in which S and the coefficients $P_{32,i}$ lie if $S = S_C$. The boundary cubic has therefore an inflection in the Bernstein–Bézier polygon unless $\beta \geq 1$, i.e., $S(n) \geq P_{31,i}(n)$.

For $s > 4$, set $A(n) =: kB(n)$, $k > 1$, and $b := \frac{1-2a}{6} < 0$. Since $\beta \geq 1$, $P_{33}(n) = P_{31}(n) + \epsilon b B(n)$, $\epsilon \geq 0$. Then

$$\begin{bmatrix} P_{03} & P_{13} & P_{23} & P_{33} \\ P_{02} & P_{12} & P_{22} & P_{32} \\ P_{01} & P_{11} & P_{21} & P_{31} \\ P_{00} & P_{10} & P_{20} & P_{30} \end{bmatrix}(n)$$

$$= \begin{bmatrix} \frac{1}{2} & b & (1+\epsilon)b & (1+\epsilon)b \\ \frac{1}{2} & b+\frac{a}{9} & (1+(1-c)\epsilon)b & (1+\epsilon)b \\ \frac{1}{2}+\frac{k}{12} & \frac{10}{36}+\frac{1-5a}{36}k & b+\frac{a}{9} & b \\ \frac{1}{2}+\frac{k}{4} & \frac{1}{2}+\frac{k}{12} & \frac{1}{2} & \frac{1}{2} \end{bmatrix} B(n).$$

We check the following four nontrivial difficult cases for monotonicity.

$$P_{22}(n) < P_{21}(n): \qquad (1-c)\epsilon b < 0 < \frac{a}{9},$$

$$P_{21}(n) < P_{20}(n): \qquad b + \frac{a}{9} = \frac{3-4a}{18} < 0 < \frac{1}{2},$$

$$P_{11}(n) < P_{10}(n): \qquad \frac{10-18+k-3k-5ak}{36} < -\frac{8+7k}{36} < 0,$$

$$P_{12}(n) < P_{11}(n): \qquad \frac{6-8a-10-k+5ak}{36} < 0.$$

The last relation is responsible for the extra condition on the normal component of the A_i and B_{ij}.

LEMMA 11.3.12. *An edge between two cells with zero cut ratios is interpolated. Planar cells with zero cut ratios are covered by a planar surface.*

Proof. Step A3 applies at two types of irregularities in the mesh. A vertex irregularity originates from a vertex of the input mesh that does not have four neighbors. Here all points A_i, B_{ij}, C_i, and hence all patches are coalesced into one input mesh point if the ratio is zero and hence the point is interpolated by the surface. A cell irregularity originates from a cell of the input mesh that is not quadrilateral. Here the points A_i, B_{ij}, and C_i are coalesced into their respective input mesh point, say L_i, and the quadratic border patches degenerate into straight lines $L_i\ M_i$ and $L_i\ M_{i-1}$ where $M_i = (L_{i+1} + L_i)/2$. Since the cubic construction averages the L_i and the L_i are coplanar, the second claim follows.

For a similar surface construction, [18] establishes that the surface lies in the convex hull of thc control mesh.

11.3.4. A remark on parametric C^1 continuity, bicubic patches, and triangular mesh cells. Choosing the extension across the boundaries of the biquadratic complex to be parametrically C^1, i.e., selecting $\phi_{i,j}$ to be the identity, leads in general to an inconsistent system of equations for bicubics. In particular, $D_1\psi_i^{\prime[1]}$ has to be at least quadratic, since $D_2D_1\psi_i^{\prime[1]}(0,0)$ is zero rather than $-2c$. This results in n additional constraints but only one of n tangent coefficients and one of n twist coefficients may be chosen freely in addition to S. Therefore, in general one cannot cover an n-sided hole with bicubics that extend a biquadratic patch complex parametrically C^1 across its boundary.

An exception occurs for $n = 3$ due to the fact that three points always lie in a plane. One can then choose the boundary curves to be the quadratic $p_{i-1}(0,t) = p_i(0,t) = M_i(1-t)^2 + (C_i + C_{i+1})t(1-t) + St^2$,

$$\phi_{ij} = \text{id} \quad \text{and} \quad \psi_i := \text{id} + t_1(1-t_2)^2 \begin{bmatrix} 2c \\ 0 \end{bmatrix},$$

and

$$\begin{aligned} S &= (C_1 + C_2 + C_3)/3, \\ P_{32,i} &= (4C_i + 4C_{i+1} + C_{i-1})/9, \\ P_{22,i} &= \frac{1}{9}(7C_i + C_{i+1} + C_{i-1}) + \frac{1}{6}C_i - \frac{1}{12}\sum_{j=0}^{2}(-1)^j(B_{j,2} + B_{j+1,1}). \end{aligned}$$

This solution can also be applied if the surrounding patch complex is bicubic resulting in the construction of [11, Figs. 4.1–4.3]. Figure 11.4 compares the distribution of the Bernstein–Bézier coefficients.

11.4. Examples

Figures 11.5 and 11.6 illustrate the flexibility of the algorithm with respect to the mesh and the blend ratio. The regular mesh regions are covered by a darker shaded biquadratic spline complex. The second object uses $\alpha = 0.15$ everywhere in the first refinement except at the mesh points labeled $\alpha = 0.0$ and $\alpha = 0.5$. This illustrates a more general rule for associating blend ratios

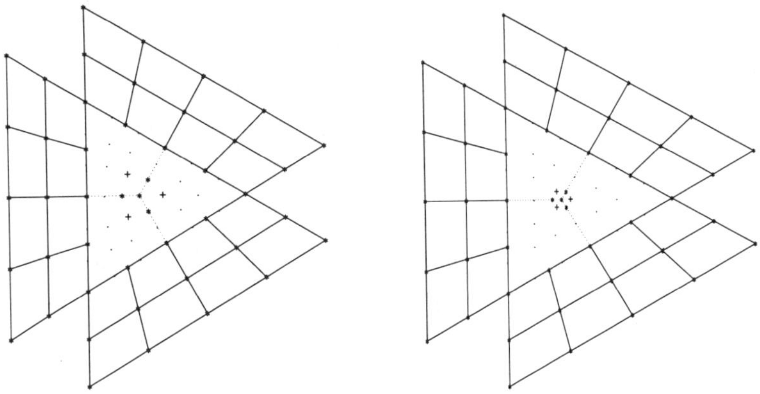

FIG. 11.4 *Coefficient distribution for* $\phi = \mathrm{id} + t_1 t_2 \begin{bmatrix} a \\ a(1-t_2) \end{bmatrix}$ *and* $\phi = \mathrm{id}$.

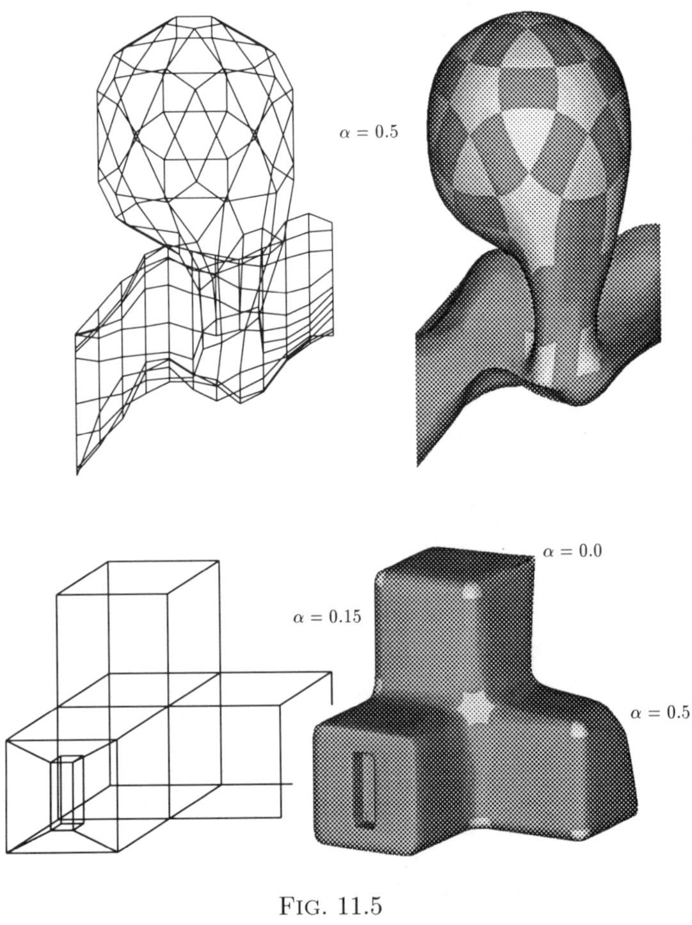

FIG. 11.5

with edges to achieve locally a larger or zero blend radius. Figure 11.6 features an asymmetric six-sided as well as irregular three-, four-, and five-sided mesh cells. The details below show a shading by the maximum of the absolute value of the two principal curvatures. The grey values range linearly from the least curvature on the object (black) to the maximal curvature (white). High curvature concentrates in the hyperbolic twist region and along edges leading into the relatively flat six-sided region.

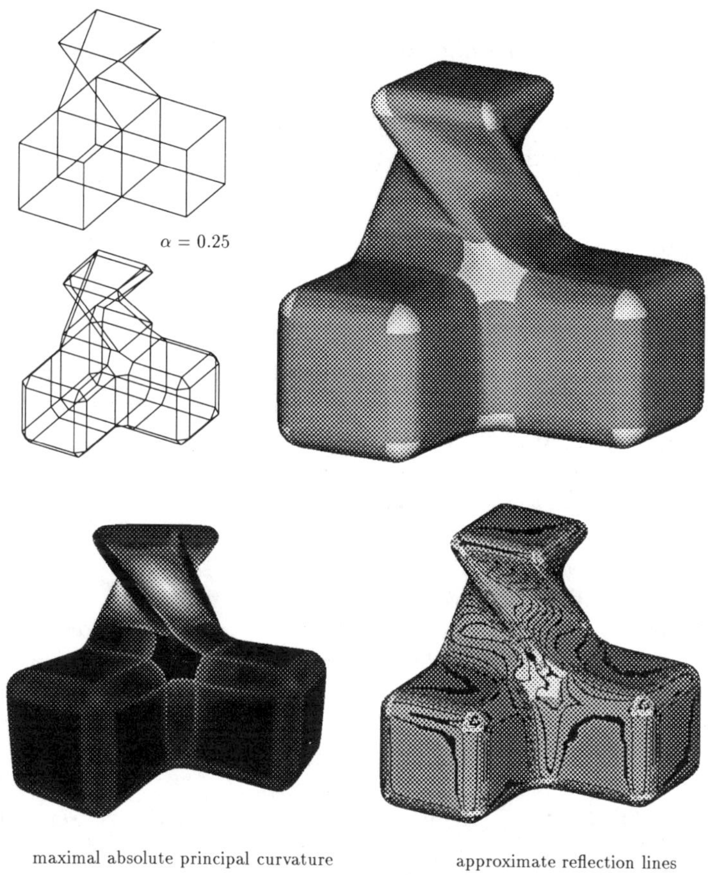

FIG. 11.6

11.5. Conclusion

The algorithm takes a mesh of points that models a surface with or without global boundary and constructs a C^1 surface that, subject to blend ratios, closely follows the mesh. Most of the surface is parametrized by a biquadratic spline whose control points are obtained by refining the input mesh. The remaining mesh regions are parametrized by bicubic patches in Bernstein–

Bézier form. The construction can be extended to rational patches and to interpolate at the vertices of the input mesh.

The algorithm uses a number of known ideas and techniques, for example, interpreting the regular mesh points as control points of a tensor-product spline and refining a mesh. However, the refinement is used for a new and different purpose. Rather than iterating to the limit, it is only applied twice. This gives the surface its rough shape and creates a new mesh with separated irregularities. The quadratic reparametrization joining the biquadratic spline complex with the bicubic patches is the key to filling an n-sided hole with tensor product patches of low degree.

The strong points of the algorithm are its low degree, tensor-product representation, and the simplicity of the construction. The algorithm defines an interpolating spline space for modeling surfaces of arbitrary topology. The low degree of the parametrization and the construction by averaging limit the potential for introducing extraneous features when smoothing the input mesh.

References

[1] A. A. Ball and D. J. T. Storry, *A matrix approach to the analysis of recursively generated B-spline surfaces*, Comput. Aided Design 18 (1986), pp. 437–442.

[2] ———, *Conditions for tangent plane continuity over recursively generated B-spline surfaces*, ACM TOG 7 (1988), pp. 83–102.

[3] ———, *Design of an n-sided surface patch from Hermite boundary data*, Comput. Aided Geom. Design 6 (1989), pp. 111–120.

[4] E. Catmull and J. Clark, *Recursively generated B-spline surfaces on arbitrary topological meshes*, Comput. Aided Design 10 (1978), pp. 350–355.

[5] H. Chiyokura and F. Kimura, *Design of Solids with Free-form Surfaces*, Computer Graphics, 17 (1983), pp. 289–298.

[6] D. Doo, *A subdivision algorithm for smoothing down irregularly shaped polyhedrons*, in Proceedings on Interactive Techniques in Computer Aided Design, Bologna, 1978, pp. 157–165.

[7] D. Doo and M. Sabin, *Behaviour of recursive division surfaces near extraordinary points*, Comput. Aided Design 10 (1978), pp. 350–355.

[8] N. Dyn, D. Levin, and D. Liu, *Interpolatory convexity preserving subdivision schemes for curves and surfaces*, preprint, 1992.

[9] T. N. T. Goodman, *Closed surfaces defined from biquadratic splines*, Constructive Approximation, 7 (1991), pp. 149–160.

[10] J. A. Gregory, *Smooth parametric surfaces and n-sided patches*, in Computation of Curves and Surfaces, W. Dahmen, M. Gasca, and C.A. Micchelli, eds., Kluwer Academic Publishers, Dordrecht, Germany, 1990, pp. 457–498.

[11] J. A. Gregory and J. Zhou, *Filling polygonal holes with bicubic patches*, TR-05-91, Brunel University,Uxbridge, UB8 3PH, England, March 1991.

[12] J. M. Hahn, *Filling polygonal holes with rectangular patches*, Theory and Practice of Geometric Modeling, W. Straßer and H.-P. Seidel, eds., Springer-Verlag, New York, 1989.

[13] K. Höllig and H. Mögerle, *G-splines*, Comput. Aided Geom. Design 7 (1989), pp. 197–207.

[14] C. Loop, *Smooth subdivision surfaces based on triangles*, Master's thesis, University of Utah, Salt Lake City, UT, 1987.

[15] ———, *A G^1 triangular spline surface of arbitrary topological type*, Comput. Aided Geom. Design 11 (1994), pp. 303–330.

[16] C. Loop and T. DeRose, *Generalized B-spline surfaces of arbitrary topology*, Proceedings of Siggraph, Association of Computing Machinery, New York, 1990.

[17] H. Mögerle, *G-splines höherer Ordnung*, thesis, Math. Inst. A, University of Stuttgart, Germany, 1992.

[18] J. Peters, *Smooth splines over irregular meshes built from few polynomial pieces of low degree*, CSD-TR-93-019, Purdue University, West Lafayette, IN, March 1993.

[19] M. Sabin, *Non-rectangular surface patches suitable for inclusion in a B-spline surface*, in P. ten Hagen, ed., Proceedings of Eurographics, 1983, North Holland, Amsterdam, 1983, pp. 57–69.

[20] R. F. Sarraga, *G^1 Interpolation of Generally Unrestricted Cubic Bézier curves*, Comput. Aided Geom. Design 4 (1987), pp. 23–40.

[21] J. J. van Wijk, *Bicubic patches for approximating non-rectangular control-point meshes*, Comput. Aided Geom. Design 3 (1984), pp. 1–13.

CHAPTER

12

A Convolution Approach to N-Sided Patches and Vertex Blending

Yan Zhao and Alyn Rockwood

12.1. Introduction and Background

Most commercial surface modelers used today employ four-sided patches joined together in what is topologically equivalent to a rectangular grid (e.g., Catia™, Geomod™, Euclid™, STRIM™, etc.). This is due primarily to the convenience of generalizing curve definitions by incorporating two orthogonal parameters into one surface. Coons' methods and the more specific tensor product approaches are examples [Far90]. However, objects that are not topologically a rectangular mesh create difficult problems for these methods (see Fig. 12.1). This creates the need for multisided or "N-sided" patches. A number of schemes for creating such patches have been developed and proposed. Objects of general topology require such patches.

Initial attempts to solve the topology problem used four-sided Coons' patches degenerated to obtain three-sided patches [Fau78]. Special case three- and five-sided patches have been developed that use similar methods [Bar73], [Far86], [Cha84]. Another approach combined three-sided and four-sided patches into one more general patch [Jen87]. A popular technique joined four-sided patches at a common midpoint to form N-sided "macropatches" [Chi83], [Var87], [Hah89], [Sar87]. Recently, the definition of homogeneous, smooth N-sided patches has been developed [Var87], [Loo89], [Blo89].

There are advantages and disadvantages to each of the above approaches. The degenerate patches make derivative computations awkward [Fau78]. Homogeneous N-sided patches are usually high degree, computationally expensive, and require special handling routines. From a practical point of view, the macropatch techniques with four-sided subpatches meeting at a common midpoint have a strong implementation advantage; they can be easily integrated into existing systems.

We adopt this four-sided approach as well. The main problems to overcome with these kinds of approaches are (1) defining smoothness among subpatches, e.g., at the midpoint where all subpatches join together and

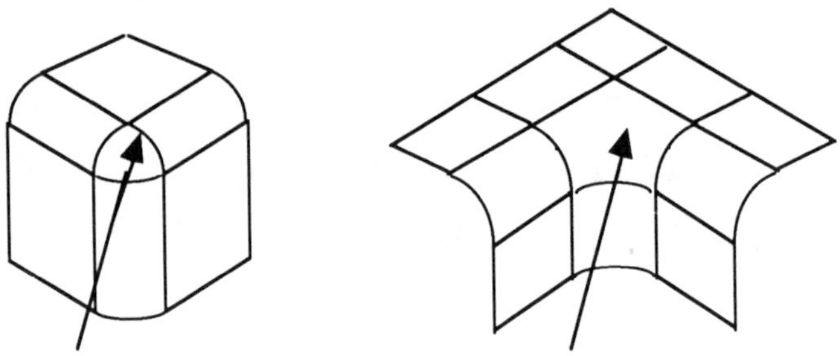

FIG. 12.1. *Problems for four-sided patches.*

between neighboring patches, and (2) controlling unwanted shape anomalies such as undulations. Most of the four-sided subpatch methods presented recently have typically depended on solving a system of equations built from constraints for continuity or derivative conditions among subpatches and between neighboring patches. One modeler, Designbase^{TM}, for example, is structured around the four-sided Gregory patch [Chi86] in this manner. Our approach, however, does not use systems of equations; instead, the emphasis is to design the control for *positional patches* and for *weight patches.* The positional patches form the *geometric primitives* and the weight patches form the *convolution kernel*; together they constitute a convolution surface.

Our approach brings several new conceptual tools to bear on the problems of smoothness and shape for the *N-sided patch.* The N-sided patch is defined by N vertices and a tangent plane for each vertex. Each of its four-sided subpatches arise naturally from a convolution surface, which interpolates to one of the "predefined" vertices and its corresponding tangent plane. The convolution kernel is a bicubic piecewise Bézier patch that convolves with biquadratic positional Bézier patches as simple geometric primitives, to form the biquintic surface. The construction of the positional Bézier patches is based on piecewise planar *control patterns.* These geometric patterns are designed on each vertex tangent plane. They control the shape of each four-sided subpatch and therefore the smoothness of the N-sided patch. With some constraints in the construction of the convolution kernel and the control patterns, the neighboring subpatches are made to meet smoothly with G^1 continuity along common boundaries and at the midpoint of the N-sided patch. This approach is especially effective for blending at vertices, as will be illustrated by several examples.

Section 12.2 describes the convolution concept followed by the description of the general approach for N-sided patch generation. Section 12.3 derives constraints for normal continuity, while §12.4 gives some construction details. Finally, some applications and examples are given in §12.5.

12.2. The N-sided Patch

Convolution surfaces incorporate the smooth blending power and easy manipulability of potential surfaces while expanding simple geometric primitives [Blo91]. A bivariate convolution surface is given by

$$S(x,y) = (W * P)(x,y) = \int_P W(u-x, v-y)P(u,v)dudv, \tag{12.1}$$

where $W(x,y)$ is the convolution kernel and $P(x,y)$ defines a geometric primitive function. Let $P(u,v)$ be a piecewise function consisting of several local parametric patches, denoted as $P_i(u,v)$, $i = 0,\dots,N-1$ for an N-sided patch. Let the kernel function $W(u,v)$ be piecewise with $W_i(u,v)$, $i = 0,\dots,N-1$ as N local parametric patches. Therefore

$$S(u,v) = \sum_{i=0}^{N-1} W_i(u,v)P_i(u,v). \tag{12.2}$$

Functionally, $W_i(u,v)$, $i = 0,\dots,N-1$ behave as weights and $S(u,v)$ as the weighted sum of the geometric patches $P_i(u,v), i = 0,\dots,N-1$. Therefore, we call $W_i(u,v)$ the *weight function*, *weight patch*, or simply *weight*, and $P_i(u,v)$ is the *positional patch*.

The N-sided patch interpolates N predefined vertices and tangent planes at each vertex. One four-sided subpatch is formed for each vertex and its tangent plane. The four-sided subpatch is generated by

$$S^{(j)}(u,v) = \frac{\sum_{i=0}^{N-1} W_i(u,v)P_i^{(j)}(u,v)}{\sum_{i=0}^{N-1} W_i(u,v)}, \qquad j = 0,\dots,N-1 \tag{12.3}$$

and the N-sided surface patch is the union of the N subpatches,

$$S(u,v) = \bigcup_{j=0}^{N-1} S^{(j)}(u,v).$$

Formula (12.3) generalizes (12.2) by dividing by the sum of the weights to make a convex combination, with the assumption that $W_i(u,v) \geq 0$. Each four-sided subpatch forms a rational surface patch. Traditional rational Bézier patches use weighted sums of control points; we use weighted sums for positional patches. In the special case, if we make

$$\sum_{i=0}^{N-1} W_i(u,v) = 1,$$

a polynomial patch results.

We define positional patches as biquadratic Bézier patches,

$$P_i^{(j)}(u,v) = \begin{bmatrix} B_0(u) & B_1(u) & B_2(u) \end{bmatrix} \begin{bmatrix} p_i^{(j)}00 & p_i^{(j)}01 & p_i^{(j)}02 \\ p_i^{(j)}10 & p_i^{(j)}11 & p_i^{(j)}12 \\ p_i^{(j)}20 & p_i^{(j)}21 & p_i^{(j)}22 \end{bmatrix} \begin{bmatrix} B_0(v) \\ B_1(v) \\ B_2(v) \end{bmatrix}, \quad (12.4)$$

where the control points in the matrix are obtained from control patterns inscribed on the tangent planes of the vertices of the N-sided patch. For example, let $N = 3$ and let the three vertices p_0, p_1, p_2 lie on the axial planes, that is, $x = 0$, $y = 0$, and $z = 0$. Assume the tangent plane for each vertex to be the axial planes as well. See Fig. 12.2.

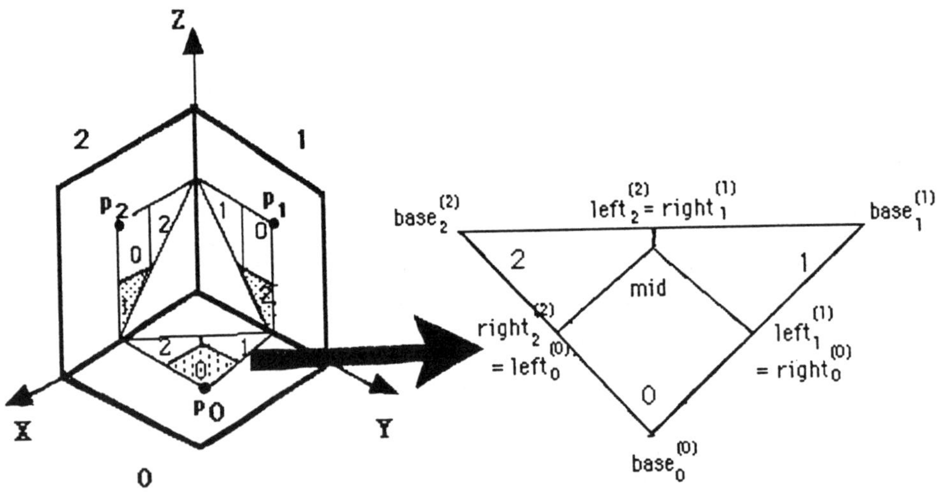

FIG. 12.2. *Vertices, tangent planes, and control patterns for a three-sided patch.*

A control pattern is constructed for each of the three given vertices on their tangent planes. Each pattern consists of three four-sided polygons. Neighboring polygons share a common edge and all polygons share a common midpoint. Denote the vertices for each polygon in the control pattern as "base," "left," "right," and "mid." Where "base" is the vertex not shared with any other polygons, "left" and "right" are shared with the left and right neighboring polygons, respectively, and "mid" is shared by all polygons. The right-hand object in Fig. 12.2 is the control pattern 0, located on the plane $z = 0$.

The vertices for each polygon in the control patterns correspond to control matrices for the positional patches by

$$\begin{bmatrix} p_i^{(j)}00 & p_i^{(j)}01 & p_i^{(j)}02 \\ p_i^{(j)}10 & p_i^{(j)}11 & p_i^{(j)}12 \\ p_i^{(j)}20 & p_i^{(j)}21 & p_i^{(j)}22 \end{bmatrix} = \begin{bmatrix} \text{base}_i^{(j)} & \text{base}_i^{(j)} & \text{right}_i^{(j)} \\ \text{base}_i^{(j)} & \text{mid}^{(j)} & \text{mid}^{(j)} \\ \text{left}_i^{(j)} & \text{mid}^{(j)} & \text{mid}^{(j)} \end{bmatrix}, \quad (12.5)$$

where the superscript $j = 0, \ldots, N-1$ indicates the control pattern and also indexes the vertex tangent plane. The subscript $i = 0, \ldots, N-1$ indexes the polygons within the jth control pattern. The N given vertices for the N-sided patch are assigned to $\text{base}_0^{(j)}$. They are interpolated in their tangent planes by the resulting N-sided patch. The other $\text{base}_i^{(j)}$, with $i \neq 0$, are on the jth tangent plane with the ith polygon that is numbered counterclockwise. The choice of the $\text{base}_i^{(j)}$ and control patterns influences the shape of the N-sided patch. The midpoints $\text{mid}^{(j)}$ can be adjusted inside the control pattern interior. The overall shape and curvature at the patch's midpoint depend strongly on the positions of the control pattern midpoints. If the midpoints in the control patterns are moved toward the joining center of the control patterns, the curvature at the midpoint of the patches is larger. The resulting surface becomes sharper at the joint. If the midpoints in the control patterns are moved away from the joining center of the control patterns, the curvature at the midpoint of the patches is smaller, and the resulting surface is flatter around the joint. We call this effect the *thumbweight* effect; the sharper patch has the larger thumbweight. All of the midpoints in the control patterns map to the same point on the N-sided patch where the subpatches join. By global G^1 continuity requirements, detailed in §12.4, the "left" and "right" points must be located in the middle of any two neighboring "base" points. Note that $\text{left}_i^{(j)} = \text{right}_{i-1}^{(j-1)}$ for $i = 0, \ldots, N-1 \pmod N$.

Consider again the configuration in Fig. 12.2, in which the three shaded polygons in the control patterns contribute to $S^{(0)}(u,v)$ as control points in the matrices of $P_i^{(0)}(u,v)$. The zeroth polygon in control pattern 0 contributes to $P_0^{(0)}(u,v)$, the first polygon in control pattern 2 contributes to $P_1^{(0)}(u,v)$, and the second polygon in control pattern 1 contributes to $P_2^{(0)}(u,v)$. In general, $P_i^{(j)}(u,v)$ in formula (12.3) is contributed by the ith polygon in control pattern $N+j-i$. As mentioned, the ordering for control patterns is counterclockwise. For each vertex of the N-sided patch, one control pattern will be built. The ordering for the four-sided polygon inside one control pattern is also from 0 to $N-1$ counterclockwise, beginning with the given vertex of the N-sided patch.

The weight functions are formed by Kth degree Bézier patches:

(12.6)

$$W_i(u,v)$$
$$= \begin{bmatrix} B_0(u) & B_1(u) & \ldots & B_K(u) \end{bmatrix} \begin{bmatrix} w_i00 & w_i01 & \ldots & w_i0K \\ w_i10 & w_i11 & \ldots & w_i1K \\ \vdots & \vdots & \vdots & \vdots \\ w_iK0 & w_iK1 & \ldots & w_iKK \end{bmatrix} \begin{bmatrix} B_0(v) \\ B_1(v) \\ \vdots \\ B_K(v) \end{bmatrix}.$$

The polynomial degree K and the control points in the *weight matrices* are determined by the continuity requirements and by keeping the polynomial degree K as low as possible. We show later that weight functions with $K = 3$ can be constructed for any N-sided patch. For $N = 4$, we can construct a weight function with $K = 1$. More detail is found in §§12.3 and 12.4.

The parameter space for the N-sided patch lies in $\mathbf{R}^2$, with $(u, v) \in [0, 1]^2$. All four-sided subpatches are mapped from the same parameter space. See Fig. 12.3.

Notice that each subpatch is defined by convex combinations of Bézier patches $P_i^{(j)}(u, v)$, which are the convex combination of control patterns. Therefore subpatches lie in the convex hull of their control patterns and so does the N-sided macropatch.

12.3. Normal Continuity for the N-sided Patch

For neighboring subpatches $S^{(j-1)}(u, 1)$ and $S^{(j)}(1, v)$, $j = 0, \ldots, N-1 \pmod N$ in Fig. 12.3, the requirements for normal (G^1) continuity are

$$S^{(j-1)}(u,1) = S^{(j)}(1,v) \tag{12.7}$$

and

$$\begin{cases} \dfrac{\dfrac{\partial S^{(j-1)}(u,v)}{\partial u} \times \dfrac{\partial S^{(j-1)}(u,v)}{\partial v}\Big|_{v=1}}{\left|\dfrac{\partial S^{(j-1)}(u,v)}{\partial u} \times \dfrac{\partial S^{(j-1)}(u,v)}{\partial v}\Big|_{v=1}\right|} \\ \qquad = \dfrac{\dfrac{\partial S^{(j)}(u,v)}{\partial u} \times \dfrac{\partial S^{(j)}(u,v)}{\partial v}\Big|_{u=1}}{\left|\dfrac{\partial S^{(j)}(u,v)}{\partial u} \times \dfrac{\partial S^{(j)}(u,v)}{\partial v}\Big|_{u=1}\right|}, & (u,v) \neq (1,1), \\ \lim\limits_{u\to 1} \dfrac{\dfrac{\partial S^{(j-1)}(u,v)}{\partial u} \times \dfrac{\partial S^{(j-1)}(u,v)}{\partial v}\Big|_{v=1}}{\left|\dfrac{\partial S^{(j-1)}(u,v)}{\partial u} \times \dfrac{\partial S^{(j-1)}(u,v)}{\partial v}\Big|_{v=1}\right|} \\ \qquad = \lim\limits_{v\to 1} \dfrac{\dfrac{\partial S^{(j)}(u,v)}{\partial u} \times \dfrac{\partial S^{(j)}(u,v)}{\partial v}\Big|_{u=1}}{\left|\dfrac{\partial S^{(j)}(u,v)}{\partial u} \times \dfrac{\partial S^{(j)}(u,v)}{\partial v}\Big|_{u=1}\right|}, & (u,v) = (1,1). \end{cases} \tag{12.8}$$

Formula (12.7) is the requirement for positional values to match along the common boundary, the C^0 requirement. Formula (12.8) is for the normal vectors of the neighboring subpatches to match across their common boundaries.

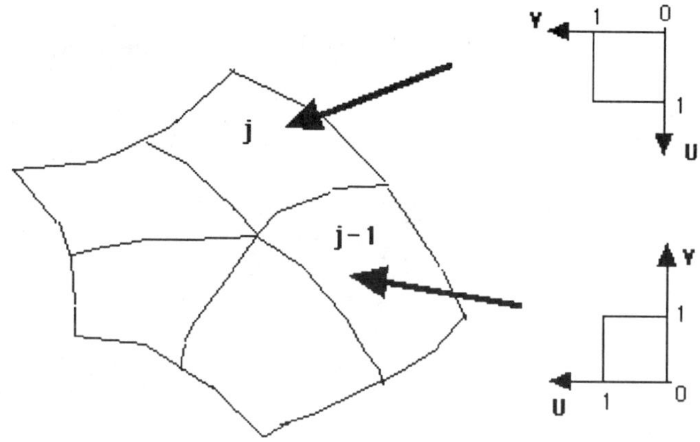

FIG. 12.3. *The map from parameter space to the N-sided patch.*

From (12.3) and (12.7) we have

$$\frac{\sum_{i=0}^{N-1} W_i(u,1)P_i^{(j-1)}(u,1)}{\sum_{i=0}^{N-1} W_i(u,1)} = \frac{\sum_{i=0}^{N-1} W_i(1,v)P_i^{(j)}(1,v)}{\sum_{i=0}^{N-1} W_i(1,v)}. \tag{12.9}$$

Note also that

(12.10)

$$\begin{aligned} P_i^{(j-1)}&(u,1) \\ &= \begin{bmatrix} B_0(u) & B_1(u) & B_2(u) \end{bmatrix} \begin{bmatrix} \text{base}_i^{(j-1)} & \text{base}_i^{(j-1)} & \text{right}_i^{(j-1)} \\ \text{base}_i^{(j-1)} & \text{mid}^{(j-1)} & \text{mid}^{(j-1)} \\ \text{left}_i^{(j-1)} & \text{mid}^{(j-1)} & \text{mid}^{(j-1)} \end{bmatrix} \begin{bmatrix} 0 \\ 0 \\ 1 \end{bmatrix} \\ &= \text{right}_i^{(j-1)} \cdot B_0(u) + \text{mid}^{(j-1)}(B_1(u) + B_2(u)) \end{aligned}$$

and

$$\begin{aligned} P_i^{(j)}(1,v) &= \begin{bmatrix} 0 & 0 & 1 \end{bmatrix} \begin{bmatrix} \text{base}_i^{(j)} & \text{base}_i^{(j)} & \text{right}_i^{(j)} \\ \text{base}_i^{(j)} & \text{mid}^{(j)} & \text{mid}^{(j)} \\ \text{left}_i^{(j)} & \text{mid}^{(j)} & \text{mid}^{(j)} \end{bmatrix} \begin{bmatrix} B_0(v) \\ B_1(v) \\ B_2(v) \end{bmatrix} \\ &= \text{left}_i^{(j)} \cdot B_0(v) + \text{mid}^{(j)}(B_1(v) + B_2(v)). \end{aligned} \tag{12.11}$$

Because $\text{left}_i^{(j)} = \text{right}_{i-1}^{(j-1)}$ for $i = 0, \ldots, N-1 \pmod N$, we deduce, for $i = 0, \ldots, N-1 \pmod N$, that

$$P_i^{(j)}(1,u) = P_{i-1}^{(j-1)}(u,1). \tag{12.12}$$

Let $W_i(1,u) = W_{i-1}(u,1)$, then the control points in the weight control point matrices satisfy

$$w_iKk = w_{i-1}kK, \qquad i = 0,\ldots,N-1, \tag{12.13}$$

where K is the degree of the weight patches, and $k = 0,\ldots,K$. Therefore, $S^{(j-1)}(u,1) = S^{(j)}(1,v)$.

In (12.8), we have

(12.14)

$$\left.\frac{\partial S^{(j-1)}(u,v)}{\partial u}\right|_{v=1} = \frac{\left[\sum_{i=0}^{N-1} W_i(u,1)\frac{\partial P_i^{(j-1)}(u,1)}{\partial u} + \sum_{i=0}^{N-1} P_i^{(j-1)}(u,1)\frac{\partial W_i(u,1)}{\partial u}\right]\sum_{i=0}^{N-1} W_i(u,1) - \sum_{i=0}^{N-1}\frac{\partial W_i(u,1)}{\partial u}\sum_{i=0}^{N-1} W_i(u,1)P_i^{(j-1)}(u,1)}{\left[\sum_{i-0}^{N-1} W_i(u,1)\right]^2},$$

(12.15)

$$\left.\frac{\partial S^{(j-1)}(u,v)}{\partial v}\right|_{v=1} = \frac{\left[\sum_{i=0}^{N-1} W_i(u,1)\left.\frac{\partial P_i^{(j-1)}(u,v)}{\partial v}\right|_{v=1} + \sum_{i=0}^{N-1} P_i^{(j-1)}(u,1)\left.\frac{\partial W_i(u,v)}{\partial v}\right|_{v=1}\right]\sum_{i=0}^{N-1} W_i(u,1) - \sum_{i=0}^{N-1}\left.\frac{\partial W_i(u,v)}{\partial v}\right|_{v=1}\sum_{i=0}^{N-1} W_i(u,1)P_i^{(j-1)}(u,1)}{\left[\sum_{i-0}^{N-1} W_i(u,1)\right]^2},$$

(12.16)

$$\left.\frac{\partial S^{(j)}(u,v)}{\partial u}\right|_{u=1} = \frac{\left[\sum_{i=0}^{N-1} W_i(1,v)\left.\frac{\partial P_i^{(j-1)}(u,v)}{\partial u}\right|_{u=1} + \sum_{i=0}^{N-1} P_i^{(j-1)}(1,v)\left.\frac{\partial W_i(u,v)}{\partial u}\right|_{u=1}\right]\sum_{i=0}^{N-1} W_i(1,v) - \sum_{i=0}^{N-1}\left.\frac{\partial W_i(u,v)}{\partial u}\right|_{u=1}\sum_{i=0}^{N-1} W_i(1,v)P_i^{(j-1)}(1,v)}{\left[\sum_{i-0}^{N-1} W_i(1,v)\right]^2},$$

(12.17)

$$\left.\frac{\partial S^{(j)}(u,v)}{\partial v}\right|_{u=1} = \frac{\left[\sum_{i=0}^{N-1} W_i(1,v)\frac{\partial P_i^{(j-1)}(1,v)}{\partial v} + \sum_{i=0}^{N-1} P_i^{(j-1)}(1,v)\frac{\partial W_i(1,v)}{\partial v}\right]\sum_{i=0}^{N-1} W_i(1,v) - \sum_{i=0}^{N-1}\frac{\partial W_i(1,v)}{\partial v}\sum_{i=0}^{N-1} W_i(1,v)P_i^{(j-1)}(1,v)}{\left[\sum_{i-0}^{N-1} W_i(1,v)\right]^2},$$

where

$$\left.\frac{\partial W_i(u,v)}{\partial u}\right|_{u=1} = K\sum_{k=0}^{K}(w_iKk - w_iK-1,k)B_k(v), \tag{12.18}$$

$$\left.\frac{\partial W_i(u,v)}{\partial u}\right|_{v=1} = \sum_{k=0}^{K} w_ikK \cdot DB_k(u), \tag{12.19}$$

$$\left.\frac{\partial W_i(u,v)}{\partial v}\right|_{u=1} = \sum_{k=0}^{K} w_iKk \cdot DB_k(v), \tag{12.20}$$

$$\left.\frac{\partial W_i(u,v)}{\partial v}\right|_{v=1} = K\sum_{k=0}^{K}(w_ikK - w_ik, K-1)B_k(u), \tag{12.21}$$

$$\left.\frac{\partial P_i^{(j)}(u,v)}{\partial u}\right|_{u=1} = 2B_0(v)\cdot(\text{left}_i^{(j)} - \text{base}_i^{(j)}), \tag{12.22}$$

$$\left.\frac{\partial P_i^{(j)}(u,v)}{\partial u}\right|_{v=1} = \text{right}_i^{(j)} DB_0(u) + \text{mid}^{(j)}(DB_1(u) + DB_2(u)), \tag{12.23}$$

$$\left.\frac{\partial P_i^{(j)}(u,v)}{\partial v}\right|_{u=1} = \text{left}_i^{(j)} DB_0(v) + \text{mid}^{(j)}(DB_1(v) + DB_2(v)), \tag{12.24}$$

$$\left.\frac{\partial P_i^{(j)}(u,v)}{\partial v}\right|_{v=1} = 2B_0(u)\cdot(\text{right}_i^{(j)} - \text{base}_i^{(j)}). \tag{12.25}$$

Here $DB_i(u)$ is the derivative of the Bernstein polynomial $B_i(u)$ of degree two. Inserting (12.18)–(12.25) into (12.14)–(12.17), and using the relationships (12.12) and (12.13), we see that formula (12.8) is satisfied by the following constraints:

$$\begin{gathered} w_iKk - w_iK-1,k = w_{i-1}k, K-1 - w_{i-1}k, K, \\ i = 0,\ldots,N-1, \quad \text{mod}(N); \quad k = 0,\ldots,K, \end{gathered} \tag{12.26}$$

$$\begin{gathered} \text{left}_i^{(j)} - \text{base}_i^{(j)} = \text{base}_{i-1}^{(j-1)} - \text{right}_{i-1}^{(j-1)}, \\ i,j = 0,\ldots,N-1, \quad \text{mod}(N). \end{gathered} \tag{12.27}$$

Hence, for the ith weight matrix, the difference of the bottom row and the next row up equals the difference of the next to the right column and the right column in the $(i-1)$th weight matrix. For the vertices of the polygons in the control patterns, the vector from the "left" point to the "base" point in the ith polygon should equal the vector from the "base" point to the "right" point in the $(i-1)$th polygon in the same control pattern. As shown in the right figure of Fig. 12.2, we should have

$$\begin{aligned}
\mathrm{left}_1 - \mathrm{base}_1 &= \mathrm{base}_0 - \mathrm{right}_0, \\
\mathrm{left}_2 - \mathrm{base}_2 &= \mathrm{base}_1 - \mathrm{right}_1, \\
\mathrm{left}_0 - \mathrm{base}_0 &= \mathrm{base}_2 - \mathrm{right}_2,
\end{aligned}$$

More detail about constructions of weight matrices and control patterns is discussed in the next section.

12.4. The Construction of Weight Functions and Control Patterns

12.4.1. Constructing weight functions. Note that in (12.6) the construction of the weight functions depends entirely upon the control points of the weight matrices. The choice of the control points must satisfy equations (12.13) and (12.26) in order to get a patch that maintains normal continuity and vertex interpolation properties. For computational and geometric simplicity, we select the polynomial degree K as low as possible. For general N-sided patches with $N \geq 3$, we can find the control points in weight matrices to construct weight function with $K = 3$. For the four-sided macropatch, we can find control points in weight matrices that make $K = 1$. Consider the general case with $K = 3$ first, that is,

(12.28)

$$W_i(u,v) = \begin{bmatrix} B_0(u) & B_1(u) & B_2(u) & B_3(u) \end{bmatrix} \begin{bmatrix} w_i00 & w_i01 & w_i02 & w_i03 \\ w_i10 & w_i11 & w_i12 & w_i13 \\ w_i20 & w_i21 & w_i22 & w_i23 \\ w_i30 & w_i31 & w_i32 & w_i33 \end{bmatrix} \begin{bmatrix} B_0(v) \\ B_1(v) \\ B_2(v) \\ B_3(v) \end{bmatrix}.$$

One reasonable choice for the weight matrices, which distribute the weights evenly, is

(12.29)

$$W_0 = \begin{bmatrix} 1.0 & 1.0 & 0.75 & 0.5 \\ 1.0 & 1.0 & 0.75 & 0.5 \\ 0.75 & 0.75 & 0.25 & 0.25 \\ 0.5 & 0.5 & 0.25 & 0.25 \end{bmatrix}, \qquad W_1 = \begin{bmatrix} 0. & 0. & 0. & 0. \\ 0. & 0. & 0. & 0. \\ 0.25 & 0.25 & 0.25 & 0.25 \\ 0.5 & 0.5 & 0.25 & 0.25 \end{bmatrix},$$

$$W_{N-1} = \begin{bmatrix} 0. & 0. & 0.25 & 0.5 \\ 0. & 0. & 0.25 & 0.5 \\ 0. & 0. & 0.25 & 0.25 \\ 0. & 0. & 0.25 & 0.25 \end{bmatrix},$$

$$W_2 = , \ldots, = W_{N-2} = \begin{bmatrix} 0. & 0. & 0. & 0. \\ 0. & 0. & 0. & 0. \\ 0. & 0. & 0.25 & 0.25 \\ 0. & 0. & 0.25 & 0.25 \end{bmatrix}.$$

This is used for the six-sided patch and four-sided patches in Figs. 12.9–12.13 and also for the eight-sided patch in Figs. 12.14 and 12.15. The four-sided patches in Figs. 12.16 and 12.17 use bilinear weight functions and bilinear positional patches. With $K = 1$, the weight matrices are

$$W_0 = \begin{bmatrix} 1.0 & 0.5 \\ 0.5 & 0.25 \end{bmatrix}, \qquad W_1 = \begin{bmatrix} 0 & 0 \\ 0.5 & 0.25 \end{bmatrix},$$

$$W_2 = \begin{bmatrix} 0 & 0 \\ 0 & 0.25 \end{bmatrix}, \qquad W_3 = \begin{bmatrix} 0 & 0.5 \\ 0 & 0.25 \end{bmatrix}. \tag{12.30}$$

Notice the way the weight matrices are built. They form a control net for global C^1 piecewise Bézier weight patches. An example for $N = 5$ is shown in Fig. 12.4.

The weight matrix in $W_0(u, v)$ forces each vertex $\text{base}_0^{(j)}$ in the positional patch $P_0^{(j)}(u, v)$, $j = 0, \ldots, N-1$, to be interpolated at $u = v = 0$ by subpatch $S^{(j)}(u, v)$, as described in (12.3). The major contribution to $S^{(j)}(u, v)$ comes from $P_0^{(j)}(u, v)$. $P_1^{(j)}(u, v)$ and $P_{N-1}^{(j)}(u, v)$, two neighboring positional patches of $P_0^{(j)}(u, v)$ that construct subpatch $S^{(j)}(u, v)$, contribute one half on their shared boundaries with $P_0^{(j)}(u, v)$ and contribute less going away from the

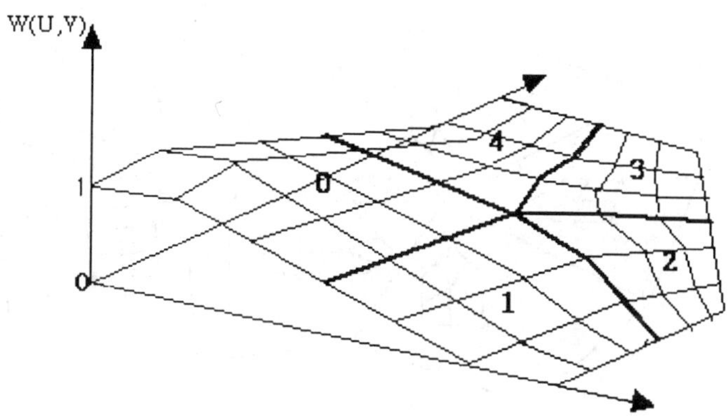

FIG. 12.4. *Control net for global C^1 piecewise weight patches of $N = 5$.*

boundaries. Other positional patches contribute one fourth to the common midpoint and the contribution is gradually reduced to zero moving from the midpoint to the far boundaries.

12.4.2. Constructing control patterns. As in §12.2 and Fig. 12.2, one control pattern is constructed for each predefined vertex of an N-sided patch. Each control patten consists of N four-sided polygons. These polygons can be degenerated into triangles, lines, or even points when necessary. The four corners of the polygons are denoted by "base," "left," "right," and "mid," where the "base" point is located at each corner of the control patterns, and "left" and "right" are shared points for a polygon meeting its left and right neighboring polygons. The "mid" point is shared by all polygons in the same control pattern. The control patterns must satisfy (12.27) to obtain global G^1 continuity. Therefore, the "left" and "right" points are selected as the midpoint between two neighboring "base" points. The "mid" points inside the interior of the control patterns are available to shape the N-sided patch. Notice that the control patterns must not overlap.

To construct two N-sided patches meeting at their common boundary, two vertices must be shared and interpolated by both N-sided patches. These two vertices form the endpoints of their common boundary curve. To make them G^1 at their common boundary, the edges of the polygons in the neighboring control patterns crossing the common boundary of the two N-sided patches must meet in a line, shown in Fig. 12.5 and Fig. 12.8. One general way to construct control patterns is called a *corner cutting control pattern*, which can be considered as a default. For each given vertex, build a rectangle with corners cut to obtain the required number of polygon sides. The "midpoint" of the control pattern is placed in the middle of the rectangle. The two polygons at the bottom along the N-sided patch boundary must be rectangular. A predefined vertex is at one corner of the bottom two polygons. That corner will be interpolated

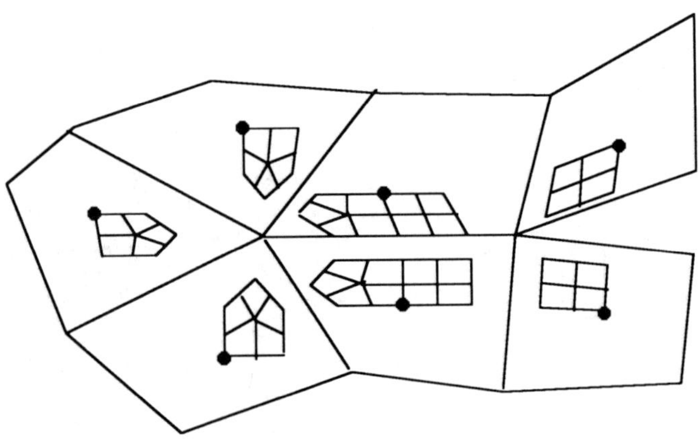

FIG. 12.5. *Two sets of control patterns for two neighboring N-sided patches.*

by the N-sided patch. Figure 12.5 shows predefined vertices denoted by "•", and control patterns for four-sided and five-sided patches. Figure 12.8 shows control patterns for four-sided and six-sided patches.

If two given vertices have the same tangent plane, the corresponding two control patterns are also in the same plane. Thus the boundary curve for the N-sided patch that uses the two vertices as its endpoints will be located in that plane as well. The patch interpolates to the vertices and their tangent plane at its boundary. If we select weight matrices as in §12.4.1, the patch's boundary curve through two neighboring vertices are actually polynomial Bézier curve. This is shown in Figs. 12.8 and 12.9, where for each six-sided or four-sided patch, two control patterns are built on one face of the cube. The given vertices are shown in Fig. 12.6. The three faces of the cube serve as three tangent planes for the vertices. It can be seen in Fig. 12.9 that the six-sided and four-sided patches interpolate to the three faces of the cube. It is also shown in Fig. 12.7 that two central patterns in each plane cause the resulting eight-sided patch to interpolate the four planes at its boundary.

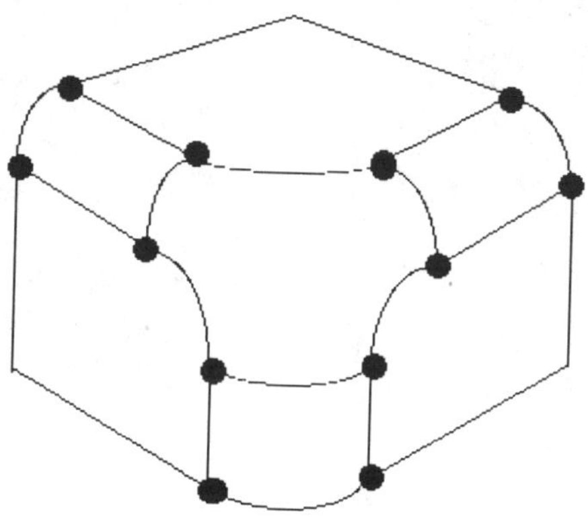

FIG. 12.6. *Suitcase corner blending problem with predefined vertices for six-sided and four-sided patches.*

12.5. Applications

12.5.1. The suitcase corner blending problem. We give a solution to the well-known suitcase corner blending problem. We use a six-sided patch for the corner and three four-sided patches for each rounded edge (see Fig. 12.6). Both the six-sided and four-sided patches are constructed with the approach described. The vertices, shown with an "•" symbol, and their tangent planes are illustrated.

More precisely, given the unit cube defined in $\mathbf{R}^3$, we round its three edges by three four-sided parametric patches with different fillet radii. The corner joint is also rounded. The control patterns for the rounded edges and corner are shown in Fig. 12.8. The resulting parametric patches are shown in Fig. 12.9. The patches are depicted by contouring in Fig. 12.10, which shows the first-

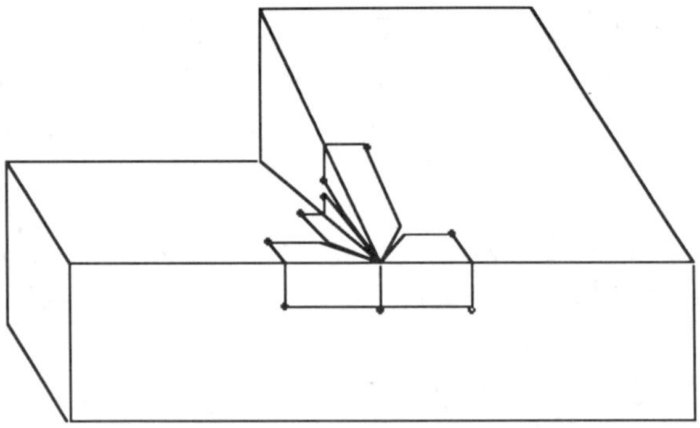

FIG. 12.7. *Object with outlines of its control patterns and vertices for blending patch at the joint.*

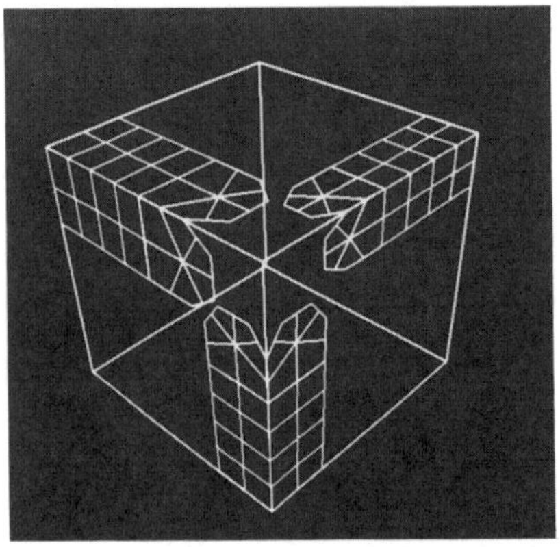

FIG. 12.8. *Control patterns for six-sided and four-sided patches in suitcase corner blends.*

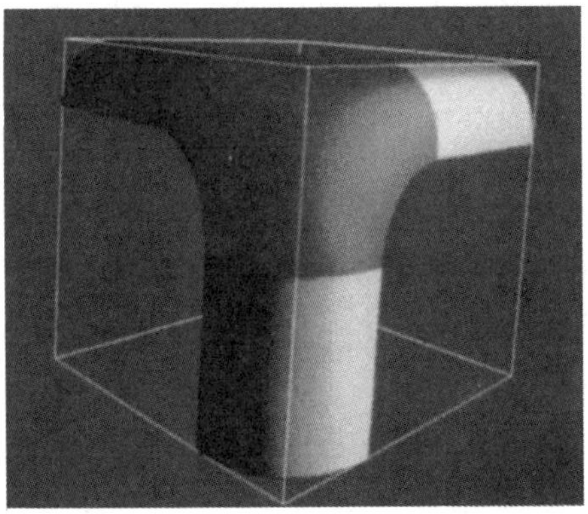

FIG. 12.9. *Six-sided and four-sided patches for suitcase corner blends.*

order geometric continuity. The patches with isophote patterns are shown in Fig. 12.11. They indicate second-order continuity. The control patterns for the three rounded edges differ only in size. For each edge to be rounded, four vertices and their four control patterns determine the rounding radius. The vertices for the six-sided patch match the vertices for the rounded edges. In the control patterns for the corner, moving the "*mid*" point in each control pattern towards the corner increases the thumbweight of the blending surface patch at the corner. Moving the "*mid*" point away from the corner causes a flattened corner surface. This is shown in Fig. 12.12.

The weight matrices for each subpatch of the four-sided patches are the same as described in (12.29), with $N = 4$. The weight matrices for each subpatch of the six-sided patch use (12.29) with $N = 6$. For surface rendering, the normal computations for each of the blending patches are computed separately

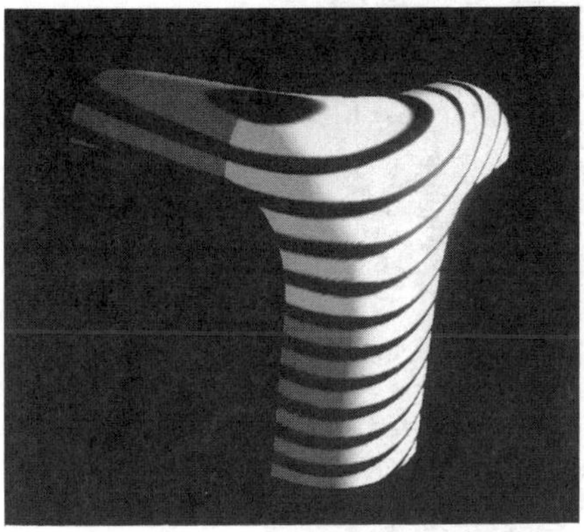

FIG. 12.10. *Contour surface for the surface in Fig.* 12.9.

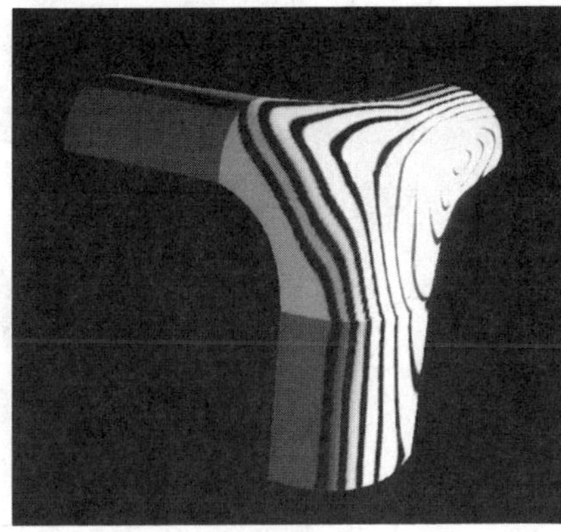

FIG. 12.11. *Isophote for the surface in Fig.* 12.9.

from their subpatches, as

$$\frac{\partial S^{(j)}(u,v)}{\partial u} \times \frac{\partial S^{(j)}(u,v)}{\partial v}.$$

The partial derivatives are obtained directly from the surface formulas (12.3)–(12.5) and (12.28). At the corner point for each subpatch, where $(u,v) = (0,0)$ and $(u,v) = (1,1)$, all partial derivatives are zero. To evaluate the normal at these points, we get approximate normals, for example,

$$\left.\frac{\Delta S^{(j)}(u,v)}{\Delta u}\right|_{u=1} \times \left.\frac{\Delta S^{(j)}(u,v)}{\Delta v}\right|_{v=1}, \tag{12.31}$$

by using backward differencing. We then average the approximate normals if more than one subpatch meets at the corner points.

The N-sided patch surface can be easily "retracted" by a transformation of its control patterns. A retracted surface for the suitcase blend in Fig. 12.9 is shown in Fig. 12.13. In addition to satisfying the corner blending problem, the N-sided macropatch is also an example for the solution of the "N-hole" problem [Loo89]. Figure 12.6 is an example for $N = 6$, the six-sided hole is shown in the middle. The left figure in Fig. 12.1 can also be described as a suitcase corner blending problem with its corner blending patch being three-sided, or as an "N-hole" problem with $N = 3$. The control patterns for the three-sided patch are shown in Fig. 12.2.

12.5.2. Blending rounded edges and a joint. Consider the object shown in Fig. 12.7. There are three boundary edges and one intersection edge leading to a common point. All edges need to be rounded, and the joint needs to be blended. We use an eight-sided patch here to construct the blending patch at the joining area. The outlines of their control patterns and the eight

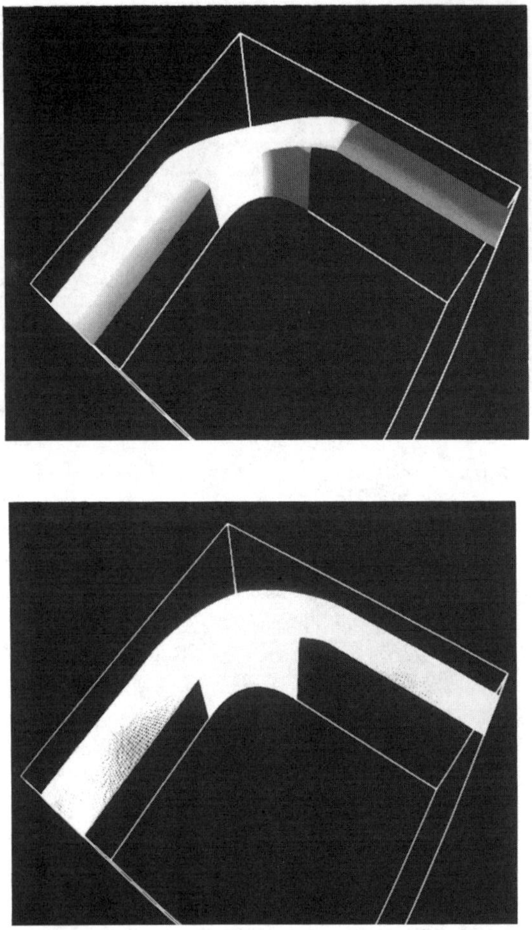

FIG. 12.12. *"Thumbweight" effects through "mid" point changes in control pattern of the six-sided patch.*

predefined vertices are shown in Fig. 12.7. The rounded edges are constructed by translational surfaces based on the boundary curves and the crossboundary tangent directions of the eight-sided blending patch. The resulting object surface is shown in Fig. 12.14. Contour lines on the constructed surface are shown in Fig. 12.15. Bicubic piecewise weight function and biquadratic positional patches are used, as described in §12.4. The corner blending patch is a biquintic eight-sided patch.

12.5.3. Blending of a rounded edge cube intersecting a plane. See Fig. 12.16. A cube is located on a plane. The two edges of the cube intersecting the plane are rounded with different radii as well as the vertical edge of the cube. A blending surface patch at the joint is needed. As before, four-sided patches are used for blending edges. A four-sided patch is also used for the blending patch at the joint. For the four-sided patches, good results can be

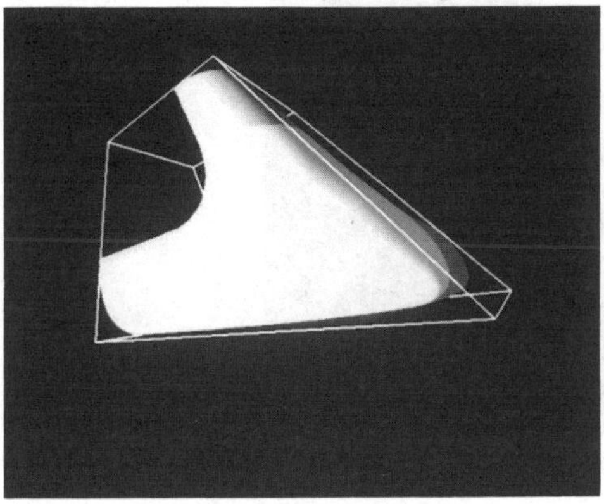

FIG. 12.13. *Retracted surface for the suitcase corner blending problem.*

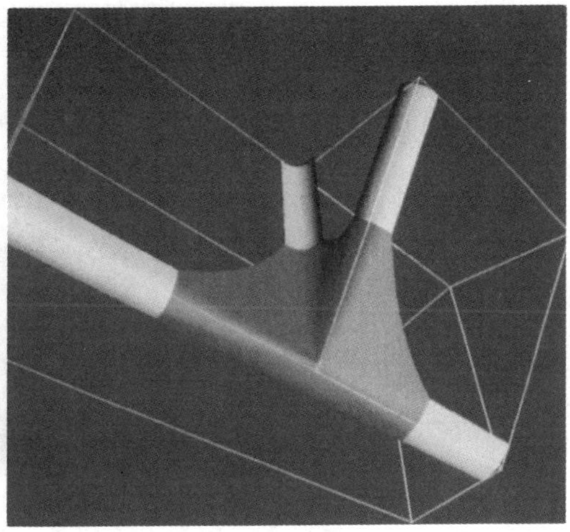

FIG. 12.14. *Blending surface of rounded edge and joint.*

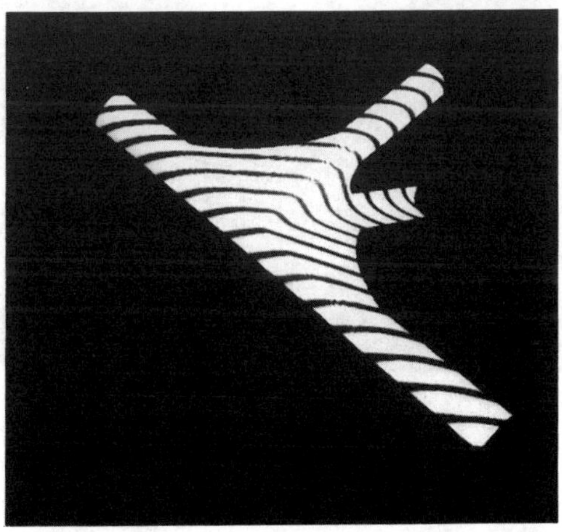

FIG. 12.15. *Contour surface for the surface in Fig.* 12.14.

expected even with bilinear weight functions and bilinear positional patches. The patch will be biquadratic.

For the jth subpatch of the four-sided macropatch, the weight matrices are as in (12.30). The positional patches are

(12.32)

$$P_i^{(j)}(u,v) = \begin{bmatrix} B_0(u) & B_1(u) \end{bmatrix} \begin{bmatrix} \text{base}_i^{(j)} & \text{right}_i^{(j)} \\ \text{left}_i^{(j)} & \text{mid}_i^{(j)} \end{bmatrix} \begin{bmatrix} B_0(v) \\ B_1(v) \end{bmatrix}, \qquad i = 1, 2, 3.$$

The constructed surface with contour lines is shown in Fig. 12.17.

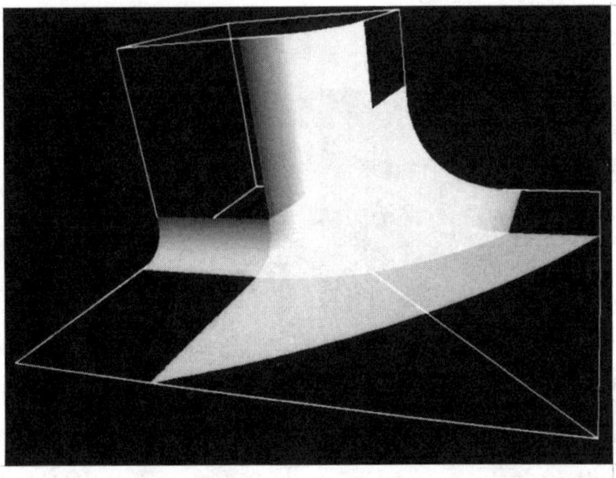

FIG. 12.16. *Blends of a rounded edge cube intersecting a plane.*

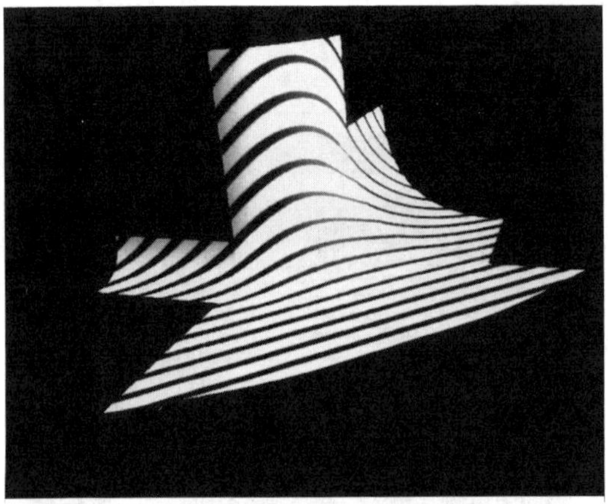

FIG. 12.17. *Contour surfaces for the surface in Fig.* 12.16.

12.6. Conclusion

Our approach to N-sided surface patches is based on convolution concept and vertex blending, where continuity and shape questions are formulated in terms of such concepts as weights and control patterns. Design constraints are seen in terms of pattern geometries, crossboundary surface information, and weight matrices. Overall shape is altered by the thumbweight facility and the location of the control patterns. Because the constructed weight functions are always positive, the resulting surface patch is in the convex hull of its control patterns. Visual interrogation of the blending surface via contouring and isophotes has confirmed the generally pleasing shape of the blends obtained from our approach.

Several examples are given. We expect to apply these ideas to many more problems such as the general vertex blend and the general N-hole problem.

Investigation into higher-order continuity convolution kernel functions should lead to weights that produce surfaces of higher-order continuity.

Acknowledgments

We appreciate support and assistance from Spatial Technology, Inc. of Boulder, Colorado, and technical conversations with Three Space, Ltd. of Cambridge, England.

References

[Bar73] R. Barnhill, G. Birkhoff, and W. Gordon, *Smooth interpolation in triangles*, J. Approx. Theory, 8 (1973), pp. 114–128.

[Blo91] J. Bloomenthal and K. Shoemake, *Convolution surfaces*, Proceedings of SIGGRAPH '91, Computer Graphics, 25 (1991), pp. 251–256.

[Blo89] M. I. G. Bloor and M. J. Wilson, *Generating blending surfaces with partial differential equations*, Comput. Aided Design, 21 (1989), pp. 165–171.

[Cat78] E. Catmull and J. Clark, *Recursive subdivision surfaces on arbitrary topological meshes*, Comput. Aided Design, 10 (1978), pp. 350–355.

[Cha84] P. Charrot and J. Gregory, *A pentagonal surface patch for computer aided geometric design*, Comput. Aided Geom. Design, 1 (1984), pp. 87–94.

[Chi83] H. Chiyokura and F. Kimura, *Design of solids with free-form surfaces*, Comput. Graphics, 17 (1983), pp. 289–298.

[Chi86] H. Chiyokura, *Solid Modeling with Design Base, Theory and Implementation*, Addison Wesley, New York, 1986.

[Doo78] D. Doo and M. Sabin, *Behavior of recursive subdivision surfaces near extraordinary points*, Comput. Aided Design, 10 (1978), pp. 356–360.

[Far86] G. Farin, *Triangular Bernstein-Bézier Patches*, Comput. Aided Geom. Design, 3 (1986), pp. 83–128.

[Far90] ——, *Curves and Surfaces for Computer Aided Geometric Design*, second ed., Academic Press, New York, 1990.

[Fau78] I. Faux and M. Pratt, *Computational Geometry for Design and Manufacture*, Ellis Horwood, Chichester, UK, 1978.

[Gre86] J. Gregory, *N-sided surface patches*, in *The Mathematics of Surfaces*, J. Gregory, ed., Clarendon Press, Oxford, UK, 1986, pp. 217–232.

[Hah89] J. Hahn, *Filling polygonal holes with rectangular patches*, in *Theory and Practice of Geometric Modeling*, W. Strasser and H. P. Seidel, eds., Springer-Verlag, New York, 1989, pp. 81–91.

[Jen87] T. Jensen, *Assembling triangular and rectangular patches and multivariate splines*, in *Geometric Modeling: Algorithms and New Trends*, G. E. Farin, ed., Society for Industrial and Applied Mathematics, Philadelphia, PA, 1987, pp. 203–220.

[Liu89] D. Liu and J. Hoschek, *GC^1 continuity conditions between adjacent rectangular and triangular Bézier surface patches*, Comput. Aided Design, 21 (1989), pp. 194–200.

[Loo89] C. Loop and T. DeRose, *A multisided generalization of Bézier surfaces*, ACM Trans. Graphics, 8 (1989), pp. 204–234.

[Sab86] M. Sabin, *Some negative results in N-sided patches*, Comput. Aided Design, 18 (1986), pp. 38–44.

[Sar87] R. Sarraga, *G^1 interpolation of generally unrestricted cubic Bézier curves*, Comput. Aided Geom. Design, 4 (1987), pp. 23–40.

[Sch46] I. Schoenberg, *Contribution to the problem of approximation of equidistant data by analytic functions*, Quart. Appl. Math., 4 (1946), pp. 45–99.

[Var87] R. Varady, *Survey and New Results in N-sided Patch Generation*, in *The Mathematics of Surfaces II*, R. Martin, ed., Clarendon Press, Oxford, 1987, pp. 203–236.

[Vee82] P. Veenman, *The design of sculptured surfaces using recursive subdivision techniques*, Proc. of the Conf. on CAD/CAM Technology in Mech. Eng., Massachusetts Institute of Technology, Cambridge, MA, 1982, pp. 54–63.

Index